Communications in Computer and Information Science 1063

Commenced Publication in 2007
Founding and Former Series Editors:
Phoebe Chen, Alfredo Cuzzocrea, Xiaoyong Du, Orhun Kara, Ting Liu,
Krishna M. Sivalingam, Dominik Ślęzak, Takashi Washio, and Xiaokang Yang

More information about this series at http://www.springer.com/series/7899

Leonid Sokolinsky · Mikhail Zymbler (Eds.)

Parallel Computational Technologies

13th International Conference, PCT 2019
Kaliningrad, Russia, April 2–4, 2019
Revised Selected Papers

 Springer

Editors
Leonid Sokolinsky 🆔
South Ural State University
Chelyabinsk, Russia

Mikhail Zymbler 🆔
South Ural State University
Chelyabinsk, Russia

ISSN 1865-0929   ISSN 1865-0937 (electronic)
Communications in Computer and Information Science
ISBN 978-3-030-28162-5   ISBN 978-3-030-28163-2 (eBook)
https://doi.org/10.1007/978-3-030-28163-2

This Springer imprint is published by the registered company Springer Nature Switzerland AG
The registered company address is: Gewerbestrasse 11, 6330 Cham, Switzerland

Preface

This volume contains a selection of the papers presented at the 13th International Scientific Conference on Parallel Computational Technologies, PCT 2019, held during April 2–4, 2019, in Kaliningrad, Russia.

The PCT series of conferences aims at providing an opportunity to discuss the future of parallel computing, as well as to report the results achieved by leading research groups in solving both scientific and practical issues using supercomputer technologies. The scope of the PCT series of conferences includes all aspects of high performance computing in science and technology such as applications, hardware and software, specialized languages, and packages.

The PCT series is organized by the Supercomputing Consortium of Russian Universities and the Federal Agency for Scientific Organizations. Originated in 2007 at the South Ural State University (Chelyabinsk, Russia), the PCT series of conferences has now become one of the most prestigious Russian scientific meetings on parallel programming and high-performance computing. PCT 2019 in Kaliningrad continued the series after Chelyabinsk (2007), St. Petersburg (2008), Nizhny Novgorod (2009), Ufa (2010), Moscow (2011), Novosibirsk (2012), Chelyabinsk (2013), Rostov-on-Don (2014), Ekaterinburg (2015), Arkhangelsk (2016), Kazan (2017), and Rostov-on-Don (2018).

Each paper submitted to the conference was scrupulously evaluated by three reviewers on the relevance to the conference topics, scientific and practical contribution, experimental evaluation of the results, and presentation quality. The Program Committee of PCT selected the 24 best papers to be included in this CCIS proceedings volume.

We would like to thank the Russian Foundation for Basic Research for their continued financial support of the PCT series of conferences, as well as respected PCT 2019 sponsors, namely platinum sponsor, Intel, gold sponsors, RSC Group and Hewlett Packard Enterprise, silver sponsors, T-Platforms and AMD, track sponsor IBS, and conference partner, Softline.

We would like to express our gratitude to every individual who contributed to the success of PCT 2019. Special thanks to the Program Committee members and the external reviewers for evaluating papers submitted to the conference. Thanks also to the Organizing Committee members and all the colleagues involved in the conference organization from Immanuel Kant Baltic Federal University, the South Ural State University, and Moscow State University. We thank the participants of PCT 2019 for sharing their research and presenting their achievements as well.

Finally, we thank Springer for publishing the proceedings of PCT 2019 in the *Communications in Computer and Information Science* series.

May 2019

Leonid Sokolinsky
Mikhail Zymbler

Organization

The 13th International Scientific Conference on Parallel Computational Technologies, PCT 2019, was organized by the Supercomputing Consortium of Russian Universities and the Ministry of Science and Higher Education of the Russian Federation.

Steering Committee

Berdyshev, V. I.	Krasovskii Institute of Mathematics and Mechanics, Yekaterinburg, Russia
Ershov, Yu. L.	United Scientific Council on Mathematics and Informatics, Novosibirsk, Russia
Minkin, V. I.	South Federal University, Rostov-on-Don, Russia
Moiseev, E. I.	Moscow State University, Russia
Savin, G. I.	Joint Supercomputer Center, Russian Academy of Sciences, Moscow, Russia
Sadovnichiy, V. A.	Moscow State University, Russia
Chetverushkin, B. N.	Keldysh Institute of Applied Mathematics, Russian Academy of Sciences, Moscow, Russia
Shokin, Yu. I.	Institute of Computational Technologies, Russian Academy of Sciences, Novosibirsk, Russia

Program Committee

Sadovnichiy, V. A. (Chair)	Moscow State University, Russia
Dongarra, J. (Co-chair)	University of Tennessee, USA
Sokolinsky, L. B. (Co-chair)	South Ural State University, Russia
Voevodin, Vl. V. (Co-chair)	Moscow State University, Russia
Zymbler, M. L. (Academic Secretary)	South Ural State University, Russia
Ablameyko, S. V.	Belarusian State University, Belarus
Afanasiev, A. P.	Institute for Systems Analysis RAS, Russia
Akimova, E. N.	Krasovskii Institute of Mathematics and Mechanics, Russia
Andrzejak, A.	Heidelberg University, Germany
Balaji, P.	Argonne National Laboratory, USA
Boldyrev, Y. Ya.	Saint-Petersburg Polytechnic University, Russia
Carretero, J.	Carlos III University of Madrid, Spain
Gazizov, R. K.	Ufa State Aviation Technical University, Russia
Gergel, V. P.	Lobachevsky State University of Nizhny Novgorod, Russia

Glinsky, B. M.	Institute of Computational Mathematics and Mathematical Geophysics SB RAS, Russia
Goryachev, V. D.	Tver State Technical University, Russia
Il'in, V. P.	Institute of Computational Mathematics and Mathematical Geophysics SB RAS, Russia
Kobayashi, H.	Tohoku University, Japan
Kunkel, J.	University of Hamburg, Germany
Labarta, J.	Barcelona Supercomputing Center, Spain
Lastovetsky, A.	University College Dublin, Ireland
Ludwig, T.	German Climate Computing Center, Germany
Lykosov, V. N.	Institute of Numerical Mathematics RAS, Russia
Mallmann, D.	Julich Supercomputing Centre, Germany
Michalewicz, M.	A*STAR Computational Resource Centre, Singapore
Malyshkin, V. E.	Institute of Computational Mathematics and Mathematical Geophysics SB RAS, Russia
Modorsky, V. Ya.	Perm Polytechnic University, Russia
Shamakina, A. V.	High Performance Computing Center in Stuttgart, Germany
Shumyatsky, P.	University of Brasilia, Brazil
Sithole, H.	Centre for High Performance Computing, South Africa
Starchenko, A. V.	Tomsk State University, Russia
Sterling, T.	Indiana University, USA
Taufer, M.	University of Delaware, USA
Turlapov, V. E.	Lobachevsky State University of Nizhny Novgorod, Russia
Wyrzykowski, R.	Czestochowa University of Technology, Poland
Yakobovskiy, M. V.	Keldysh Institute of Applied Mathematics RAS, Russia
Yamazaki, Y.	Federal University of Pelotas, Brazil

Organizing Committee

Erokhin, G. N. (Chair)	Baltic Federal University, Russia
Demin, M. V. (Co-chair)	Baltic Federal University, Russia
Belova, A. V. (Secretary)	Baltic Federal University, Russia
Antonov, A. S.	Moscow State University, Russia
Antonova, A. P.	Moscow State University, Russia
Bardina, M. G.	South Ural State University, Russia
Danilin, A. N.	Baltic Federal University, Russia
Filatova, V. M.	Baltic Federal University, Russia
Kraeva, Ya. A.	South Ural State University, Russia
Nikitenko, D. A.	Moscow State University, Russia
Ostrovskikh, N. O.	South Ural State University, Russia
Pestov, L. N.	Baltic Federal University, Russia
Samusev, I. G.	Baltic Federal University, Russia
Sobolev, S. I.	Moscow State University, Russia

Voevodin, Vad. V.   Moscow State University, Russia
Yurov, A. V.   Baltic Federal University, Russia
Zymbler, M. L.   South Ural State University, Russia

Contents

Supercomputer Simulation

High Performance Architectures, Tools and Technologies

HPC Software for Massive Analysis of the Parallel Efficiency of Applications

Pavel Shvets⊙, Vadim Voevodin(✉)⊙, and Sergey Zhumatiy⊙

Research Computing Center of Lomonosov Moscow State University,
Leninskie Gory, 1, bld. 4, Moscow, Russia
shvets.pavel.srcc@gmail.com,{vadim,serg}@parallel.ru

Abstract. Efficiency is a major weakness in modern supercomputers. Low efficiency of user applications is one of the main reasons for that. There are many software tools for analyzing and improving the performance of parallel applications. However, supercomputer users often do not have sufficient knowledge and skills to apply these tools correctly in their specific case. Moreover, users often do not know that their applications work inefficiently.

The main goal of our project is to help any HPC user to detect performance flaws in their applications and find out how to deal with them. To this end, we plan to develop an open-source software solution that performs automatic massive analysis of all jobs running on a supercomputer to identify those with efficiency issues and helps users to conduct a detailed analysis of an individual program (using existing software tools) to identify and eliminate the root causes of the loss of efficiency.

Keywords: Supercomputing · Efficiency analysis · Parallel program · Massive analysis · Application performance

1 Introduction

Many modern supercomputers are used inefficiently. This means that a huge part of the computational resources provided by a supercomputer system is often underutilized or even idle. This is not due to low demand for such resources. On the contrary, supercomputer users often have to wait for the availability of resources. For example, on the Lomonosov-1 supercomputer, the average wait time for a job to start is more than 10 h (since the beginning of 2018).

The main reason for supercomputer underutilization is the low efficiency of applications that are executed on it [1]. There are two major reasons for this. The first is that the HPC area is becoming more and more large-scale [2]. At the Supercomputer Center of Moscow State University, the number of users has grown sixfold in the last five years (up to 3000); a similar situation can be seen in many other supercomputer centers. This happens because of a growing demand for computationally-intensive experiments in many research areas. In turn, this leads to a growing number of new users that are usually skilled experts in their

© Springer Nature Switzerland AG 2019
L. Sokolinsky and M. Zymbler (Eds.): PCT 2019, CCIS 1063, pp. 3–18, 2019.
https://doi.org/10.1007/978-3-030-28163-2_1

respective scientific areas but are not experienced enough in developing efficient parallel applications (even generally in parallel computing). The second reason is the overwhelming complexity of modern supercomputer systems. Not only users but also system administrators do not know all the manifold peculiarities of hardware components which can potentially influence the performance of an application running on a supercomputer. Our practice shows that users often must know a lot about supercomputer architecture (especially architecture of computing nodes) if they want to achieve high efficiency.

This problem has been topical for many years. Many useful software tools and approaches have been developed for performance analysis and optimization of parallel programs (profilers, trace analysis tools, debuggers, emulators). There are numerous examples of successful application of these tools in practice. But our experience of many years at the Research Computing Center of Moscow State University (RCC MSU), providing support for the largest supercomputers in Russia, shows that, unfortunately, such tools are seldom actually used. There are different reasons for that:

- Users often do not know about many useful analysis tools.
- Users often do not know which tool they should start with.
- Users often do not know how to interpret the results given by analysis tools.
- Users often do not even know that there are some performance issues in their applications.

Consequently, a solution is needed to help users deal with these issues. For this purpose, we started the project that will be described in this paper. The main goal of this project is to assist any HPC user in identifying performance issues in their applications and dealing with them (if there are any). With this intention mind, we are currently developing an open-source software complex that will cover all jobs executed on a supercomputer and will be useful for users of any level of knowledge in performance analysis. Our current research is focused on the MSU supercomputer center but our future plans include making this solution portable to other supercomputer centers as well.

The main contribution of this work is a detailed description of the holistic approach proposed at the RCC MSU for mass performance analysis of parallel applications in the supercomputer job flow.

The paper is organized as follows. In Sect. 2, we describe our own previous experience in optimizing the efficiency of supercomputer usage and outline various approaches suggested by other researchers to deal with the same problem. Section 3 is devoted to a detailed description of our approach, consisting of three main parts: primary automatic analysis of the overall supercomputer job flow Sect. 3.1, detailed analysis of one particular application Sect. 3.2, and complex analysis of the overall supercomputer behavior Sect. 3.3, with the third part being based on the first two. A short description of the current state of the complex and examples of its use in practice are given in Sect. 4. Section 5 includes our conclusions and plans for further development and improvement of our solution.

2   Background and Related Work

The main issue discussed in the project is low efficiency of parallel program execution. There is a large number of software solutions that have achieved certain results in performance analysis and optimization of a particular application (e.g., debugging, profiling, and tracing tools), but these solutions are aimed at studying one particular efficiency aspect, namely the optimality of cache memory usage, efficiency of MPI data transfer, etc. There are only a few studies intended to perform a more global analysis aimed at integrating knowledge about different efficiency aspects. Such integration can help users to understand what type of performance issue occurred in their particular case and how they can overcome it. This is the main concern of our project.

One of the biggest and most interesting projects aimed at achieving this goal is POP CoE (Performance Optimization and Productivity Center of Excellence) [3,4]. Within the framework of this project, a team of experts performs detailed analyses and optimization of real-world applications provided by users from different subject areas. Several levels of application analysis are available. Performance Audit is a primary service aimed at an initial analysis and evaluation of program performance. Performance Plan is a more detailed service for detecting root causes as well as determining possible ways for their elimination. The last level is Proof-of-Concept, which offers an assessment of the possible benefits of the suggested optimization. This overall concept of helping users in detecting and overcoming root causes of performance degradation is close to the one proposed in our project. But there is a major difference—POP CoE analysis is performed manually by experts, they do not develop any software for automating this process, which is what we plan to do in our project. POP approach is, therefore, much more focused on each particular application, while massive analysis and automation are unavailable in this case. It should be noted, however, that POP project results have influenced the development of the overall project architecture we propose in the present paper.

The Score-P Package [5] is another solution designed to help users analyze program efficiency from different points of view. This package integrates several performance analysis tools (Scalasca, Periscope, Vampir, etc.) and provides unified access for profiling, tracing, and analyzing applications. This approach provides assistance in performing a uniform analysis of both serial and parallel programs with different parallel programming models (MPI, OpenMP, or both). However, Score-P offers integration only on the technical side. This means that it eases the process of using the mentioned tools but it does not help users in choosing an appropriate software tool or interpreting analysis results.

TAU Performance System toolkit [6] is a similar software solution which assists in tracing and profiling parallel applications and helps users to study different aspects of parallel program behavior through a unified approach. But similarly to Score-P, it cannot be employed to carry out a more global analysis for detecting different types of performance issues since it is unable to integrate accumulated knowledge on different aspects of program performance.

One more solution that is worth mentioning is Valgrind [7], a suite of tools aimed at a holistic analysis of memory usage in a program. The Valgrind core emulates program execution, allowing to collect the full trace of memory references. Similarly to Score-P and TAU, Valgrind has a unified interface for a number of modules which can be used to study different features of memory usage, such as cache subsystem performance (via Cachegrind), heap usage (via Massif), call stack (via Callgrind), and others.

The Intel Parallel Studio XE package [8] provides significantly wider functionality. The package incorporates, in particular, Intel Advisor, a tool designed for optimizing vectorization and threading, Intel VTune Amplifier, for performance profiling, Intel Inspector, for memory usage errors detection, Intel Trace Analyzer and Collector, for tracing parallel applications, along with a number of other useful tools. However, the user must decide himself which tool to use in each particular case, how to integrate the results from several tools in this package, and how to correctly interpret the results obtained.

At the RCC MSU, we have acquired a rich experience in analyzing and optimizing the performance of real-life applications (see, for example, [9] or [10]) and have developed a significant portfolio of software tools for the study of different aspects of program efficiency (e.g., the JobDigest system for generating job efficiency reports [11], or a system for detecting abnormally inefficient applications [12]). At the same time, we have accumulated a vast knowledge base on possible root causes of efficiency degradation and ways for their elimination, as well as best practices for the proper use of existing analysis tools (including those mentioned above) and correct interpretation of their results.

At the moment, there is no software available that can solve the global problem of helping users in detecting performance issues in parallel programs and tackling their root causes. Nevertheless, the aforementioned projects are capable of solving specific subproblems, so they could be integrated into the overall solution. This is planned to be done within the framework of the current project.

3   Software Complex for Massive Analysis of the Efficiency of Applications

The main goal of our project is to develop a software solution to help any HPC user in detecting performance issues in their job, finding their root causes, and determining possible ways for their elimination. To this end, we are currently developing three subsystems in the framework of the overall solution. The first subsystem (job primary analysis) is responsible for the primary massive analysis of the supercomputer overall job flow. Its main goal is to detect possible performance issues and notify users about them. The next subsystem (job detailed analysis) is aimed at the detailed analysis of a particular application, it should assist users in determining and eliminating the root causes of performance issues detected at the primary analysis stage. The third subsystem (supercomputer analysis) is intended for system administrators and management, it should respond for the detection of performance issues at the overall supercomputer system level and study the general efficiency of different super-computer software and hardware components (such as software packages, file system, resource manager, overall job flow) and their interaction.

The general architecture of the proposed solution is shown in Fig. 1. The overall working algorithm starts in the job primary analysis subsystem with a constant primary analysis of the supercomputer job flow (the top left part of the figure). This subsystem should collect different kinds of input data: low-level monitoring data, information about the job itself, tags (classes assigned to jobs by other performance analysis tools), etc. The collected input data are then processed (in the S1 analysis module), making it possible to obtain primary analysis results for every job in the job flow. These results can be shown to the user (front-end module on the left) and are also stored in the knowledge base, depicted in the center of Fig. 1.

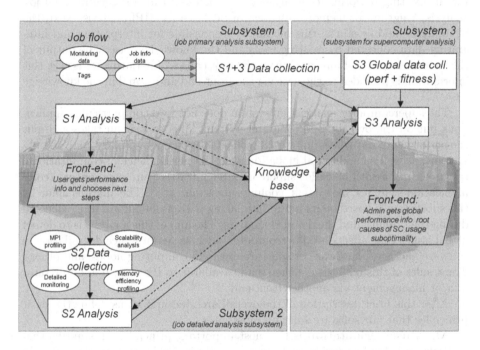

Fig. 1. The general architecture of the proposed software complex

The concept of knowledge base is very important in our solution. Its main task is to store every piece of useful information produced by any subsystem in the past. The knowledge base works as the main interaction point between differ-ent subsystems. Indeed, the detailed analysis is always based on primary results, whereas the analysis of the overall supercomputer behavior is based on results from both subsystems, namely the primary and the detailed analyses. Moreover, storing the information from past analysis results enables us to request and use it in the future. For instance, when analyzing a new program, we can use this database to find similar, already studied situations, and thus use the experience gained earlier. The dashed arrows in Fig. 1, connecting the knowledge base with the analysis modules, portray this idea, i.e. the use of past experience for conducting a performance analysis of either the current job or the supercomputer state.

The potential performance issues detected in each user job by the primary analysis subsystem are shown to the user via the front-end (a web page). If the information provided is not enough, then the user can start a detailed analysis of a particular application using the next subsystem and based on issues detected earlier. We plan to implement various types of detailed analysis, such as MPI profiling and tracing, memory efficiency and correctness analysis, network activity analysis, etc. When the user chooses an appropriate analysis type, the data collection module starts. Our plan is to incorporate into this module various existing performance analysis tools, such as Vtune for general serial job analysis, Valgrind for memory research, or Scalasca for MPI analysis. Next, the S2 analysis module starts; this module is responsible for interpreting and integrating data collected by existing tools and the knowledge base. The output of this module consists of information on the root causes of performance issues; this information is provided to the user via the front-end and stored in the knowledge base.

The supercomputer analysis subsystem does not focus on particular applications but on the overall supercomputer system. This subsystem takes as input the performance data on running jobs (collected by the job primary analysis subsystem), as well as additional global data on the performance of computational resources and health of major supercomputer infrastructure components (communication switches, file system, cooling subsystem, etc.). The analysis made by this subsystem is based on its input data, along with useful information collected during the primary and the detailed analysis. The aim of this analysis is to detect global issues that affect the supercomputer, such as performance degradation of software packages, hardware or software errors in computing nodes, and others. The results of this analysis are then made available to the system administrator and/or management via another front-end.

After this brief description of the overall architecture, let us define the basic principles that our software solution must meet.

Mass user orientation is the most important principle. This means that our software should provide performance analysis results to a user on all jobs she/he is running on the supercomputer. Moreover, our software should be useful to users with various levels of proficiency: no additional training should be required for basic software usage; at the same time, performance data should be available with any level of detail if needed. This principle also implies that our software should support role separation, i.e. supercomputer users should be allowed to view only results on their own jobs; leaders of supercomputer projects should be able to analyze the jobs of every user in their projects; system administrators should be granted the right to view all the results collected by our software solution.

The next principle is the **integration of available software tools**. As it was stated earlier, there exist many software tools that can be used for analyzing particular features of program behavior, especially at the detailed analysis stage. It is not our aim in this project to compete with existing performance analysis tools; on the contrary, we want to complement them and teach supercomputer users how to use them. Moreover, we are interested in integrating with as many existing tools as possible since it would enable us to make a more holistic analysis of applications.

We would like our solution to be useful to the whole community, so **portability** is also one of our basic principles. Currently, during the development process, we have to focus on one supercomputer (namely the Lomonosov-2 system). However, we are interested in the involvement of other supercomputer centers as well, so all subsystems are planned to be as portable as possible. It should be noted that some components are definitely going to be system-dependent; this will be described in more detail during the description of the three software subsystems.

The following subsections are devoted to the description of each subsystem of the overall solution.

3.1   Primary Massive Analysis of Supercomputer Job Flow

This subsection describes the first part of the complex we are developing, namely the primary massive analysis of the supercomputer job flow. The main goal of massive analysis is to automatically scan through all jobs running on the supercomputer, detect jobs with possible performance issues (such as inefficient MPI usage, bad memory locality, underutilization of resources, etc.), and notify users about them. This can (and must) be done automatically, because: (1) it is impossible for the system administrator to manually perform such analysis on a daily basis; (2) at this initial stage, users are not yet aware about any performance issues in their jobs, so no user interaction is involved.

The detection of possible performance issues is conducted by using a set of static rules. Each rule determines one particular performance issue but it also defines assumptions on its possible root causes and recommendations for their elimination (manually developed by the authors of this paper). At this primary analysis stage, we do not have any information about the internal structure of the jobs being analyzed, so only a rough evaluation of possible problems can be made. This stage can thus be seen as a starting point for the detailed analysis, which is made by a different subsystem, described in Subsect. 3.2.

The input data used in these rules is mainly performance data provided by the monitoring system. For each job, we collect values describing CPU utilization (CPU user/system/iowait/idle load, number of active processes, rate of instructions retired), memory usage (rate of L1/L2/L3 cache-misses per second, rate of load/store operations), and Infiniband usage (number of bytes/packets sent/received per second). This set of input data was manually chosen as the most informative one for conducting performance analyses, based on our opinion. Since this stage is primary and is not aimed at a detailed fine analysis of each application, only integral values (average, minimum, maximum) are currently collected for each job. In the future, we plan to enhance these input data with more detailed information, such as time series of values during job execution or more "intellectual" data provided by other performance analysis tools as, for example, anomaly detection software [12].

We have currently developed a set of 25 rules which are implemented and used on the Lomonosov-2 supercomputer. They are divided into nine groups, each covering a specific type of performance issue: inefficient size of MPI packets, incorrect job behavior or operation mode, hanged jobs, significant disbalance of workload, and others.

Three examples of particular rules are given below to explain the corresponding concept. The first rule catches underutilization of compute node resources. More specifically, it detects a quite common situation, namely only one active process per node is detected along with a low usage of memory, GPU, and communication network (a major resource underutilization is detected). In this case, the assumption is that the user explicitly specified the allocation of one MPI process per node during job launch, even though it is not needed under such conditions. Moreover, this leads to a significant waste of computational resources. Therefore, our recommendation in this situation is to specify more processes per node to fully utilize the resources available.

One type of memory-related issue is considered in the next rule example. This rule is triggered if low job activity is detected together with highly intensive memory usage and low memory locality. This behavior usually means that memory usage is organized inefficiently and probably takes too much time, leading to CPU being idle while waiting for memory access to be executed. In this case, we recommend to perform an analysis of memory usage efficiency (applying such tools as Valgrind).

The last example describes a rule for detection of general inefficiency problems. The rule is triggered if the execution rate of CPU instructions is low and, at the same time, no high external activity (GPU usage or data transfer over Infiniband) is detected. That means that there is definitely a significant underutilization of resources but no signs of possible root causes for that are found at this stage. Nevertheless, this performance issue should be reported to the user, so that she/he can make attempts to investigate it by using, for instance, our detailed analysis subsystem. Under such conditions, only general recommendations, such as the use of holistic program analyzers (e.g., Intel Vtune Amplifier), can be given.

It should be noted that we have a strict criterion for each rule, which allows to automatically perform rule checking. Each rule developed for the primary massive analysis is described according to the following detailed structure:

- *Criterion for rule triggering*. This is the core of the rule which formally describes what input data on job behavior triggers the specified rule. It is implemented as a Python function.
- *Description of detected performance issue*. A text description of the performance issue detected by this rule; it is shown to the user.

- *Description of possible root causes.* A text that describes the possible reasons for this issue; it is shown to the user.
- *Recommendations on further analysis.* Information on possible ways to overcome the cause of the issue or analyze it in detail. This includes a text description for the user as well as machine data that is automatically analyzed within other parts of the proposed overall solution.
- *Confidence level.* This value represents our confidence that the specified performance issue can be accurately detected by the suggested criterion. It takes a value from 1 (not sure at all) to 5 (completely sure).
- *Criticality level.* This value indicates how critical the detected performance issue is. It takes a value from 1 (the performance issue is unimportant or no issues were found) to 5 (incorrect job behavior, we can automatically cancel it).
- *Group name and priority.* Each rule belongs to a certain group according to the type of performance issue it describes. Within each group, a priority is specified for each rule; the bigger the priority value, the more critical the performance issue detected. Only rules with the biggest priority values within a group are shown to the user.

Such a detailed description enables us to create flexible rule management. In particular, this allows for rule suppression (displaying only the most important rules to prevent the output to be littered with superfluous information), convenient rule filtering, as well as custom notification of users and/or administrators in critical situations.

An important issue is to verify the correctness and completeness of the proposed rules. The main difficulty here lies in the fact that there are no strict criteria for determining inefficiency problems. These criteria have therefore to be defined heuristically and tested on real-life data. For these purpose, we study the results of primary analyses of real jobs; several rules have already been tuned based on the information obtained.

A pilot version of this subsystem has been implemented and is currently under evaluation on the Lomonosov-2 supercomputer. A more detailed description and other examples of using primary massive analysis is given in [13].

3.2 Detailed Analysis of Particular Applications

The primary analysis helps in obtaining basic information on a job performance. However, this information is far from sufficient for users to clearly determine the root causes of performance degradation in their programs and what optimization steps need to be taken to overcome them. We designed the next subsystem of the overall software solution to carry out a more detailed analysis of individual jobs.

The goal of this subsystem is to obtain detailed information about specific features of the behavior of a job: memory or communication network usage, utilization of computing resources, scalability, etc. Obtaining information on each individual feature is almost an independent process; this means that the implementation of different detailed analysis types can be developed in parallel. As it was stated earlier, we plan to actively use various external tools at this stage.

A detailed analysis should always be done in close collaboration with the user. That is, nothing is automatically collected for the job at this stage, unlike the primary analysis. The implementation of interaction with the user is the major and probably most difficult part of this subsystem. Currently, we plan this interaction to be implemented as follows:

- During the data collection step, we will provide either scripts for the automated execution of external tools for the chosen application analysis or user guides on which tools to use and how to collect data with them.
- During the data analysis step, we will either provide scripts for an automated analysis of the output of external tools (e.g., parsing output files) or ask the user questions that will help us acquire useful information about the job behavior.

It should be noted that, unlike the primary analysis stage which is designed to work with users of any level of proficiency, this stage requires a deeper knowledge in parallel programming and program efficiency analysis.

There are different types of detailed analysis that we are planning to implement. One of them is the **analysis of job execution dynamics**. This type of analysis is designed to trace the dynamics of changes in performance characteristics during a job execution; for example, the user can see how the CPU load or, let us say, the number of MPI bytes sent per second changes. For this purpose, tools like JobDigest [11] or monitoring tools like Nagios or Zabbix can be used. This type of analysis does not require any manipulations with the program: it is only necessary to collect and store monitoring data for each job.

A method for performing this type of detailed analysis is currently under active development. Data collection is already available since it is currently based on the JobDigest approach which requires only monitoring data already being collected on the Lomonosov-2 supercomputer. The data analysis step in this case is planned to be implemented via a set of rules similar to the ones described in Subsect. 3.1: they contain descriptions of potential issues, assumptions on their root causes, and possible recommendations for their elimination. In this case, however, it is difficult to check the triggering of the rules automatically, so it will be formed as a set of questions to the user. At this moment, we have already developed 13 rules concerning the distribution of MPI activity between nodes, different behavior of the CPU load on nodes, visible correlations between dynamic characteristics, and others.

MPI profiling is another example of detailed analysis. This type is designed to obtain basic information about MPI programs. Using MPI program profiling tools, the user can find out what proportion of the total job execution time it took to perform MPI operations, which types of MPI operations turned out to be the most expensive, whether there is an imbalance between processes, etc. This type of analysis is planned to be used in almost all cases where the primary analysis has revealed performance issues correlated with inefficient execution of MPI operations.

The first version of this type of analysis is planned to be implemented using the mpiP tool [14], a popular light-weight software for MPI profiling. We are planning to provide a simple user manual at the data collection step with instructions on how to link a program with the mpiP tool and execute the instrumented program. The data analysis step is supposed to be performed automatically, without the user participation, since mpiP provides a convenient output file which can be easily parsed and analyzed.

Low-level analysis is a type of detailed analysis that can be useful for experienced users who want to explore all the fine peculiarities of job behavior. It aims to study low-level software and hardware sensors, available, for example, via the PAPI library [15]. Such an analysis can be useful to investigate low-level root causes of idle computational resources. For this purpose, methods as the top-down approach [16] can be used. This approach helps to identify hardware bottlenecks preventing a program from being more efficient: big rates of branch mispredictions, DRAM bandwidth, CPU port utilization, etc.

We are also planning to implement other types of detailed analysis, such as MPI tracing, analysis of memory usage efficiency or correctness, I/O analysis, scalability analysis, general serial job analysis, and others.

It is worth noting the importance of collecting feedback from users at this stage. The data analysis step will be very complex, so in some cases we will not be able to perform a complete root-cause investigation. Thus, when providing each new result, we will need users to determine whether it is correct. Without user feedback, it will be very difficult to make the detailed analysis step accurate enough.

This subsystem is currently under development; several methods (such as analysis of execution dynamics, detailed monitoring, and MPI profiling) are planned to be implemented in the beginning of the year 2019.

3.3   Analysis of Supercomputer Behavior

The goal of the supercomputer analysis subsystem is to obtain comprehensive information about the performance of the entire supercomputer. Within this subsystem, we are planning to implement various "smart screens", i.e. separate web pages for displaying diverse features of the supercomputer general state. For example, a separate "smart screen" will be devoted to the reliability and availability of the supercomputer: it will show where and when system failures occurred and how the supercomputer efficiency was impacted. Another example is a general "smart screen" regarding the efficiency of use of software packages installed on the supercomputer. Do users need to upgrade a particular package to achieve a better performance? How does the performance of a package change for different users? Which packages show low efficiency in general? This "smart screen" will be designed to answer all these questions.

An important "smart screen" will be devoted to the integration and analysis of useful information obtained during the primary and the detailed analyses.

Up to this point, these results have been mentioned only as a means to study a particular job. However, the same data can be very useful if one wants to assess the situation on the supercomputer as a whole. In particular, the study of combined results of both the primary and the detailed analysis subsystems for a job is expected to provide answers to the following questions:

- What are the performance issues most often encountered in practice and what are their root causes?
- How much does the average job performance increase after performing an optimization procedure based on primary and/or detailed analysis results?
- How much do the failures in the supercomputer infrastructure affect the performance of user jobs? Which failures have the greatest impact?

Within this subsystem, we are planning to detect performance issues and their root causes but in terms of the whole supercomputer. Let us give a specific example of a situation that we would like to discover. Assume that according to the results of the primary and the detailed analyses, it was found that the number of jobs having performance issues among those using a specific software package significantly increased starting at a certain time X. Furthermore, studying the information on the state of the supercomputer, it was also discovered that the software package was updated at that time X. This suggests that the package was somehow incorrectly configured during the upgrade process, and this caused a performance degradation throughout the entire supercomputer job flow. In such cases, system administrators should be notified in order to check the correctness of the package configuration.

It should be noted that this subsystem is currently the least developed since it heavily relies on the implementation of the other two subsystems.

4   Developing the Implementation on the Lomonosov-2 Supercomputer

Currently, the primary analysis subsystem has been implemented within the overall software complex and is being evaluated on the Lomonosov-2 supercomputer. It can be accessed by any user of the Lomonosov-2 via the Octoshell system [17] under the "Efficiency" tab (available only in Russian language). The main page provides basic information about jobs executed by the current user (see Fig. 2). Each row corresponds to a particular job; the values on the right show the average values of the most important performance characteristics: CPU and GPU load, number of active processes (load average), number of instructions executed per second (IPC), and rate of bytes sent/received using MPI (the color indicates how big the value is).

Potential performance issues found for each job are shown in the "Detected issues" column and are specified as icons, one for each group of rules (see Subsect. 3.1). Clicking on the leftmost column will expand a brief description of these issues. A more detailed description of the performance behavior (detected issues,

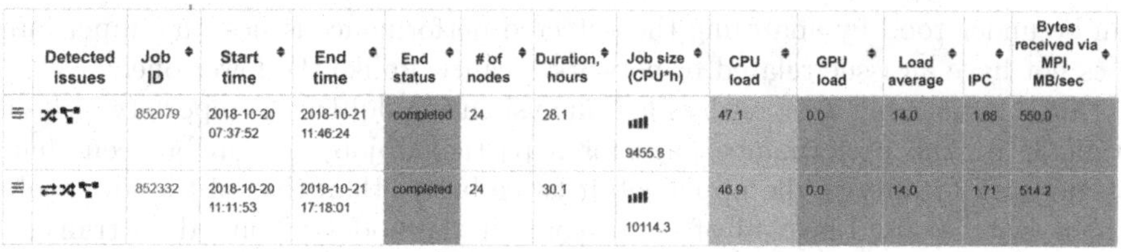

Detected issues	Job ID	Start time	End time	End status	# of nodes	Duration, hours	Job size (CPU*h)	CPU load	GPU load	Load average	IPC	Bytes received via MPI, MB/sec
≡ ✕⌄⌃	852079	2018-10-20 07:37:52	2018-10-21 11:46:24	completed	24	28.1	▮▮▮ 9455.8	47.1	0.0	14.0	1.66	550.0
≡ ⇄✕⌄⌃	852332	2018-10-20 11:11:53	2018-10-21 17:18:01	completed	24	30.1	▮▮▮ 10114.3	46.9	0.0	14.0	1.71	514.2

Fig. 2. Screenshot of a job list with primary analysis results obtained for real-life jobs

dynamic characteristics, JobDigest report) for each job is also available; it can be accessed by clicking on the corresponding number in the "Job ID" column.

The brief description of the performance issues for the lower job is provided in Fig. 3. The upper issue is correlated with network locality, which evaluates the distance between computing nodes allocated to this job (in terms of the number of switches used for connecting them). In this case, MPI usage intensity is high while network locality is low, which could lead to noticeable program slowdown due to high latency of MPI data transfer. The middle issue indicates that the job is running on the supercomputer partition intended for GPU-based jobs but the job itself does not utilize GPU. The lower issue notifies that the program generates too small MPI packets, although MPI activity is high. In this case, data transfer overheads can be significant.

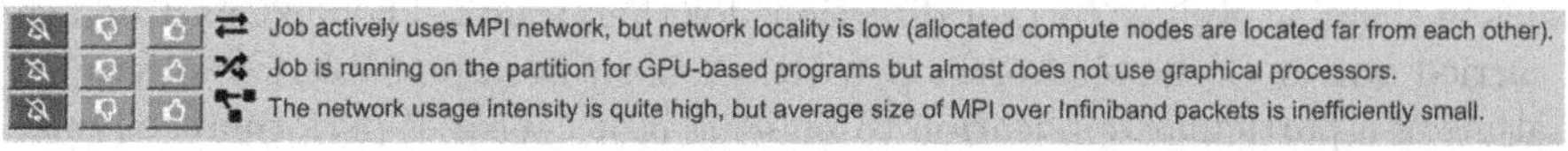

✕	🔇	👍	⇄	Job actively uses MPI network, but network locality is low (allocated compute nodes are located far from each other).
✕	🔇	👍	✕	Job is running on the partition for GPU-based programs but almost does not use graphical processors.
✕	🔇	👍	⌄⌃	The network usage intensity is quite high, but average size of MPI over Infiniband packets is inefficiently small.

Fig. 3. A brief description of detected performance issues for the lower job shown in Fig. 2

We encourage the user to provide feedback on each issue and adjust the notifications. This is even more important for us because initially we cannot determine the most comprehensive and correct criteria for rule triggering and need feedback from users in order to tune them. The "crossed bell" button hides this performance issue for every job, so it will not be shown until the user explicitly marks it as visible again (this can be done on a page with full descriptions of all possible primary performance issues). The "thumbs up" button indicates that the user agrees that the issue applies to her/his job. The "thumbs down" button, on the contrary, shows that the user disagrees, and allows her/him to specify a reason for that. Such information is particularly valuable for us since it helps to fine-tune our primary rules.

There is another interesting fact that is worth noting. Figure 2 shows two identical launches of a program (same input data, number of processes, and launch parameters). It is known that this program always executes with almost the same wall time. In this case, however, the execution time is somewhat different: 30 h for the lower job versus 28 for the upper one. The probable root cause can be understood by analyzing the detected performance issues: the upper job does not have an issue related to network locality, unlike the lower one.

After performing a more detailed analysis using JobDigest report (which is available on the performance page for a particular job), it can be seen that the network locality of the upper job is twice better than that of the lower job (three switches used instead of six to connect the nodes). Thus, data transfer over the MPI network is performed longer on average, which is confirmed by a slightly reduced MPI data transfer rate (the right column in Fig. 2). It should be noted that this situation is consistently repeated: launches of this program on the allocated nodes with good network locality are performed two to three hours faster than launches with low network locality.

More details on the software implementation of the job primary analysis subsystem, as well as other results obtained in practice, can be found in [13].

5   Conclusions and Further Work

The paper describes our project aimed at assisting supercomputer users in detecting performance issues in their parallel applications, determining their root causes and eliminating them. For that purpose, we are currently creating an open-source software complex which consists of three parts. The first part performs an automatic primary massive analysis of every job running on the supercomputer, so as to obtain a general understanding of potential performance issues in them, as well as further directions for their study. The next part, which is carried out in close collaboration with the user, is responsible for a detailed analysis of a particular application to find the root causes of performance issues detected. This part includes, for instance, a study of the job execution dynamics, a scalability analysis or MPI profiling performed using existing performance analysis tools. The last part is aimed at assisting system administrators and management of supercomputer centers in the analysis and optimization of the overall supercomputer performance. To that end, we plan to integrate the results obtained during the primary and the detailed analysis concerning different applications, as well as to detect root causes of performance degradation at the level of the entire supercomputer.

Our plans for the near future include the implementation of different types of detailed analysis, which will greatly enhance the functionality of the software complex we are currently developing. We want to focus firstly on the analysis of execution dynamics, detailed monitoring, and MPI profiling. The next step is to develop new primary rules, in particular by adding new types of performance data as input to the analysis. Another direction for further work is to start the creation of a knowledge base of potential performance issues and their root causes, so that we can use the accumulated experience to produce more accurate analyses in the future.

Acknowledgments. The results described in this paper were achieved at Lomonosov Moscow State University with the financial support of the Russian Science Foundation (agreement No. 17-71-20114). The research was carried out on the HPC equipment of the shared research facilities at Lomonosov Moscow State University and was supported through the project RFMEFI62117X0011.

References

1. Voevodin, V., Voevodin, V.: Efficiency of exascale supercomputer centers and supercomputing education. In: Gitler, I., Klapp, J. (eds.) ISUM 2015. CCIS, vol. 595, pp. 14–23. Springer, Cham (2016). https://doi.org/10.1007/978-3-319-32243-8_2
2. Joseph, E., Conway, S.: Major trends in the worldwide HPC market. Technical report (2017). https://hpcuserforum.com/presentations/stuttgart2017/IDC-update-HLRS.pdf
3. Bridgwater, S.: Performance optimisation and productivity centre of excellence. In: 2016 International Conference on High Performance Computing & Simulation (HPCS), pp. 1033–1034. IEEE (2016). https://doi.org/10.1109/HPCSim.2016.7568454
4. Performance Optimisation and Productivity—A Centre of Excellence in Computing Applications. https://pop-coe.eu/
5. Knüpfer, A., et al.: Score-P: a joint performance measurement run-time infrastructure for Periscope, Scalasca, TAU, and Vampir. In: Brunst, H., Müller, M., Nagel, W., Resch, M. (eds.) Tools for High Performance Computing 2011, pp. 79–91. Springer, Heidelberg (2012). https://doi.org/10.1007/978-3-642-31476-6_7
6. Shende, S.S., Malony, A.D.: The TAU parallel performance system. Int. J. High Perform. Comput. Appl. **20**(2), 287–311 (2006)
7. Nethercote, N., Seward, J.: Valgrind: a framework for heavyweight dynamic binary instrumentation. SIGPLAN Not. **42**(6), 89–100 (2007). https://doi.org/10.1145/1273442.1250746
8. Intel Parallel Studio XE. https://software.intel.com/en-us/parallel-studio-xe
9. Neytcheva, M., et al.: Multidimensional performance and scalability analysis for diverse applications based on system monitoring data. In: Wyrzykowski, R., Dongarra, J., Deelman, E., Karczewski, K. (eds.) PPAM 2017. LNCS, vol. 10777, pp. 417–431. Springer, Cham (2018). https://doi.org/10.1007/978-3-319-78024-5_37
10. Afanasyev, I.V., et al.: Developing efficient implementations of Bellman-Ford and forward-backward graph algorithms for NEC SX-ACE. Supercomput. Front. Innov. **5**(3), 65–69 (2018)
11. Nikitenko, D., et al.: JobDigest detailed system monitoring-based supercomputer application behavior analysis. In: Voevodin, V., Sobolev, S. (eds.) Russian Supercomputing Days, RuSCDays 2017, vol. 793, pp. 516–529. Springer, Cham (2017). https://doi.org/10.1007/978-3-319-71255-0_42
12. Shaykhislamov, D., Voevodin, V.: An approach for dynamic detection of inefficient supercomputer applications. Proc. Comput. Sci. **136**, 35–43 (2018)
13. Shvets, P., Voevodin, V., Zhumatiy, S.: Primary automatic analysis of the entire flow of supercomputer applications. In: Proceedings of the 4rd Ural Workshop on Parallel, Distributed, and Cloud Computing for Young Scientists, CEUR Workshop Proceedings, vol. 2281, pp. 20–32 (2018)
14. Vetter, J., Chambreau, C.: mpiP: Lightweight, scalable MPI profiling (2005)
15. Browne, S., Dongarra, J., Garner, N., Ho, G., Mucci, P.: A portable programming interface for performance evaluation on modern processors. Int. J. High Perform. Comput. Appl. **14**(3), 189–204 (2000)

16. Tuning Applications Using a Top-down Microarchitecture Analysis Method. https://software.intel.com/en-us/vtune-amplifier-help-tuning-applications-using-a-top-down-microarchitecture-analysis-method
17. Nikitenko, D., Voevodin, V., Zhumatiy, S.: Resolving frontier problems of mastering large-scale supercomputer complexes. In: Proceedings of the ACM International Conference on Computing Frontiers - CF 2016, pp. 349–352. ACM Press, New York (2016). https://doi.org/10.1145/2903150.2903481

Evolution of the Octoshell HPC Center Management System

Dmitry Nikitenko$^{(\boxtimes)}$ ⓘ, Sergey Zhumatiy ⓘ, Andrei Paokin ⓘ,
Vadim Voevodin ⓘ, and Vladimir Voevodin ⓘ

Research Computing Center of Lomonosov Moscow State University,
Moscow, Russia
{dan,serg,vadim,voevodin}@parallel.ru, andrejpaokin@yandex.ru

Abstract. Managing and administering an HPC center is a real challenge. The Supercomputing Center at Moscow State University is the largest in Russia, with petascale machines running in the interest of thousands of users engaged in hundreds of research projects. Since its early stages of development, the Octoshell system has been instrumental in mastering the equipment and tackling technical issues. The more problems are solved, the more new ideas arise and chances for improvement are discovered. In this paper, we discuss major enhancements recently made to Octoshell, including the integration of new modules (such as extended object description, license management, hardware accounting, and others), and outline plans for the near future.

Keywords: Supercomputer center · Resource management ·
HPC center workflow · Workflow automation · Resource utilization

1 Introduction and Background

Making a supercomputer center run efficiently is always a challenging task. The difficulties that arise depend on the scale of the center, its workflow and regulation peculiarities, and many other factors. As the number of research projects and/or users grows, system holders usually have to move from a stack of standalone tools, such as ticketing or accounting systems, to an integrated environment, something that provides possibility to fix diverse management issues. Some centers employ special toolkits provided by system vendors or purchase commercial tools, but any functionally rich solution is really costly, hence the high demand for open-source software.

Octoshell is an open-source software toolkit originally developed to tackle the whole spectrum of problems related to HPC center management [1,2]. It was designed in a way to support various workflows, but its basic functionality was planned according to the needs of the Supercomputing Center at Lomonosov Moscow State University [3,4]. This center is the largest in Russia and features two top-level systems, Lomonosov and Lomonosov-2 [5,6], both ranked in the

ⓒ Springer Nature Switzerland AG 2019
L. Sokolinsky and M. Zymbler (Eds.): PCT 2019, CCIS 1063, pp. 19–33, 2019.
https://doi.org/10.1007/978-3-030-28163-2_2

TOP50 [7,8] (rating of the most powerful supercomputers in the CIS) and the TOP500 [9] lists.

At present, all users of the MSU HPC center use Octoshell in their everyday work and study. In numbers, it is over 3000 users all over the world, representing hundreds of companies collaborating in several hundreds of research projects.

The first version of Octoshell focused mainly on accounting: managing workgroups, synchronizing user accounts between clusters, and helpdesk. One year later, some optimization changes were made to the engine and a major version of the system (v2) was released [2]. This time, some new features were developed, in particular annual report expertise, and multiple interface improvements were introduced.

The more experience we gain running the system, the more opportunities for development we find.

In this paper, we discuss recent advancements in Octoshell and plans for the near future.

Section 2 offers a general overview of recent challenges and appropriate methods to deal with them. In Sect. 3, we describe the architectural organization of the system. Section 4 concludes the paper and outlines upcoming goals.

2   Recent HPC Center Management Challenges and Their Solutions

In this section, we give more details on the most recently introduced modules and functionality that determine the development of Octoshell v2.0 to Octoshell v2.5. There have been many bug fixes and logic upgrades, but we would like to draw attention to the following major improvements which significantly extended Octoshell functionality:

- software license accounting;
- hardware accounting;
- resource utilization statistics;
- application efficiency analysis;
- computing cost estimation;
- extended object description;
- multilanguage support.

Below, we give a general idea on the functionality and implementation principles of these modules.

2.1   Software License Accounting

Accounting for installed software packages was implemented in the software management module. This module intends to organize all information about software packages deployed on supercomputers and manage the access to packages, including commercial ones. The most important functionality points are:

– accounting for all versions of software packages;
– metadata organization (install path, source-code path, installation instructions, support service contact information, etc.);
– access management for users (logins);
– detailed license information (type, cost, lifetime, etc.);
– role-dependent levels of access, ability to hide some software information from users;
– accounting feature for special software, not needed for regular users (system software).

Any software package may have one or more versions which can be installed on different supercomputers of the HPC center. A certain version of a package can be active or inactive, and can also be marked as "service". Service versions of packages are usually unavailable to regular users.

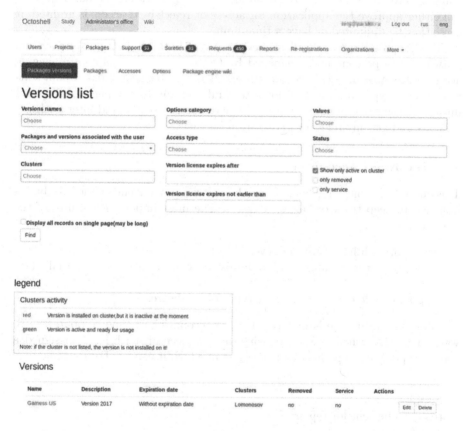

Fig. 1. A screenshot of the Octoshell software license module

Each version of the package has a name, a description, attributes as "service", "deleted", as well as a list of supercomputer systems that it is installed on, the license type, and the validity period of the license.

The type and validity period of a license can assist administrators in tracking whether a renewal of the package is necessary and in providing access to the package only for those users who are allowed under the terms of the license (for example, students, members of certain research projects).

Also, any attributes in the form of a "key-value" can be attached to a version of the package. The list of keys is common to all versions of all packages, so it can be used for extending information about all installed software. For example, one can attach to each version of a package a link to its documentation, contact information of the person who is responsible for the installation or that of an expert in the package, etc.

To gain access to commercial software, an Octoshell user only needs to file a request for access on the page of the desired package version. The application is automatically converted into a ticket in the support module of Octoshell, and the administrator receives the corresponding notification. Later, the administrator can either approve the application for access or reject it. If access is granted, it can either be unlimited or have a time limit.

The proposed approach allows accounting of both applications and system software on supercomputers managed by Octoshell. It also helps in managing access to them by different groups of users. It is possible to link data about software license type and any additional text information to any package. Figure 1 illustrates the corresponding interface, where one can search and filter packages versions and get any relevant information on them.

2.2 Hardware Accounting

This module is somewhat similar to the software management module but is designed to keep track of hardware items. The most important features of the module are:

– history of each hardware item event;
– automated state change and administrator notification on special state changes;
– support for life cycles of different types of hardware.

This special module is intended for the accounting of supercomputer hardware in the Octoshell system, in which every piece of hardware is assigned a special type. Each type has a set of states in which it can be. For example:

– works;
– disabled;
– ready to be sent for repair;
– sent for repair;
– repaired but not installed;
– disassembled;
– cannot be used anymore.

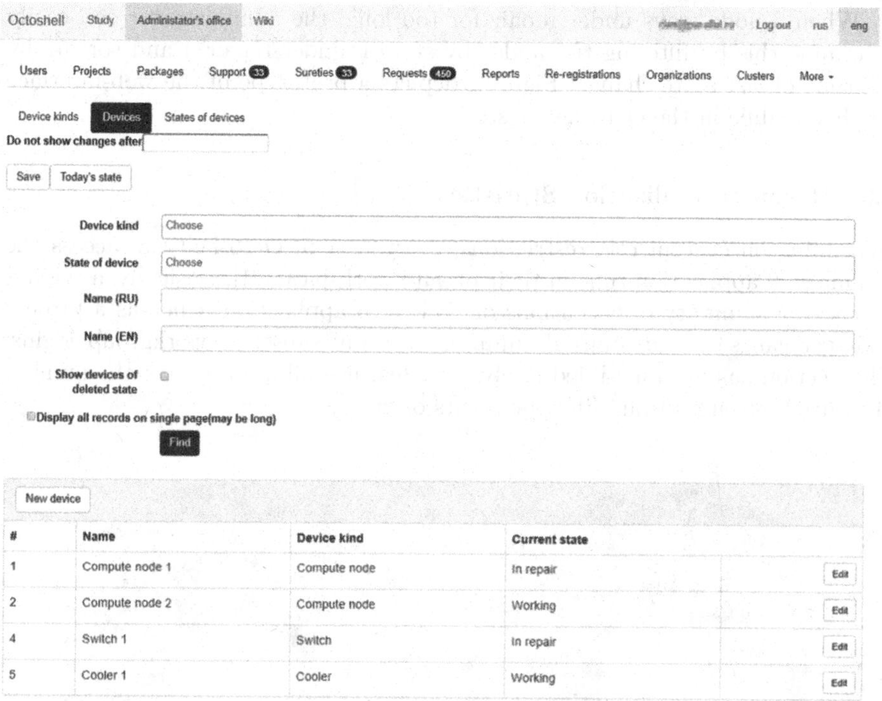

Fig. 2. A screenshot of the Octoshell hardware accounting module

The set of states can be different for different types of hardware. When a piece of hardware is changed from one state to another, the time of the change is saved along with an administrator's comment. The administrator can view the history of state changes for any specific hardware item or for all the items, over a certain time interval.

Every piece of hardware has a name, description, type, and state history. In this module, data can be imported from a file to simplify the creation of the initial hardware list. If necessary, the set of states of any type of hardware can be extended at any time but it cannot be reduced.

This module does not provide access for ordinary users; it is available only to administrators. Supercomputer centers use various monitoring systems, both well-known and proprietary (see, for instance, [10]). For example, for closer integration with the supercomputer, monitoring systems can change the state of the hardware through Octoshell using an external API. If a hardware memory error is detected in one of the computing nodes, then the node is transferred through the Octoshell API to "failed" state with the comment "memory bank 1 failed". After this, the administrator receives an e-mail notification and, if the node is under warranty, they can timely apply for hardware support from the manufacturer, providing the node's serial number which can be taken from the node description.

When a node stays under repair for too long, the administrator can easily determine this by filtering the nodes by state ("under repair") and sorting by the time of last state change. Figure 2 depicts a prototype of the web-interface for this module in the Octoshell system.

2.3 Resource Utilization Statistics

Depending on custom role restrictions, every user of Octoshell can access the summary of application runs in their private workspace [11]. Basically, a regular user sees a summary of executions of their own application, whereas a project manager can see a customized summary for any subset of workgroup logins. This section has been modified lately. New features allow observing the number of utilized resources from different points of view.

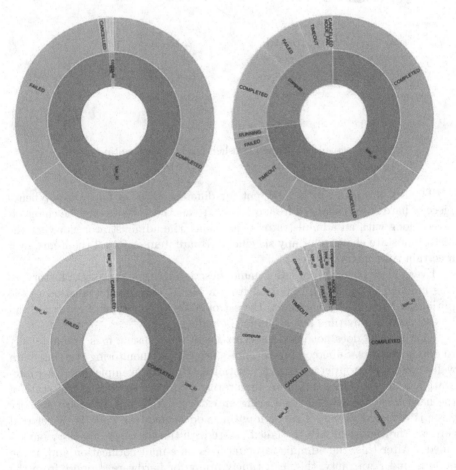

Fig. 3. Enhanced visualization of resource utilization in Octoshell

The basic hierarchical presentation from the level of systems to the level of job states, passing by the level of partitions, is now complemented with job-state-oriented statistics, which proves useful in estimating the number of resources spent in testing or debugging (or even wasted in failed-job states) and comparing it with the number of resources utilized in normally executed jobs.

In Fig. 3, we see various screenshots of an interactive visualization prototype complementary to the regular tablewise representations. The example provided in Fig. 3 corresponds to the data given below in Fig. 6.

Full-working resource utilization statistics are already available in developer mode to administrators. Regular users and research project managers will be able to benefit from it by the beginning of the next annual reregistration, which is scheduled for the end of January, 2019.

2.4   Application Efficiency Analysis

This module provides the user with information on the performance of their jobs. Currently, the module is only available in Russian. The main page offers a list of jobs (executed by the user) with the most general performance description (see Fig. 4). This includes integral values of the most important performance characteristics, such as CPU and GPU load, load average (number of active processes per node), IPC (instructions per cycle), and MPI usage intensity. The color indicates how large the value of each characteristic is (green is high, yellow is medium, red is low). By using this data, the user can roughly estimate the utilization of different types of computational resources or perform a general comparison of consecutive program launches. This is the renewed approach to the analysis of integral and dynamic job characteristics described in [15, 16].

The most insightful information on this page is given in the leftmost column. The icons show potential performance issues detected for each job by the efficiency analysis software developed at the Research Computing Center of Moscow State University [12–14]. A performance issue can be correlated to the efficiency of MPI usage, memory locality, and many other aspects of program behavior.

The process of detecting performance issues is organized as follows. The efficiency analysis software employs a set of 25 static rules. Each rule is aimed at detecting one particular performance issue which can potentially degrade the efficiency of the studied job. A rule is defined by a description of a specific issue (and also strict criteria for rule triggering), assumptions on possible root causes, as well as our recommendations on how these causes can be eliminated. Each job executed on the supercomputer is independently checked to ensure its compliance with all rules, and the list of applied rules is shown to the user. These 25 rules are currently divided into nine main groups according to the type of performance issue.

There are, for instance, groups of rules intended to detect significant workload imbalance, inefficient size of MPI packets, or underutilization of particular types of computational resources. It should be noted that, without knowing the internal structure of a job, it is almost impossible to detect clearly the existence or impact of any performance issue, so this efficiency analysis module can only make hypotheses about issues that can potentially occur in a job.

Detected issues	Job ID	Start time	End time	End status	# of nodes	Duration, hours	Job size (CPU*h)	CPU load	GPU load	Load average	IPC	Bytes received via MPI, MB/sec	Bytes sent via MPI, MB/sec
≡ ⌄⇄✕	882682	2018-11-25 12:11:29	2018-11-26 19:46:08	completed	3	31.0	1326.3	47.0	8.0	14.0	1.82	221.2	221.3
≡ ⇄⊘	882905	2018-11-26 12:20:46	2018-11-26 16:08:35	completed	4	3.8	212.6	2.8	999.6	1.0	1.52	0.0	0.0
≡ ⌄⇄✕	882948	2018-11-26 16:08:40	2018-11-26 19:49:38	completed	15	3.7	773.4	47.0	0.0	14.0	1.95	609.2	609.2

Fig. 4. The main page of the module, showing a job list with basic performance information

A detailed description of detected performance issues for each job can be found in the individual job page (this page can be accessed by clicking on the job ID in the second column to the left in Fig. 4). Figure 5 portrays the main part of this page, describing the last job from the list given in Fig. 4.

At the top of Fig. 5, we can see a description of three rules triggered for this job, which means that three issues were detected. The first rule states that MPI packets are too small in this job, and this, together with high intensity of use of the communication network, can lead to significant overheads, which is the root cause for this performance issue. In this case, it is recommended to perform an MPI usage profiling, with the assistance of performance analysis tools as mpiP or Scalasca, for instance. It should be noted that this module is planned to help users in conducting such kind of analysis but this functionality is currently under development.

The second rule detects low network locality. This issue occurs if a job actively uses the MPI network, but the nodes allocated to the job are located far from each other, which leads to high data transfer latency. In fact, this problem is not the user's fault and arises because of the non-optimal work of the resource manager, so there is very little the user can do about it (they can explicitly allocate nodes or try to optimize the MPI part). This information is, however, useful since it helps users to understand why their programs do not perform better.

The third rule indicates that the user has chosen an improper supercomputer partition for his job: it is executed on a partition for GPU-based applications but does not utilize GPUs at all.

The individual job page includes not only information about detected performance issues. It also provides a JobDigest report [17] which shows a detailed description of job performance characteristics changing throughout execution. A mini-JobDigest showing the most important characteristics (the same provided in the main page; see Fig. 4) can be seen at the bottom of Fig. 5. A full JobDigest can be accessed via the link "Full description". According to these data, we

Description	Supposition	Recommendation
The network usage intensity is quite high, but average size of MPI over Infiniband packets is inefficiently small.	The overhead of MPI messages transfer can be significant.	Use profiling tools for analysis of MPI programs.
Job actively uses MPI network, but network locality is low (allocated compute nodes are located far from each other).	A suboptimal set of compute nodes was selected by the resource manager for job execution. It is recommended to run the job on a better set of nodes (if necessary, explicitly specify the nodes to run the job on), or try to optimize the MPI part of the program.	Use profiling tools for analysis of MPI programs.
Job is running in the partition for GPU-based programs but almost do not use graphical processors.	The partition for this job is selected incorrectly.	It is recommended to change the partition for this job.

Task behavior dynamics
Full description

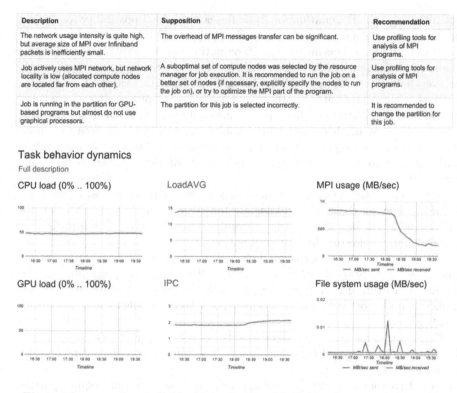

Fig. 5. The page of an individual job, showing detailed performance information

can see, for example, that GPU activity is zero all the time, while MPI usage intensity significantly decreases in the second half of the job execution (although it still remains high, at the level of 200–300 MB/s). Such information helps to gain deeper insights into the program behavior: it enables the user to evaluate and compare the performance of different stages of their program, which is impossible with integral values.

2.5   Computing Cost Estimation

One of the most important issues, albeit usually put aside when working at shared research facilities, such as the one at Lomonosov Moscow State University, is computing cost. The reason is quite simple. Most users get access for free and do not worry about efficiency, the number of utilized resources, or running costs.

The bad thing is that somebody else pays for that, the system holder in most cases. In some occasions, when facilities are built through financial support from the federal budget, strict restrictions are even established by the law, requiring that the system holder compensate the expenses. The good thing is that some foundations declared special computing-related expenses in cost sheets of supported projects. This means that those users who do big science funded by some institutions (as the Foundation for Basic Research) have a possibility to pay just for the computing resources they actually need.

At the same time, this means that if computing resources can be provided on a commercial basis, inefficient computing becomes more than ever naturally costly for everybody, both the system holder and the user. In Octoshell, each HPC system has some attributes related to availability of resources. It was a natural idea to extend the attribute list with a computing-cost item which would show, for example, how much it costs to the holder to provide a core hour or a node hour of a given system type. Of course, these figures must take into consideration hardware amortization, energy costs, rental costs, salaries, software licenses, and even insurance.

There are many known methodologies to estimate such costs (see [18]). Following one of those methodologies, the average node-hour cost for the Lomonosov-2 can be estimated as 21 rubles approximately. This rate is used in the example below as the cost assumption.

There is a section in Octoshell dedicated to statistics of application runs in every user's private space, where one can see all job states, appropriate numbers of runs, and utilized resources. Regular users can see the details on their own application runs; project managers can see the activity details for each workgroup member.

Extending these statistical data with the cost of executed jobs can perfectly illustrate the cost of bad decisions and weak debugging, as shown in the table given in Fig. 6.

Note that these are just preliminary data, and each position requires further careful analysis. In particular, a significant part of the applications within the CAN-CELED and TIMEOUT categories are actually normally completed tasks, which reflects the specific features of the work of the corresponding research groups.

As time goes by, this might turn into a billing system, but we believe that, at present, it is much more important to inform the users and induce them to reduce computing costs.

2.6   Extended Object Description

Many entities are created as a result of managing a supercomputer center. In many situations, the description of entities of various kinds are successively extended by various users. The Comments engine developed for Octoshell allows users with granted access to extend the description of entities in a convenient manner. Three types of extra information are supported:

- **Tags**, used for linking objects of different classes. For example, the project, the manager of the project, and their reports can be marked with the "top papers" tag. This information can help the project to be allocated more quotas of resources for its execution.
- **Comments**, used for addition of plain text. For example, references to scientific papers, opinions about projects and users.
- **Files**, used for adding scientific papers, plots, and so on.

Partition	State	Jobs	Corehours	Nodehours	Cost (RUB)
low_io	TOTAL	61,327	2,081,460.06	74,337.86	1,561,095.05
	COMPLETED	40,581	968,530.01	34,590.36	726,397.51
	FAILED	19,985	62,352.50	2,226.88	46,764.38
	CANCELLED	551	711,285.72	25,403.06	533,464.29
	TIMEOUT	194	297,321.65	10,618.63	222,991.24
	RUNNING	10	39,900.00	1425.00	29,925.00
	NODE_FAIL	6	2,097.19	74.90	1572.8925
compute	TOTAL	361	759,124.73	27,111.60	569,343.55
	COMPLETED	182	405,695.01	14,489.11	304,271.26
	CANCELLED	111	195,049.57	6,966.06	146,287.18
	FAILED	51	18,890.46	674.66	14,167.85
	TIMEOUT	15	117,376.22	4,192.01	88,032.17
	RUNNING	1	998.36	35.66	748.77
	NODE_FAIL	1	21,115.11	754.11	15,836.33
test	TOTAL	438	970.35	34.66	727.76
	COMPLETED	169	218.86	7.82	164.15
	FAILED	150	74.32	2.65	55.74
	CANCELLED	88	186.02	6.64	139.52
	TIMEOUT	31	491.14	17.54	368.36

Fig. 6. Resource-utilization costs for a period

Each extended information unit bears the common name of *note*. Users must have privileges based on their groups to create, read, and modify notes.

If a note has a context, then the access to the note is based on the links between the context and user groups (*context rules*); otherwise, links are established between the entity (or its class) that the note is attached to and user groups (*object rules*).

An object rule is linked with a class of objects or with an object and, according to its character, can be either a permission or a prohibition. An object rule may belong or not to a user group; the negative case corresponds to the default object rule definition which is required for all objects that notes can be attached to. Thus, default rules can be assigned for extra information management and then redefined for some groups of users.

Context rules can only be permissions. Users should be granted the special object rule *creating with context* to be allowed to create notes for a given object. Choosing a place in the web interface and editing permissions for notes are the only things required. Each user can work with all notes they have access to in a special separate web interface.

The Comments engine significantly eases the supercomputer center management, especially during large-scale processes as reregistration, making it possible for Octoshell users to process huge amounts of data in short periods of time.

2.7 Multilanguage Support

At present, the number of users of the Octoshell system deployed in the HPC center at Lomonosov Moscow State University exceeds 3000. Many of them collaborate with workgroups from all over the world. This means that it is a common and natural requirement to provide access to Octoshell for English-speaking users.

When the first version of Octoshell was implemented, the developer thought of that, but there was very little time to run, so the support of other languages stayed in an "optional feature" state. At that stage, there was just a set of locales predefined, but only the Russian one was configured and, what is more, many messages, titles, and interface blocks were hard-coded. Unfortunately, just after three years of running, the developers succeeded to find the resources to clean the code and fill in the English locale.

In general, the modification was nothing special, but, at the same time, it was possible to improve the system's usability by enriching many of the messages with hyperlinks to the objects mentioned in them.

For example, if a project member is notified that their project has been blocked, then the received message is partially highlighted, thereby providing links to the project description, details of the reason for the blocking (e.g., delayed reregistration or problems with surety documents).

All in all, there were several hundreds of messages translated and enriched. It is essential that now English-speaking users do not experience troubles with registration and basic operations within the Octoshell system and, therefore, have a more positive experience accessing the resources of the HPC center and collaborating.

3   Advanced Octoshell Architecture

After all the above, we can explain the overall Octoshell architecture by the scheme given in Fig. 7.

CORE and BASE APPLICATION together with the AUTHENTICATION and FACE modules can be used as a minimum configuration featuring basic functionality. Other modules can be used if needed. The general concept of the Octoshell workflow, the development motivation, as well as an overview of related works are considered in [1,2].

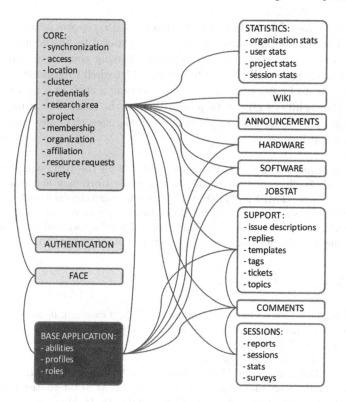

Fig. 7. Octoshell core and modules interaction

4   Conclusions and Future Work

The years of experience using Octoshell for managing the largest HPC center in Russia (at Lomonosov Moscow State University) have proved that it was a beneficial decision to develop this system. We have met new challenges, one after another, by modifying existing modules or extending the functionality.

Recent advancements include management of software packages and licensing, hardware support, extended role-sensitive description of any objects in the system, support for resource utilization accounting, means for the analysis of application run efficiency, and others.

With multilanguage support, all these features have become available to a wide range of users, including foreign experts and researchers.

When considering future plans, we focus on two major improvements (besides further enhancements to existing modules):

- the implementation of virtualization techniques that can allow for more flexible development of new modules;
- the extension of statistical sections by introducing interactive user-friendly tools.

There are some ideas to incorporate Octotron [19] reports into the Octoshell workflow. The team of Octoshell developers is grateful to all the users for their valuable feedback on the system. We hope that further comments, remarks, and suggestions from users will help us to improve Octoshell even more.

Octoshell is no longer just a tool: it has become a shell or a framework that integrates the results of numerous jointly but, at the same time, independently developed projects aimed at responding to challenges that any contemporary HPC center can face.

Acknowledgements. The research was carried out on the HPC equipment of the shared research facilities at Lomonosov Moscow State University and was supported through the project RFMEFI62117X0011. The results were obtained with the financial support of the Russian Foundation for Basic Research (grants No. 18-29-03230 and No. 17-07-00719). The results described in Subsect. 2.5 were obtained with the financial support of the Russian Science Foundation (agreement No. 17-71-20114).

References

1. Nikitenko, D.A., Voevodin, Vl.V., Zhumatiy, S.A.: Octoshell: large supercomputer complex administration system. In: Russian Supercomputing Days International Conference. CEUR Workshop Proceedings, Moscow, Russia, 28–29 September 2015, vol. 1482, pp. 69–83 (2015)
2. Nikitenko, D.A., Voevodin, Vl.V., Zhumatiy, S.A.: Resolving frontier problems of mastering large-scale supercomputer complexes. In: ACM International Conference on Computing Frontiers, CF 2016, Como, Italy, 16–18 May 2016, pp. 349–352. ACM, New York (2016). https://doi.org/10.1145/2903150.2903481
3. High Performance Computing in Moscow State University High Performance Computing in Moscow State University. http://hpc.msu.ru
4. Shared Resources MSU HPC center. http://www.parallel.ru/cluster
5. Voevodin, V.V., et al.: Practice of "Lomonosov" supercomputer. Open Syst. J. **7**, 36–39 (2012)
6. Sadovnichy, V., Tikhonravov, A., Voevodin, Vl., Opanasenko, V.: "Lomonosov": supercomputing at Moscow State University. In: Contemporary High Performance Computing: From Petascale toward Exascale (Chapman & Hall/CRC Computational Science), pp. 283–307. CRC Press, Boca Raton (2013)
7. Top50 most productive supercomputers in Russia and CIS. http://top50. supercomputers.ru
8. Nikitenko, D., Zheltkov, A.: The Top50 list vivification in the evolution of HPC rankings. In: Sokolinsky, L., Zymbler, M. (eds.) PCT 2017. CCIS, vol. 753, pp. 14–26. Springer, Cham (2017). https://doi.org/10.1007/978-3-319-67035-5_2
9. Top500 Supercomputer Sites. http://top500.org
10. Safonov, A., Kostenetskiy, P., Borodulin, K., Melekhin, F.: A monitoring system for supercomputers of SUSU. In: Russian Supercomputing Days International Conference. CEUR Workshop Proceedings, Moscow, Russian Federation, 28–29 September 2015, vol. 1482, pp. 662–666 (2015)
11. Nikitenko, D., Shvets, P., Voevodin, V., Zhumatiy, S.: Role-dependent resource utilization analysis for large HPC centers. Commun. Comput. Inf. Sci. **910**, 47–61 (2018). https://doi.org/10.1007/978-3-319-99673-8_4

12. Voevodin, Vl.V., Voevodin, Vad.V., Shaikhislamov, D.I., Nikitenko, D.A.: Data mining method for anomaly detection in the supercomputer task flow: numerical Computations: Theory and Algorithms. In: The 2nd International Conference and Summer School. AIP Conference Proceedings, Pizzo calabro, Italy, 20–24 June 2016, vol. 1776, pp. 090015-1–090015-4 (2016). https://doi.org/10.1063/1.4965379

13. Shaykhislamov, D., Voevodin, V.: An approach for dynamic detection of inefficient supercomputer applications. Proc. Comput. Sci. **136**, 35–43 (2018). https://doi.org/10.1016/j.procs.2018.08.235. ISSN 1877–0509

14. Shvets, P., Voevodin, V., Zhumatiy, S.: Primary automatic analysis of the entire flow of supercomputer applications. In: Proceedings of the 4th Ural Workshop on Parallel, Distributed, and Cloud Computing for Young Scientists. CEUR Workshop Proceedings, Yekaterinburg, Russia, vol. 2281, pp. 20–32 (2018)

15. Nikitenko, D.A., et al.: Supercomputer application integral characteristics analysis for the whole queued job collection of large-scale HPC systems. In: 10th Annual International Scientific Conference on Parallel Computing Technologies, PCT 2016. CEUR Workshop Proceedings, Arkhangelsk, Russian Federation, 29–31 March 2016, vol. 1576, pp. 20–30 (2016)

16. Nikitenko, D., Stefanov, K., Zhumatiy, S., Voevodin, V., Teplov, A., Shvets, P.: System monitoring-based holistic resource utilization analysis for every user of a large HPC center. In: Carretero, J., et al. (eds.) ICA3PP 2016. LNCS, vol. 10049, pp. 305–318. Springer, Cham (2016). https://doi.org/10.1007/978-3-319-49956-7_24

17. Nikitenko, D., et al.: JobDigest - detailed system monitoring-based supercomputer application behavior analysis. In: Voevodin, V., Sobolev, S. (eds.) Supercomputing. RuSCDays 2017. CCIS, vol. 793, pp. 516–529. Springer, Cham (2017). https://doi.org/10.1007/978-3-319-71255-0_42

18. Belkina, Yu., Nikitenko, D.: Computing cost and accounting challenges for Octoshell management system. In: Proceedings of the 4th Ural Workshop on Parallel, Distributed, and Cloud Computing for Young Scientists. CEUR Workshop Proceedings, Yekaterinburg, Russia, vol. 2281, pp. 146–158 (2018)

19. Antonov, A., et al.: An approach for ensuring reliable functioning of a supercomputer based on a formal model. In: Wyrzykowski, R., Deelman, E., Dongarra, J., Karczewski, K., Kitowski, J., Wiatr, K. (eds.) PPAM 2015. LNCS, vol. 9573, pp. 12–22. Springer, Cham (2016). https://doi.org/10.1007/978-3-319-32149-3_2

Numerical Algorithms for HPC Systems and Fault Tolerance

Boris N. Chetverushkin[1], Mikhail V. Yakobovskiy[1], Marina A. Kornilina[1(✉)], and Alena V. Semenova[2]

[1] Keldysh Institute of Applied Mathematics, Russian Academy of Sciences, Moscow, Russia
office@keldysh.ru, {lira,mary}@imamod.ru
[2] Moscow Institute of Physics and Technology, Dolgoprudny, Russia
alena.semenova@phystech.edu

Abstract. We discuss in the paper the influence exerted by applied mathematics techniques on the progress of the architecture and performance of computing systems. We focus on computational algorithms aimed at solving the problem of fault tolerance which is topical for future exaFLOPS and super-exaFLOPS systems.

Keywords: Supercomputer · Fault tolerance · Hyperbolic systems of equations · Control point

1 Introduction

It is evident that the architecture of current and forthcoming computer systems greatly influences the progress of algorithms for solving applied problems. An example of this is the trend of development of parallel computer systems based on distributed memory architecture, which started in the 1980s and ultimately led to a significant change in how parallel algorithms are viewed. Minimization of interprocessor communication became one of the key requirements placed upon algorithms.

Further progress in computer technology was closely tied to the emergence of multi-core processors and the use of graphics processing units (GPUs) as accelerators, which in turn led to tougher requirements imposed upon algorithms. Since then, both logical simplicity and efficiency of algorithms came to the fore. However, the creation of algorithms with a good combination of both properties is a rather tricky task.

A well-known and successful example of such technique is the one that is based on explicit schemes for parabolic equations. Explicit schemes are logically simple and very easy to implement on various multiprocessor architectures. However, the strong stability condition,

$$\Delta t \le h^2, \tag{1}$$

This work was supported by the Russian Foundation for Basic Research (project No. 17-07-01604-a).

hinders the practical use of explicit schemes on detailed spatial grids since the time step Δt must be unreasonably small (here, Δt is the permissible time step, while h is the characteristic step of the spatial grid).

We will consider in this paper the opposite scenario, namely the influence of computational algorithms on the architecture of multiprocessor systems.

2   Hybrid Computer Systems

After their appearance in the late 1990s, GPUs rapidly developed but their use was mainly limited to the industry of computer games. High-performance computing at a relatively low price and low energy consumption were their main advantages. However, considering the challenging problems that supercomputers faced in the 2000s, these advantages were canceled out by the fact that graphics cards were not efficient in branching algorithms. It should be noted that the logical simplicity of algorithms was one of the factors that contributed to the success of multiprocessor computations regardless of the type of computer architecture. The situation required much effort to develop logically simple and, at the same time, efficient algorithms. Such algorithms provided the background that stimulated the use of GPUs for non-graphical computations in the solution of important practical problems, including scientific applications.

K-100 was the first massive computer system in Russia that deployed GPUs as accelerators. The K-100 was built at the Keldysh Institute of Applied Mathematics in 2010 and featured a peak performance of 107 teraFLOPS. Each node of the K-100 is equipped with two six-core processors Intel Xeon X5670 and three graphics accelerators nVidia Fermi C2050 (each having 448 CUDA cores). The processors are connected through MVS-express and Infiniband [1,2] high-speed networks which control communication and data transfer.

Algorithms that are efficient when running on the K-100 [1,3] have two main components: a small number of operations governing the general structure and the logic of calculations (executed on uniform CPU cores of general-purpose processors) and a set of uniform arithmetic operations executed on GPUs over a large set of uniform data.

At the Keldysh Institute, the K-100 is used for a wide spectrum of tasks, such as simulations of gas dynamics, multidimensional radiative transfer, and molecular dynamics. Scientists in Russia and abroad are currently trying to expand the scope of problems that can be solved with the aid of such heterogeneous systems.

The possibility of using such accelerators to solve practically important problems has been a strong incentive for new improvements in GPU technology. GPUs show a rapidly increasing performance while featuring a relatively low energy consumption.

The development of graphics cards and, especially, the expansion of their scope of use, owing to newly created algorithms, have succeeded in creating the widespread opinion that future computing systems capable of delivering a performance of several dozens of petaFLOPS and even more will be based on a heterogeneous architecture.

It should be noted that the successful development of a new marketable component base is impossible without taking into account prior experience in mathematical modeling and recent progress in computational algorithms. The most successful strategy consists in the development of a new processor simultaneously with models of rather complicated tasks on its functional prototype.

In the next section, we will explain how computational algorithms can substantially help in solving the problem of fault tolerance which is essential for exascale-class systems.

3 Achieving Fault Tolerance Through Hyperbolization of Continuum Mechanics Equations

Exascale computers (or, in other words, computers capable of performing at least 10^{18} floating point operations per second, i.e. 1 ExaFLOPS) are expected to appear in the near future, most likely in the mid-2020s. They will consist of billions of independent processor cores. Such prediction is based on current technological trends in computer components without taking into consideration possible scientific and engineering breakthroughs that may radically change the situation. Naturally, the probability of processor failure for an exascale system is very high. Experts estimate that failure of at least one processor will occur every 20–25 min in one-exaFLOPS systems. For performances exceeding 1 ExaFLOPS, the time of failure-free operation of the entire system will shorten even further [4–7].

To recover from a failure, the following routine might be applied. For a long-running evolutionary (nonstationary) application, the intermediate state of the program is regularly saved at specific time intervals δ. This technique is called checkpointing. In case of a processor failure, the program execution is suspended, the failed processor is replaced with a reserve processor, the program rolls back to its latest checkpoint and then resumes execution.

Given today's rates of failures, 1 ExaFLOPS seems to be the critical barrier of computer performance. Indeed, the typical time for data storing and reading in practical applications on exascale supercomputers can be estimated as several dozens of seconds and more. Thus, the supercomputer will actually run for nothing, over and over rolling back to the same checkpoint and replacing the failed processors.

Note that the replacement process also requires some time which is logarithmically dependent on the number of computing processors but independent of the number of modeling time steps since the last checkpoint. Therefore, the time it takes to replace a processor does not have a significant effect on the efficiency of computations and is not a crucial factor for supercomputing.

Surprisingly, the way out of this deadlock can be found thanks to computational algorithms. The appropriate method is based on the technique of hyperbolization of continuum-mechanics equations suggested in [8,9].

Let us write down the continuum-mechanics equations in the form (see [10])

$$\frac{\partial Q_i}{\partial t} = \operatorname{div} \overrightarrow{S_i}, \tag{2}$$

where Q_i is a vector of main variables (density, momentum, and total energy) and $\vec{S_i}$ is the flux vector of the i-th component, which causes Q_i to change. Note that (2) is a parabolic system in the case of the Navier–Stokes equations.

By using connections between kinetic and gas-dynamics approaches to the description of continuous media, we can write the following alternative system instead of (2):

$$\frac{\partial Q_i}{\partial t} + \frac{\tau}{2}\frac{\partial^2 Q_i}{\partial t^2} = \operatorname{div} S_i^* \tag{3}$$

In this case, the gas-dynamics continuity equation for the system given in (3) takes the form

$$\frac{\partial \rho}{\partial t} + \frac{\tau}{2}\frac{\partial^2 \rho}{\partial t^2} = \frac{\partial}{\partial x_k}\left[-\rho U_k + \frac{\tau}{2}\frac{\partial}{\partial x_p}\left(\rho U_k U_p + p\delta_{kp}\right)\right], \tag{4}$$

where ρ is density, U is velocity, p is gas-dynamics pressure, and τ is the characteristic time between collisions of gas molecules.

It has been proven (see [8, 11, 12]) that the hyperbolized system given by (3) for the description of viscous heat-conducting gas flows differs from the corresponding Navier–Stokes equations by terms of the second order in the Knudsen number $O(\mathrm{Kn}^2)$. The system is called the quasi-gasdynamic (QGD) system of equations [8].

The perfect agreement between the results of numerical simulations based on the Navier–Stokes equations and those based on the QGD system is confirmed by numerous computational experiments. Furthermore, the QGD approach was extended to the simulation of magneto-gasdynamic phenomena in astrophysics, which led to fundamentally new results obtained by using more than 10^9 spatial nodes in a 3D mesh [13, 14].

QGD system (3) and (4) is hyperbolic [15]. This fact enables us to construct logically simple and efficient algorithms well adapted for implementation on extra massively parallel computer systems.

Below, we present a concept of fault-tolerant technique with accelerated recomputations in the failed-processor domain. With the aid of this technique, processor failures will not slow down the execution of the entire program. We will demonstrate how the proposed technique can be implemented. In this regard, we will consider the simulation of the following one-dimensional wave equation (5) with a source $F(x,t)$:

$$\frac{\partial^2 u}{\partial x^2} - \frac{1}{C}\frac{\partial^2 u}{\partial t^2} = F(x,t). \tag{5}$$

The one-dimensional wave equation is hyperbolic. From the theory of partial differential equations (PDEs), we know that the properties of hyperbolic PDEs sharply differ from those of parabolic PDEs. Indeed, parabolic PDEs have infinite domains of influence and dependence, which implies that all points of space feel the disturbance at once; in the case of hyperbolic PDEs, these domains are finite. The domain of dependence for the solution of Eq. (5) is bounded by two characteristics, $x = x_0 - Ct$ and $x = x_1 + Ct$, as shown in Fig. 1. The numerical solution of PDE (5) at a time t_0 in the computational domain n_0 of a failed processor CPU2 depends on all the values in its domain of dependence n_1 at a previous time $t_0 - \delta$.

In contrast to the solution of a parabolic equation, the solution of a hyperbolic equation at a time t depends on the values of the function at an earlier time $t - \delta$, not on the whole domain of the function but only on the values of the function in a small spatial domain. This domain of influence for the solution of Eq. (5) is determined by two characteristics, $x - Ct$ and $x + Ct$. The simple process of replacing a failed computing processor in this case consists in the following: if a processor fails at some model time step t_0, then the code execution stops in the entire domain, the processor is replaced with a working one, and after a rollback and recovery from the checkpointed data related to time $t_0 - \delta$, the code is resumed [16].

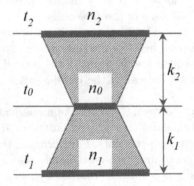

Fig. 1. The domain of fast computations

In accordance with the principle of "geometric parallelism"[1] and the method of characteristics, it is easy to determine the domain of dependence n_1 which at time $t_0 - \delta$ exerts influence on the solution at time t_0 within the computational domain n_0 of the failed processor (see Fig. 1). The domain n_1 is larger than domain n_0 (the computational domain of the failed processor). Therefore, to execute the computations in the domain n_1 at time $t_0 - \delta$, we should increase the number of processors. However, the number of additional processors that is required to perform computations in the domain n_1 from time $t_0 - \delta$ to time t_0 and then to time $t_0 + \delta$, is much less than the total number of processors involved in the computations by the principle of geometric parallelism [16–18].

---

[1] Principle of "geometric parallelism", also referred to as domain decomposition, is a parallelization method widely used in solving problems of mathematical physics. As specified by this technique, the computational domain is divided into subdomains, then each processor executes the same code on its own individual subdomain and exchanges boundary data with neighboring processors.

Let us write the lower estimate of the number of additional processors p_d, without taking into account the costs for processor replacing and data redistribution among processors:

$$p_d > \frac{1}{d+1} \frac{1}{k_2} \sum_{j=1}^{2} k_j \sum_{i=0}^{d} \alpha_j^i, \tag{6}$$

where d $= 1$, 2, 3 is the spatial dimension of the problem, $\alpha_j = 1 + 2\gamma \frac{k_j}{n_0}$, $\gamma = \frac{C\Delta t}{h}$ is the Courant number, k_1 is the number of model time steps since the last checkpointing, k_2 is the number of model time steps to the end of recalculation, and n_0^d is the number of mesh nodes per processor at the initial stage of computations, before the first failure occurs.

We can describe the proposed strategy as follows. Throughout the computations, each working processor periodically saves the local checkpoint, namely it copies the data for restoration of computations to another working processor. Figure 2 shows an example of computations on three processors (CPU1, CPU2, CPU3). The processors CPU4 and CPU5 are reserved and, initially, do not perform computations while the other processors are running the program. CPU1 periodically saves the local checkpoint (CH1) both in its own RAM and in the RAM of CPU2. The other working processors do the same operation. If CPU2 fails, it can be replaced by reserved processors which are idle. The number of required processors is given by (6). In Fig. 2, the processor CPU2 is replaced by two processors. The added processors, CPU4 and CPU5, read the latest checkpoint CH2 from the RAM of CP3 and share the relevant workload evenly. Then, the added processors recover the data that was lost due to the failed processor. While CPU4 and CPU5 recompute the lost data, all remaining processors keep on computing. Therefore, the added processors should proceed with the calculations until they catch up with the others at time step $t_0 + \delta$. The total overhead Δ consists of the time required for replacing the failed processor with reserve ones, reconstructing the virtual topology of the computational network, reading the data from the checkpoint, and partitioning the workload among the added processors. Evidently, the overhead has no correlation with the amount of lost computation results for the time steps from $t_0 - \delta$ to t_0.

After the reconciliation procedure for adjusting local time steps (as far as each processor obtained computation results for time $t_0 + \delta$), the processor CPU5 passes its data to processor CPU4, which continues the computation instead of the failed processor CPU2. Then the processor CPU5 returns to the reserve pool of processors and does not take part in computations until the next failure.

It is important that, during the procedure of recovery of local data lost by the failed processor CPU2, resumption of computations using extra processors, and adjustment of the local time steps on working processors (CPU1 and CPU3) and extra processors (CPU4 and CPU5), the program continues running on the other surviving processors (they are not shown in the scheme but there can be any number of them). These processors have meanwhile been computing further on from modeling time step t_0 towards time step $t_0 + \delta$. If we increase the number of

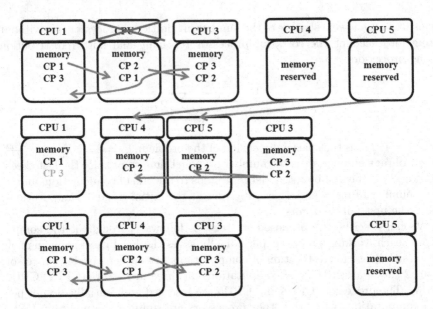

Fig. 2. General resiliency strategy

processors in the domain of dependence n_1 of the processor CPU2 in accordance
with (6) and repartition the computational load, then the recovery time will
be shortened by approximately a factor of the number of additional processors.
This succeeds for scalable parallel applications provided that the particular job
of any processor is computational-intensive enough to be repartitioned. Thus,
increasing the number of processors will reduce the number of computational
points in each subdomain, which in turn will accelerate the computations in
these subdomains and hence in the whole domain of the failed processor (Fig. 2).
This makes it possible to overcome the lag and catch up by time $t_0 + \delta$ with
the other processors which have not stopped the computations. Note that the
additional processors can be picked from the reserve pool provided at the start
of the application.

Thus, since most processors continue computing in regular mode, the proposed method keeps the running time almost unchanged regardless of the failure.
Of course, the total running time of fault-tolerant code is somewhat longer compared with that of a code that does not support fault tolerance. Nevertheless, the
corresponding overhead is regular and does not depend on the number of failed
processors. Furthermore, the suggested approach ensures the execution of computations even in case of multiple failures either simultaneous or overlapping.
A necessary condition for continuous work is a sufficient distance (in terms of
virtual computational network) between faulty processors. We should note that,
even though the probability of regular failure for massively parallel systems is
close to 1, the probability of simultaneous failures of two nearby processors is
rather low, since it is much less than the probability of failure of one processor,
which is relatively small in teraFLOPS systems today.

Let us make two important remarks concerning the suggested approach to the problem of fault tolerance.

Note 1. This approach, which is based on the features of computational algorithms, does by no means obviate the need for quality improvements in processors and other components of modern computing systems. The solution of the fault-tolerance problem can be achieved through a combination of both approaches.

Note 2. The proposed approach to the fault-tolerance problem in case of hyperbolic systems is somewhat evident. What is less obvious is the process of hyperbolization, including the determination of fluxes S_i^* in (3). It is necessary to search for other ways to solve the problem of fault tolerance by means of computational algorithms. This would make it possible to extend the class of applications that can be computed on future exascale systems.

4 Numerical Results

To simplify the computations and enable the use of explicit difference schemes, we formally set $\gamma = 1$. This assumption affects only the number of cells of the computational grid that are involved when moving from one time step to the next one at the stage of recalculation and does not impose any additional conditions upon the step Δt. Thus, the size of the calculation domain decreases at each subsequent time step (in the time interval $t_1 \leq t \leq t_0$) and increases (in the time interval $t_0 \leq t \leq t_2$) by exactly one cell for each edge of the computation domain. Accordingly, in the one-dimensional case, the estimate (6) for $\gamma = 1$ takes the form

$$p_1 > \frac{n_0(k_1 + k_2) + k_1^2 + k_2^2}{n_0 k_2}. \tag{7}$$

We developed a program to test the proposed algorithm. The program analyzes the dependence between changes in processing time and the number of reserve processors. Two cases were considered. In the first case, it was assumed that a single failure of one processor occurred in the course of the computation of (5), then computations were restored and continued for the specified model time. In the second case, all the processors were faultless. The dependence of the corresponding processing time versus the number of reserve processors is shown in Fig. 3. The computations were carried out with $n_0 = 10^5$, $k_1 = 9{\cdot}10^4$, $k_2 = 4{\cdot}10^4$, $p_{\text{work}} = 10$ (total number of working processors), and $K = 2 \cdot 10^5$ (number of time steps). The results showed that, if the number of reserve processors (7) is enough, the processing time depends insignificantly on processor failures. There is, however, some unavoidable time loss due to the cost of recalculation control and checkpoint readout.

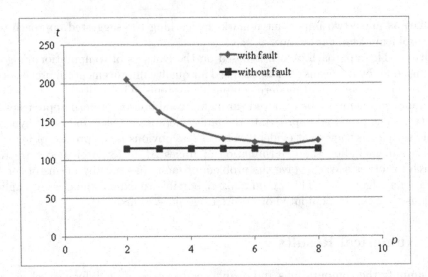

Fig. 3. Processing time t versus number of reserve processors p with a single failure (blue line) and without failure (red line) (Color figure online)

5   Conclusions

Since the 1950s (the time of the first computers), progress in computational mathematics has been inspired mostly by achievements in computer facilities. It would suffice to remember in this connection multiple algorithms that were proposed within a few years in the 1940s and 1950s. Indeed, most of them were motivated by developments in computer technology independently achieved in the USSR and the USA during that period. However, the reverse influence of computational algorithms on the development of computer technology in those days was negligibly small.

In this paper, we made an attempt to look at the problem of the mutual influence of computational algorithms and computer technology from a different angle. We expect that the emergence and evolution of novel computational tools (including their technical implementation and the introduction of new physical ideas underlying them) and also the development of new approaches and concepts in applied mathematics will determine the future progress in computing systems. It is quite probable that these fundamentally new areas of numerical mathematics will make extensive use of basic knowledge from related fields of physics and mechanics.

References

1. Davydov, A.A., Lacis, A.O., Lutsky, A.E., Smoliyanov, Yu.P., Chetverushkin, B.N., Shilnikov, E.V.: The hybrid supercomputer MVS-express. Doklady Math. **82**(2), 816–819 (2017). https://doi.org/10.1134/S1064562410050364
2. Gibridnyi vychislitel'nyi klaster k-100 [hybrid computer cluster k-100]. www.kiam.ru/MVS/resourses/k100.html
3. Gorobets, A.V., Sukov, S.A., Zhelezniakov, A.O., Bogdanov, P.B., Chetverushkin, B.N.: Rasshirenie dvukhurovnevogo rasparallelivaniia MPI+openmp posredstvom opencl dlia gazodinamicheskikh raschetov na geterogennykh sistemakh [extension of two-level parallelization MPI+openmp through opencl for hydrodynamic computations on heterogeneous systems]. Vestnik Iuzhno-Ural'skogo gosudarstvennogo universiteta **9**, 76–86 (2011)
4. Bland, W.: Int. J. High Perform. Comput. Appl. **27**(3), 244–254 (2013)
5. Cappello, F.: Int. J. High Perform. Comput. Appl. **23**(3), 212–226 (2009)
6. Cappello, F., Geist, A., Gropp, W., Kale, S., Kramer, B., Snir, M.: Int. J. High Perform. Comput. Appl. **1**(1), 1–28 (2014)
7. Snir, M., et al.: Int. J. High Perform. Comput. Appl. **28**(2), 129–173 (2014)
8. Chetverushkin, B.N.: Kineticheskie skhemy i kvazigazodinamicheskaia sistema uravnenii [Kinetic schemes and quasi-gasdynamic system of equations]. Max Press, Moscow (2004)
9. Chetverushkin, B.N.: Resolution limits of continuous media mode and their mathematical formulations. Math. Models Comput. Simul. **5**, 266–279 (2013). https://doi.org/10.1134/S2070048213030034
10. Sedov, L.I.: Metody podobiia i razmernosti v mekhanike. [Similarity and dimensionality methods in mechanics]. Nauka, Moscow (1977)
11. Zlotnik, A.A., Chetverushkin, B.N.: Entropy balance for theone-dimensional hyperbolic quasi-gasdynamic system of equations. Doklady Math. **95**(3), 276–281 (2017). https://doi.org/10.1134/S106456241703005X
12. Chetverushkin, B.N., Zlotnik, A.A.: On some properties of multidimensional hyperbolic quasi-gasdynamic systems of equations. Russ. J. Math. Phys. **24**, 299–309 (2017). https://doi.org/10.1134/S1061920817030037
13. D'Aschenzo, N., Chetverushkin, B.N., Saveliev, V.I.: On an algorithm for solving parabolic and elliptic equations. Comp. Math. Math. Phys. **55**(8), 1290–1297 (2015). https://doi.org/10.1134/S0965542515080035
14. Chetverushkin, B., D'Ascenzo, N., Saveliev, A., Saveliev, V.: A kinetic model for magnetogasdynamics. Math. Mod. Comp. Simul. **9**(5), 544–553 (2017). https://doi.org/10.1134/S2070048217050039
15. Zlotnik, A.A., Chetverushkin, B.N.: O parabolichnosti kvazigazo-dinamicheskoi sistemy uravnenii, ee giperbolicheskoi 2-go poriadka modifikatsii i ustoichivosti malykh vozmushchenii dlia nikh [on parabolicity of quasi-gasdynamic system of equations, its 2nd order hyperbolic modification, and stability of small perturbations for them]. Zhurnal vychislitel'noi matematiki i matematicheskoi fiziki, pp. 445–472 (2008)
16. Chetverushkin, B.N., Yakobovskiy, M.V.: Vychislitel'nye algoritmy i otkazoustoichivost' giperekzaflopsnykh vychislitel'nykh sistem [computational algorithms and fault-tolerance of hyper-exaflops computing systems]. Doklady Akademii nauk [Proc. Russ. Acad. Sci.] **472**(1), 1–5 (2017). https://doi.org/10.7868/S0869565217010042

17. Bondarenko, A.A., Yakobovskiy, M.V.: Modelirovanie otkazov v vysoko-proizvoditel'nykh vychislitel'nykh sistemakh v ramkakh standarta mpi i ego rasshireniia ulfm. [simulation of failures in high-performance computing systems within the mpi standard and its ulfm extension]. Vestnik Iuzhno-Ural'skogo gosudarstvennogo universiteta **4**(3), 5–12 (2015). www.mathnet.ru/links/76f8ebac383f6603304d157a36123407/vyurv1.pdf
18. Bondarenko, A.A., Kornilina, M.A., Yakobovskiy, M.V.: Fault tolerant algorithm for HPC. In: Supercomputing in Scientific and Industrial Problems KIAM RAS, p. 8 (2016). www.kiam.ru/SSIP/SSIP_abstracts.pdf#page=9

The Set@l Programming Language and Its Application for Coding Gaussian Elimination

Ilya I. Levin[1] , Aleksey I. Dordopulo[2]([envelope]) , Ivan V. Pisarenko[2] ,
and Andrey K. Melnikov[3]

[1] Academy for Engineering and Technology,
Institute of Computer Technologies and Information Security,
Southern Federal University, Taganrog, Russia
iilevin@sfedu.ru
[2] Supercomputers and Neurocomputers Research Center, Taganrog, Russia
{dordopulo,pisarenko}@superevm.ru
[3] "InformInvestGroup" CJSC, Moscow, Russia
ak@iigroup.ru

Abstract. The development of new heterogeneous architectures for computer systems is an advanced research direction in the field of supercomputer engineering. At the same time, software porting between different heterogeneous architectures requires significant code revision as a consequence of the architectural limitations of the programming languages available. To solve this problem, we propose Set@l, a new architecture-independent programming language which extends some ideas of the COLAMO and SETL programming languages. In Set@l, an algorithm and its parallelization are described by separated modules of the program. This is determined by the application of the set-theoretic code view and the aspect-oriented programming paradigm. The source code in Set@l is architecture-independent and represents the solution to a computational problem as an information graph specified in terms of set theory and relational calculus. Aspects adapt a universal algorithm to the architecture and configuration of a particular computer system by means of partition and classification of sets. In contrast to SETL, Set@l offers different classification criteria for collections, particularly in accordance with the type of parallelism. Sets of indefinite type of parallelism are denoted as implicit ones. In the paper, we describe the fundamentals of architecture-independent programming in the Set@l language and consider an example in which Set@l is applied for coding Gaussian elimination.

Keywords: Heterogeneous computer systems ·
Architecture-independent programming · Aspect-oriented approach ·
Set@l programming language

L. Sokolinsky and M. Zymbler (Eds.): PCT 2019, CCIS 1063, pp. 45–57, 2019.
https://doi.org/10.1007/978-3-030-28163-2_4

1   Introduction

The combination of microprocessors with other computational devices is one of the architectural solutions providing increase in computer systems' performance [1–3]. FPGA-based (field-programmable-gate-array-based) hybrid reconfigurable computers [4] are the most promising class of heterogeneous systems, owing to their design and technological features and on account of their capabilities of architectural adaptation to the structure of a particular computational problem [5]. However, the software for such computer systems should combine descriptions of parallel, pipeline, and procedural computations [6,7]. This requirement leads to complications in software development as well as overhead time and costs. The variety of computational architectures used in hybrid computer systems considerably complicates software porting due to the lack of effective methods and tools for architecture-independent parallel programming. In currently available programming languages, the mathematical sense of an applied problem and the decomposition of its solution are described by means of indivisible fragments of code. For that reason, any changes in parallelization features resulting from the implementation of an algorithm on a computer system with a different architecture require the development of a new application. Traditional approaches to the solution of the architectural-limitation problem have many shortcomings and rely on specialized translation algorithms (the Pifagor language of functional programming [8]) or on the fixed parallelization model, which is designed according to the principles of procedural programming (the OpenCL standard [9]).

The main programming problems of FPGA-based reconfigurable computers are solved in COLAMO, a high-level programming language [10–12]. In COLAMO code, parallelization is described in implicit form through the declaration of access types for arrays and the indexing of their elements. However, COLAMO is oriented to the structural and procedural organization of computations. This makes it impossible to port parallel COLAMO applications to computer systems with different architectures. Set@l is an architecture-independent programming language which extends the ideas contained in COLAMO. The language is based on the set-theoretical representation of source code introduced in SETL [13] and has mechanisms for partitioning and typing of sets. To describe an algorithm and its parallelization as separate program modules, Set@l uses the aspect-oriented programming (AOP) paradigm [14,15]. Accordingly, software porting consists in the generation of aspects covering the architecture and configuration of a computer system. Aspects describe partitioning and typing for the key collections of an algorithm. The source code, which specifies the mathematical sense of the solution of a problem, remains unchanged for all architectures.

In the paper we review the basic theoretical principles of the architecture-independent programming in the Set@l language and give an example of its utilization for the description of the forward Gaussian elimination.

2   Programming Principles and Syntax Elements of the Set@l Language

In conventional programming languages for high-performance computer systems, an algorithm and its parallelization are described by an indivisible code (tangling in terms of AOP) and are distributed in the text of a program (scattering according to AOP). These features result in architectural limitation of software. Since the problems of code tangling and scattering [16] can be solved within the framework of AOP technology, it is used in the Set@l language. According to AOP principles, the parallelization of computations is specified by separate modules of a program which are called aspects [17]. A universal algorithm for the solution of an applied problem is described by a source code which is independent from the architectural features of the computer system selected for its implementation. Thus, a program in Set@l contains the architecturally independent source code and also aspects aimed at the decomposition of an applied problem with regard to the architecture and configuration of a computer system. A translator-preprocessor analyzes user's markup and forms a virtual program, in which all features of the operation and iteration decomposition are weaved into the code. As a result, computations can be implemented on a certain architecture.

As noted above, the source code and aspects are the basic units of each program in Set@l (see Table 1). The source code describes the solution of an applied problem as an information graph in the language of the set theory and relational calculus [18] and forms a unified architecture-independent representation of an algorithm. The aspects, oriented to adaptation of an algorithm to the architecture and configuration of a specified computer system, part the initial sets of processing vertices and data of iterations into special collections with their own processing types and partitions. These collections can be variously combined, and owing to this, we can describe various forms of parallelism and change them by making combinations. A program can contain an arbitrary number of aspects, which describe different features of the partition and typing of basic collections of a computational problem.

Modules of the Set@l program are divided into sections, which are declared with the help of the syntax elements given in Table 1.

For the description of connections between the source code and its aspects, an interface section is included in each program module (see Table 1). The interface section contains declarations of input data (**input**), output data (**output**), undefined parameters (**extern**), and references to aspects-sources and aspects-targets. Other sections of a program are declared like the interface section, but they do not contain the lists of parameters and can have various names. For example, in the main code it is reasonable to allocate the section of data initialization and a main computational section.

Figure 1 shows the diagram, which demonstrates the structure of an aspect-oriented program written in the Set@l language and intended to solve tasks of linear algebra. To describe the parallelizing of an algorithm, one should divide the collections of rows I, columns J, and iterations K into subcollections with types depending on the architecture of a computer system. Within the source

Table 1. Syntax elements used for the description of modules and sections of a Set@l program

Module/section	Description format

Source code	`program(<name>):`
	`  <source code>;`
	`end(<name>);`

Aspect	`aspect(<name>):`
	`  <aspect code>;`
	`end(<name>);`

Interface section	`interface:`
	`  <input parameters>:input(<source modules>);`
	`  <output parameters>:output(<target modules>);`
	`  <list of undefined parameters>:extern;`
	`end(interface);`

Other sections	`<name of section>:`
	`  <code of section>;`
	`end(<name of section>);`

code, the partitions of I, J, and K are not specified. It means that the description of an algorithm is architecture-independent. The aspect of processing method specifies the way of partition of a matrix (for example, by rows, by columns or by iterations) by the declaration of generalized collections with undefined type and dimensions, marked as $[\![I]\!]$, $[\![J]\!]$, and $[\![K]\!]$ in Fig. 1. Using the parameters of the task size from the source code and certain parameters of the computer system configuration from the configuration aspect, the architectural aspect transforms or defines the type of collections according to the features of the computer system's architecture for efficient parallel-pipeline processing. As a result, the final partition of the information graph of a problem is formed. The total partition is defined by collections $\{I\}$, $\langle J \rangle$, and $\overrightarrow{\{K\}}$ with their processing types, and can be substituted into the source code for the implementation of the required variant of parallelization.

In contrast to other programming languages based on the set theory (e.g., SETL), the Set@l language provides the classification of collections by various criteria. In the context of an architectural independence, the fundamental criterion of sets typing is the parallelism of their elements during processing. Table 2 contains the collection classification by the types of parallelism, and the corresponding description formats in the Set@l language. The basic types of collections are a set (parallel-independent processing), a tuple (sequential processing), a pipeline tuple (pipeline processing) and a tuple of processing by iterations (parallel-dependent processing). In some aspects, the type of collections cannot be defined with certainty due to the lack of information about the architecture of the computer system chosen for the task solution. Such collections are marked as implicit (`imp`), and in other aspects their type is declared by the keyword type, as follows:

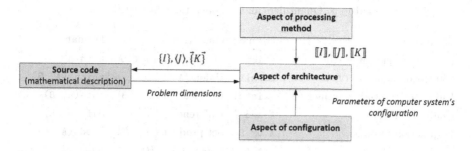

Fig. 1. Diagram of interaction between the source code and the aspects of the Set@l program for a task of linear algebra

Table 2. Types of collections classified by parallelism in Set@l

Type of collection	Processing type	Marking	Format of description
Set	Parallel-independent	$\{1, 2, \ldots, p\}$	`set(1...p)`
Tuple	Sequential	$[1, 2, \ldots, p]$	`seq(1...p)`
Pipeline tuple	Pipeline	$\langle 1, 2, \ldots, p \rangle$	`pipe(1...p)`
Set of iteration processing	Parallel-dependent	$\overrightarrow{\{1,2,\ldots,p\}}$	`conc(1...p)`
Implicit collection	Type is defined in other aspect	$[\![1, 2, \ldots, p]\!]$	`imp(1...p)`

```
type(<name of collection>)='<type of parallelism>';
```

Marking types by parallelism is only one of the variants of collection classification that are possible in the Set@l language. If it is necessary, one can add other variants of typing (e.g., classification by synchronization or by capacity) and, accordingly, increase the flexibility and adaptation capability of architecture-independent programs.

Syntax elements used in the Set@l language for the description of the basic mathematical operations of set theory are given in Table 3. For the partition of collections, we use structures formed according to the rules of relational calculus as follows:

```
A=<type of collection>(<variable>|<predicate>);
```

The descriptions of loops are made with the help of the universal quantifier marked by the keyword **forall**:

```
(forall <variable-iterator> in <set>|<predicate>):
    <loop body>;
end(forall);
```

Table 3. Quantifiers and set operations in Set@l

Operation	Format	Operation	Format
Existential quantifier (∃)	exists	Membership (a ∈ A)	a in A
Universal quantifier (∀)	forall	Union (A ∪ B)	union (A, B)
Unique existential quantifier (∃!)	exists!	Intersection (A ∩ B)	int (a, B)
Conjunction ($A \& B$)	A and B	Difference ($A \setminus B$)	dif (A, B)
Disjunction ($A \vee B$)	A or B	Set product ($A \times B$)	prod (A, B)
Inversion (!A)	not A	Subset ($A \subset B$)	subs (A, B)

In the Set@l language, the collection classification proposed by Vopenka within the alternative set theory [19] is introduced. The collection types given in Table 4 can describe the majority of computational algorithms.

As a default, all collections in the Set@l language are considered as clearly defined sets. However, some algorithms cannot be described in an architecture-independent form without the application of other collection types. An algorithm of this kind is the solution of a system of linear algebraic equations (SLAE) by the Jacobi iterative method [20]. In some cases, the collection of Jacobi iterations represents an indefinite semiset.

Table 4. Types of collections classified by indefiniteness in Set@l

Type of collection	Description	Marking	Keyword
Set	Completely defined collection of elements	{ }	set
Semiset	Partly defined collection of elements	{ ?	sm
		? }	
		{ ? ¿ }	
Class	Completely undefined collection of elements with type specified in another aspect	? ¿	cls

In addition to the basic classification of collections by parallelism and indefiniteness, developers can introduce their own set-typing attributes. The following syntactic construction is used for the declaration of new collection types:

```
attribute <name> (<set or element>):
        <attribute description>;
    end(<attribute name>);
```

An attribute assigned to a set or its element specifies a processing method or relation between the objects of the program. Multiple attributes form a collection in the same way as other objects of the Set@l programming language. By using user-defined set attributes, we can describe complex algorithms and features of their implementation in a compact set-theoretical way. Thus, the distinctive features of the Set@l language are the aspect-oriented structure of programs and the representation of a computational problem as a system of collections rather than as data arrays specified according to some type of parallelism. Thanks to partition and multicriteria typing of collections, it is possible to describe a computational algorithm regardless of parallelism and other aspects of its implementation.

3   Description of Forward Gaussian Elimination in the Set@l Language

The solution of SLAE is a fundamental applied problem in linear algebra. The Gaussian method [20] is the classic direct method for solving SLAE. It consists in a sequential exclusion of unknown variables by equivalent transformations of the equations of the system (forward elimination) and then finding all unknown variables starting with the last one (backward substitution). In the paper, we consider the forward Gaussian elimination which results in the transformation of the original SLAE until it is in row echelon (trapezoidal) form. It is assumed that the initial SLAE matrix is diagonally dominant.

To solve the problems of linear algebra, the following processing methods are used: by rows, by columns, by cells and by iterations. If a matrix is processed by rows, the collection of the matrix rows is divided into N blocks $BL(i)$, and each block has s rows (see Fig. 2-a). If a matrix is processed by columns, the collection of the matrix columns is divided into M blocks $BC(j)$, and each block has c columns (see Fig. 2-b). If a matrix is processed by cells, it is divided into $N \times M$ cells, and each cell has s rows and c columns (see Fig. 2-c). If a matrix is processed by iterations, the collection of iterations is divided into T blocks $BI(k)$ with ni iterations in each block (see Fig. 2-d).

The source code of the Set@l program for performing the forward Gaussian method is given in Fig. 3. The program module containing the source code consists of three sections: the interface (interface, lines 2–5), the data preparation block (data preparation, lines 6–11), and the main computational block (computing, lines 12–22).

The interface section interconnects the source code with the architectural aspect (architecture). The collections of rows, columns, and iterations (I, J, and K) are declared as the input parameters, while the numbers of rows n, columns m, and iterations p are declared as the output parameters.

The data preparation section is required for the declaration of the task size and for the input of the matrix that corresponds to the initial form of the SLAE. The matrix is represented as an extended matrix supplemented by the column vector of constant terms. It is modified at each iteration of the forward Gaussian method. The following format of element indexing is used in the code for the extended matrix:

```
a(<iteration>,<row>,<column>).
```

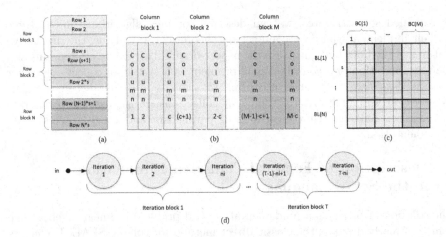

Fig. 2. Processing methods used for the parallelization of linear algebra tasks: by rows (a), by columns (b), by cells (c), by iterations (d)

Asterisk symbol "*" denotes the whole range of the corresponding index (see line 10 in Fig. 3).

The computational section is the main part of the source code. It contains the algorithm of the applied problem described in a universal mathematical language without declaring any types of parallelization. Figure 3 shows the algorithm of the forward Gaussian elimination described in terms of set theory and relational calculus (lines 13–20).

The aspect of the processing method for the forward Gaussian elimination is given in Fig. 4. The parameters of the processing method, s, N, c, M, ni, and T, are determined by the architecture and configuration of the computer system. They are defined in the architectural aspect which is reflected in the interface section of the program unit considered here (line 3). After partitioning the sets by a certain processing method, they are passed to the architectural aspect, where the type of parallelization is defined (line 4). The partition of collections I, J, and K (lines 7–9, set formation section) makes it possible to describe each processing method shown in Fig. 3. For example, if processing is done by rows, then the set I is divided into N blocks of s concurrently processed rows. The sets J and K are not divided; this fact is described by setting $M = 1$, $c = m$, $T = 1$, $ni = p$.

Figure 5 shows the code of the architectural aspect for the forward Gaussian method written in Set@l. It describes the parallelization of the problem solution for a reconfigurable computer system with an independent FPGA field (RCS) and a multiprocessor architecture (MP). The types of the sets I, J, K, and their subsets are declared in the architectural aspect of the program. Moreover, the aspect defines the parameters of the processing methods that depend on the configuration of the computer system.

```
(1)     program(Gauss_forward):
(2)        interface:
(3)           I,J,K: input(architecture);
(4)           n,m,p: output(architecture);
(5)        end(interface);
(6)        data preparation:
(7)           n=<number of rows in matrix>;
(8)           m=n+1;          // number of columns in matrix;
(9)           p=n;            // number of iterations;
(10)          a(1,*,*)=<initial extended matrix of SLAE>;
(11)       end(data preparation);
(12)       computing:
(13)          (forall k in K):
(14)             (forall seq(i,j) in prod(I,J)|i>=(k+1)):
(15)                a(k+1,i,j)=a(k,i,j)-a(k,i,k)/a(k,k,k)*a(k,k,j);
(16)             end (forall);
(17)             (forall seq(i,j) in prod(I,J)|i<(k+1)):
(18)                a(k+1,i,j)=a(k,i,j);
(19)             end (forall);
(20)          end(forall);
(21)       end(computing);
(22)    end(Gauss_forward);
```

Fig. 3. The source code of the forward Gaussian elimination in Set@l

```
(1) aspect(processing):
(2)     interface:
(3)        s,N,c,M,ni,T: input(architecture);
(4)        I,J,K: output(architecture);
(5)     end(interface);
(6)     set formation:
(7)        I=imp(BL(i)|BL(i)=imp((i-1)*s+1 … i*s) and i in seq(1…N));
(8)        J=imp(BC(j)|BC(j)=imp((j-1)*c+1 … j*c) and j in seq(1…M));
(9)        K=imp(BI(k)|BI(k)=imp((k-1)*ni+1 … k*ni) and k in seq(1…T));
(10)    end(set formation);
(11) end(processing);
```

Fig. 4. The aspect of the processing method for the forward Gaussian elimination program in Set@l

For the reconfigurable computer system we use a complex processing by rows and iterations (see Fig. 6). The number of concurrently processed rows s is defined by the number of available channels for distributed memory controllers (DMC), marked as K_krp (see Fig. 5, left column, line 10) and declared in the configuration aspect. If we know s and the total number of rows in a matrix n, we can calculate the number of row blocks N (see line 11). When the set I of row numbers is formed, then parallelization types are declared (lines 12 and 13). Parallelization by columns is not used, so the set of columns J does not contain subsets, and is described as a pipeline tuple which is equivalent to a block with all columns of the set (lines 15–18). Since the declaration of the parallelization type for a block of lines BC has no sense in this case, then the keyword null (line 18) is used instead of the certain type. The number of concurrently performed iterations ni is calculated by the formula in line 20.

```
(1) aspect(architecture):
(2)      interface:
(3)          I,J,K: input(processing);
(4)          R,R0,K_krp,q1,q2,architecture_type: input(configuration);
(5)          n,m,p: input(Gauss_forward);
(6)          I,J,K: output(Gauss_forward);
(7)      end(interface);
(8)      case(architecture_type='RCS'):  |  case(architecture_type='MP'):
(9)                    // Rows (I):
(10)         s=K_krp;                     |      s=q1;
(11)         N=n/s;                       |      N=n/s;
(12)         type(I)='pipe';              |      type(I)='seq';
(13)         type(BL(i))='par';           |      type(BL(i))='par';
(14)                   // Columns (J):
(15)         c=m;                         |      c=q2;
(16)         M=1;                         |      M=m/c;
(17)         type(J)='pipe';              |      type(J)='seq';
(18)         type(BC(j))='null';          |      type(BC(j))='par';
(19)                   // Iterations (K):
(20)         ni=min( p, floor(R/s/R0) );  |      ni=p;
(21)         T=p/ni;                      |      T=1;
(22)         type(K)='pipe';              |      type(K)='seq';
(23)         type(BI(k))='conc';          |      type(BI(k))='null';
(24)      end(case);                      |   end(case);
(25) end(architecture);
```

Fig. 5. The architectural aspect of the forward Gaussian elimination program in Set@l

The formula takes into account the total available computational resource R, the resource $R0$ required for the implementation of the minimal basic subgraph, the parallelization by s rows and limitation of ni by the total number of iterations p in the case of excess amount of available computational resource.

The functions `min()` and `floor()` are used for calculation of a minimum value and for rounding downward to the nearest whole number, respectively.

The number T of parallel-dependent blocks of iterations (line 21) is calculated as a ratio of the total number p of matrix processing iterations to the number of iterations ni in the block. It is worth noting that, in contrast to processing by rows, processing by iterations is parallel-dependent, so, each iteration block $BI(k)$ is declared as the `conc` type (line 23).

For computer systems with a multiprocessor architecture we use the method of processing by cells (see Fig. 7). The processor set is represented as a rectangular matrix with the dimension of $q1 \times q2$. The elements of each cell are processed concurrently by different processors (lines 13 and 18 in Fig. 5, right column), so, the cell size equals to the size of the processor matrix, i.e. $s = q1$ and $c = q2$ (lines 10 and 15). The number of cells, which form a matrix $N \times M$, is calculated using the matrix size parameters n and m (lines 11 and 16). Within one iteration, cells are processed sequentially (lines 12 and 17). Collection of all iterations forms the tuple, which is not divided into blocks and contains p sequential elements (lines 20–23).

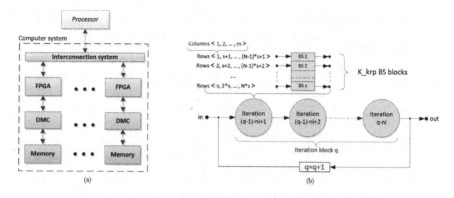

Fig. 6. The architecture of a reconfigurable computer system with an independent FPGA field (a) and parallelization of linear algebra problems by rows and iterations (b)

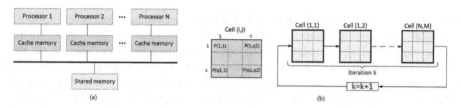

Fig. 7. The architecture of a multiprocessor computer system (a), and parallelization of linear algebra algorithms by cells (b)

The code of the configuration aspect for computer systems with reconfigurable and multiprocessor architectures is given in Fig. 8.

In the configuration aspect we define the value of the **architecture_type** variable (line 5, Fig. 8), which describes the type of the computer system architecture. According to the architecture, the configuration parameters (lines 6 and 11, Fig. 8) and branches of the architectural aspect (line 8, Fig. 5) are initialized using the case statement. Before the loading of the configuration aspect, uninitialized parameters in the source code and other aspects are considered as indefinite variables.

The considered example of the program for the forward Gaussian elimination demonstrates the key features of the Set@l language syntax. The features are required for the aspect-oriented structure of the program and for set-theoretic representation of the source code.

```
(1) aspect(configuration):
(2)     interface:
(3)         R,R0,K_krp,q1,q2,architecture_type: output(architecture);
(4)     end(interface);
(5)     architecture_type=<type of computer system architecture>;
(6)     case(architecture_type='RCS'):
(7)         R=<available computational resource>;
(8)         R0=<resource required for implementation of minimal basic
                subgraph>;
(9)         K_krp=<number of DMC in computer system>;
(10)    end(case);
(11)    case(architecture_type='MP'):
(12)        q1=<number of rows in matrix>;
(13)        q2=<number of columns in matrix>;
(14)    end(case);
(15)end(configuration);
```

Fig. 8. The configuration aspect of the Set@l program

4   Conclusions

In contrast to other programming tools for high-performance computer systems, the Set@l language represents an algorithm as sets and relations between them, rather than as data and instructions with strictly defined parallelism. If we declare various partitions of sets and classify them according to different features, we can describe an algorithm in an architecture-independent form and adapt it to the architecture and configuration of a computer system by means of aspects. Thanks to the basic mechanism of set typing by parallelism, the Set@l language makes it possible to remove architectural limitations that are typical of traditional programming languages. The description of various aspects of parallelization as independent program modules is done by creating collections with an undefined processing type which is specified in other components of the program. The Set@l language gives fundamentally new possibilities for efficient software porting to various architectures of computer systems, including hybrid and reconfigurable ones. This paper summarizes the development of Set@l theoretical principles. We thoroughly considered the distinctive features of the language and described its fundamental syntax elements. We are currently working on the development of a Set@l translator.

References

1. Bourzac, K.: Supercomputing poised for a massive speed boost. Nature **551**, 554–556 (2017). https://doi.org/10.1038/d41586-017-07523-y
2. Mittal, S., Vetter, J.: A survey of CPU-GPU heterogeneous computing techniques. ACM Comput. Surv. **47**(4), Article no. 69 (2015). https://doi.org/10.1145/2788396
3. Loshchukhina, T.E., Dorofeev, V.A.: Solving the applied tasks using the heterogeneous computing systems. Bull. Tomsk Polytech. Univ. **323**(5), 165–170 (2013). (in Russian)
4. Ebrahimi, A., Zandsalimy, M.: Evaluation of FPGA hardware as a new approach for accelerating the numerical solution of CFD problems. IEEE Access **5**, 9717–9727 (2017). https://doi.org/10.1109/ACCESS.2017.2705434

5. Dordopulo, A.I., Levin, I.I.: Resource-independent programming of hybrid reconfigurable computer systems. In: Proceedings of Russian Supercomputing Days 2017, pp. 714–723. MSU Publishing, Moscow (2017). (in Russian)
6. Andreyev, S.S., et al.: Hybrid supercomputer K-100: what next? Inf. Technol. Comput. Syst. **2**, 29–35 (2012). (in Russian)
7. Daga, M., Tschirhart, Z.S., Freitag, C.: Exploring parallel programming models for heterogeneous computing systems. In: Proceedings of 2015 IEEE International Symposium on Workload Characterization, pp. 98–107 (2015). https://doi.org/10.1109/IISWC.2015.16
8. Legalov, A.I.: Functional language for creation of architecture-independent parallel programs. Comput. Technol. **10**(1), 71–89 (2005). (in Russian)
9. OpenCL: The open standard for parallel programming of heterogeneous systems. https://www.khronos.org/opencl/
10. Kalyaev, I.A., Levin, I.I., Semernikov, E.A., Shmoilov, V.I.: Reconfigurable Multipipeline Computing Structures. Nova Science Publishers, New York (2012)
11. Dordopulo, A.I., Levin, I.I., Kalyaev, I.A., Gudkov, V.A., Gulenok, A.A.: Programming of hybrid computer systems in the programming language COLAMO. Izvestiya SFedU. Eng. Sci. (11), 39–54 (2016). https://doi.org/10.18522/2311-3103-2016-11-3954. (in Russian)
12. Kalyaev, I.A., Dordopulo, A.I., Levin, I.I., Gudkov, V.A., Gulenok, A.A.: Programming technology for hybrid computer systems. Comput. Technol. **21**(3), 33–44 (2016). (in Russian)
13. Dewar, R.: SETL and the evolution of programming. In: Davis, M., Schonberg, E. (eds.) From Linear Operators to Computational Biology: Essays in Memory of Jacob T. Schwartz, pp. 39–46. Springer, London (2013). https://doi.org/10.1007/978-1-4471-4282-9_4
14. Kurdi, H.A.: Review on aspect oriented programming. Int. J. Adv. Comput. Sci. Appl. **4**(9), 22–27 (2013)
15. Rebelo, H., Leavens, G.T.: Aspect-oriented programming reloaded. In: SBLP 2017 Proceedings of the 21st Brazilian Symposium on Programming Languages, Article no. 10 (2017). https://doi.org/10.1145/3125374
16. The AspectJ Development Environment Guide. https://www.eclipse.org/aspectj/doc/released/devguide/index.html
17. Aspect oriented programming with Spring. Core technologies. https://docs.spring.io/spring/docs/3.0.x/spring-framework-reference/html/aop.html
18. Haggarty, R.: Discrete Mathematics for Computing. Pearson Education, Harlow (2002)
19. Vopenka, P.: Introduction to Mathematics in Alternative Set Theory. Alfa, Bratislava (1989). (in Czech)
20. Dahlquist, G., Bjork, A.: Numerical Methods. Dover Publications, New York (2003)

Using Empirical Data for Scalability Analysis of Parallel Applications

Pavel Valkov$^{(\boxtimes)}$, Kristina Kazmina, and Alexander Antonov

Lomonosov Moscow State University, Moscow, Russia
pas-valkov@yandex.ru, kp.kazmina@gmail.com, asa@parallel.ru

Abstract. Scalability is the most important characteristic describing the execution of a parallel program. Scalability is a quantity that depends on many factors: the values of the input parameters, the algorithm used and its implementation details, the capabilities of hardware resources, and others. We must evaluate and compare the scalability of applications to determine their optimal execution parameters on the resources available. However, there often arises a need to assess scalability on large configurations of computing systems to which access may be limited. Therefore, an important task is to predict scalability based on the results obtained on small configurations. The concept of scalability metric was introduced for this purpose. In this article, we study the applicability of the metric and consider some approaches to predicting the scalability of applications on large supercomputer configurations. We tested the applicability of the proposed approach on the Lomonosov and Lomonosov-2 supercomputers at Lomonosov Moscow State University.

Keywords: Scalability · Metric · Prediction · HPL · NPB

1 Introduction

At present, parallel computing is widely used in many fields of science. Supercomputer systems are extremely popular resources; it is, therefore, important to use them rationally. Nonetheless, parallel applications running on supercomputers can have many drawbacks that cause the efficiency of systems to drop. The reasons for this can be either an unsuccessfully selected algorithm for solving a problem or a suboptimal allocation of resources. The research of such a distinctive characteristic of parallel applications as scalability can help in dealing with these issues.

By analyzing the scalability of parallel programs, one can improve the efficiency of use of computing resources and assess the quality of parallel applications to detect bottlenecks and characterize the system used. In the general case, the dynamics of scalability characteristics of a parallel application on a real computer system is nontrivial, hence the considerable difficulties associated with purely analytical approaches to its research.

© Springer Nature Switzerland AG 2019
L. Sokolinsky and M. Zymbler (Eds.): PCT 2019, CCIS 1063, pp. 58–73, 2019.
https://doi.org/10.1007/978-3-030-28163-2_5

The goal of this paper is to devise universal methods that resort on empirical data for the study of the scalability of parallel applications. The first method, which we introduced in [1], describes a technique for the prediction of scalability of parallel applications in the case of large supercomputer configurations; the second one presents an approach to the comparative analysis of different applications. The universality of both methods lies in the absence of strict restrictions upon the computing system and the application launched on it; moreover, the source code of the application does not go through any modifications, and there is no requirement for its availability. The applicability of this approach has been verified through the use of various benchmarks, such as Linpack [2], NPB BT, SP, and LU [3], on the Lomonosov [4] and Lomonosov-2 [5] supercomputers.

2   Related Work

The concept of a multidimensional scalability metric that can be used to compare applications was introduced and studied in [6], where scalability was defined as a parallel-program property that characterizes the dependence between the entire set of dynamic characteristics of the program and the set of launch parameters. Other definitions, as, for example, those suggested in [7–9], are special cases of the one given above. An important characteristic that makes this definition different from others and, at the same time, makes it more general is that it involves multiple launch parameters, and not just determines the dependence of scalability on a single parameter.

There is no unambiguous interpretation of what scalability prediction is. If we consider the generalized definition of scalability from [6], then predicting scalability often means predicting the dependence between parallel-program execution time and some parameters (as in [10–13]). Other sources consider the prediction of a program's execution performance [14–16], the performance per dollar [17], the energy consumption and its ratio to performance [18,19], and others.

There are studies of both strong and weak scaling [10], and also of wide scaling [20]. The results from several specific executions can be used in two manners to predict scalability on the whole space of possible executions: first, the values for large processor configurations can be predicted by extrapolation on the basis of data collected for small processor configurations [10,11,20], and second, the values for all the parameter space can be predicted by interpolation on the basis of data collected for points uniformly selected in the parameter space [12,16,18,19].

Some approaches rely on separation of computation and communication, on the use of stalled cycles [11], on the representation of program's characteristics as those obtained during the execution of benchmarks [14], on techniques allowing to extract from the entire set of parameters only those that significantly affect the result [16], and others. We should emphasize that if the predicted value depends on several parameters, then the extrapolation of these parameters separately gives a much better result than the extrapolation of some dependent unifying value [10,11].

The approximation can be achieved through various means: statistical regression methods, neural networks, partial or full simulation of program execution. Linear regression can provide results with an error not exceeding the error for other approaches [10,14,17,18,20]. The choice of nonlinear functions to characterize the scalability of applications [11,15,16] theoretically enables to describe more complex and heterogeneous behaviors. The use of neural networks and other methods of machine learning [12,13,16,18,19] provides an opportunity to identify unobvious dependencies between the studied values and the selected parameters, and this can improve the approximation accuracy. However, this approach has some drawbacks: the implicit type of the results does not allow for any analysis based on the specific form of the obtained model; moreover, correct training is only possible if a relatively large test set is provided, which is especially difficult in terms of computing on real supercomputers.

The approach based on the simulation of application execution is the most difficult to implement and requires a considerable amount of information about the program structure. Nevertheless, there are examples of implementation of this approach: modeling can be performed using Markov models [21] or partial deterministic replay [22].

The scalability prediction approaches we have listed here do not have universality in a general sense, they imply severe constraints on applications or specific computing systems and may require knowledge of the implemented algorithm or the ability to edit the program code.

3   Scalability Prediction

We developed our scalability prediction method according to the versatility requirements that had been set. This means that there are no explicit constraints on the computing systems or the program, and considering the hypothetical absence of information about the source code of the program itself and the impossibility to modify it, only the executable file can be available.

When building a model under such restrictions, it is necessary to resort to some kind of empirical data that can help to take into account program-related information acquired without violating the declared requirements. We suggest the use of dynamic characteristics obtained throughout several specific executions of the application (such executions are called *test executions* or *test runs*).

3.1   Conducting Test Executions

Usually, a task that requires a small number of nodes of a supercomputer can be executed faster than one requiring a large number of nodes. Moreover, large supercomputer configurations may be temporarily unavailable or accessible only to a limited privileged set of users. In light of this, the actual task of predicting scalability using the results of test executions of an application can be formulated as the task of using the results of several executions of this application on small configurations to predict its performance characteristics on large configurations of a computing system.

In this section, we use the application execution time as a dynamic characteristic of runs. Also, due to issues described later in Sect. 4, we consider not only the execution time itself but the minimum execution time for each fixed set of parameters on a specific configuration. In this text, sometimes this minimum will be referred to as the *execution time*.

3.2   The Extrapolation Model

The method described in this article for predicting scalability is based on an extrapolation model, that is, a model that describes the dependence between the values of a dynamic characteristic $T(q)$ of an application launched on q processes and the number of processes q. The results of k test runs on $p_1, \ldots, p_k$ processes are used to determine the parameters of the function $T(q)$. Thus, by regression on the set of points $(p_i, T(p_i))$, $i = 1, \ldots, k$, one can determine the coefficients. It is also possible to consider a generalized model describing the dependence of $T(q, inp_1, \ldots, inp_m)$ on the number of processes q for runs with input parameters $inp_1, \ldots, inp_m$.

The main problem is choosing the form of the function $T(q)$. Several assumptions can be made based on general information about the dynamics of the application execution time as the number of processes increases. The function should enable the description of instances of well-parallelized applications (when the execution time decreases as the number of processes increases) and cases of poor parallelization (when the execution time at some point begins to increase as the number of processes increases, even if the time was initially decreasing). At the same time, given a hypothetical infinite growth of the computing system configuration, the computation time per process, even for a well-parallelized application, will be, at some point, small compared with the overhead required for initialization, synchronization, and communications on a large number of processes. Usually, due to this overhead, the increase in the resulting acceleration drops at first and then becomes negative.

This means, in particular, that the derivative of the extrapolation function $T(q)$ is negative when the execution time decreases and becomes positive if the execution time at some instant starts to increase. For example, the function

$$T(q) = aq + bq^{-1} + cq^{-1/2} \tag{1}$$

can have such properties. Its derivative is

$$T^{(1)}(q) = a - \frac{2b + cq^{\frac{1}{2}}}{2q^2} \tag{2}$$

The selection of the coefficients of the function $T(q)$ can be carried out by regression using the method of least squares and the Levenberg–Marquardt algorithm [23, 24].

3.3 Prediction Error

The prediction accuracy of the model was estimated by the relative error. The execution time of an efficiently written parallel program can significantly decrease as the number of processes it is running on increases. Therefore, the data used to build and test a prediction model based on the execution time of the application may show a significant variation in some cases. At the same time, many methods used to select the regression coefficients resort to minimizing the absolute error, rather than the relative error. The absolute error of the approximation for the computation time on a small number of processes can be large with a small relative error, and conversely, the absolute error on a large number of processes can be small with a large relative error. The task posed in this section implies a greater desired prediction accuracy for a larger number of processes. This raises the problem of constructing a prediction model with a better fit on small configurations.

Actually, setting such priorities manually while fitting the model is not always necessary: practical results show that, for example, the shape of the functions chosen for approximation may prevent this. This problem can also be solved by replacing the prediction of the execution time with that of another characteristic which has a smaller dispersion of values, by excluding the results of some launches on the smallest configurations, as well as by using regression methods with weights.

3.4 Summary of the Prediction Process

The prediction process for the chosen type of extrapolation model can be divided into several steps. At the first step, multiple test runs of an application are performed on small configurations, including identical runs, and information about dynamic characteristics of the program operation is collected. At the second step, the collected data is processed: the minimum execution time for identical executions is selected, and data is unified. At the third stage, the data obtained at the previous step are used to select the parameters of the prediction model; as a result, we obtain a model with fixed parameter values. At the fourth step, the prediction of the application execution time for the given target configuration is constructed using the final model.

It is worth noting that the general form of the prediction model is, in this case, the same for all applications, but the parameters of the model are selected separately for each application considering a specific set of its test executions.

To check the correctness of the predictions, several identical runs are performed for each target configuration, and for each set of identical runs a minimum execution time is selected. These minimum execution times on the target configurations are used to calculate the relative error for the model.

4 Evaluation of the Scalability Prediction

We tested the applicability of the suggested method on the supercomputers Lomonosov and Lomonosov-2 for HPL, the Linpack benchmark, BT, SP, and

LU from NPB 3.3.1 using the OpenMPI library version 1.10.7. We studied the strong scalability of these applications for various fixed matrix sizes.

The computing nodes of the Lomonosov supercomputer are equipped with two Intel Xeon X5570 processors (2.93 GHz) and connect with each other through an InfiniBand 4x QDR communication network. The compilation was made with mpicc. The computing nodes of the Lomonosov-2 supercomputer include an Intel Haswell-EP E5-2697v3 processor (2.6 GHz) and are connected via an Infiniband FDR communication network. For compiling, we used mpicc and mpif77.

Throughout the research of HPL scalability, multiple test runs for data collection were performed on configurations with numbers of processes equal to $4n$, $n = 1, \ldots, 32$, and several runs were performed on configurations with $144, 256, 400, 576, 784$, and 1024 processes (all the sizes of the configurations are square integers to avoid imbalance which is possible due to changes in the aspect ratio of the process grid) with fixed matrix sizes equal to $16\,384, 25\,600$, and $50\,000$.

The research of the BT and SP benchmarks involved test runs on configurations with numbers of processes equal to n^2, $n = 2, \ldots, 11$, and also some executions on configurations with $144, 256, 400, 576, 784$, and 1024 processes. All BT runs were conducted with fixed matrix sizes equal to 64^3 and 162^3, corresponding to tasks of classes A and C for NPB. SP executions were performed for tasks of class C. It should be noted that runs on $576, 784$, and 1024 processes could not be executed for class A tasks, thereby making predictions in this case impossible.

The results of reruns on the same configuration showed the presence of considerable variations in runtime values. This is illustrated in Fig. 1, which shows all the runs, including identical ones, for the Linpack test on the Lomonosov supercomputer with a matrix size of $25\,600$.

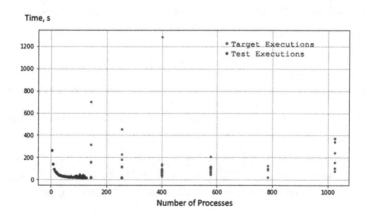

Fig. 1. Linpack execution time on the Lomonosov supercomputer. The matrix size is $25\,600$.

The maximum variation in execution time for identical runs in some cases reaches a value of 4188%. Such a deviation in execution time can be explained by changes in the communication network load, as well as by a possible computational imbalance on different nodes, potential hardware failures, and different distribution of processes over computational nodes.

Given the dispersion of the execution time, we decided to choose the minimum execution time for identical experiments. Predictions were also made for the corresponding minima of the execution time. On configurations used to construct the prediction model (up to 128 processes), we took into account the results of 3 or 4 identical test runs. On target configurations for which we made predictions (144 to 1024 processes), we took into account the results of 3 to 25 identical program executions.

To make predictions more accurate, the configuration size for test executions in Sect. 4.1 starts with 16 processes.

4.1   Testing Results for the Extrapolation Model

We used the function $T(q) = aq + bq^{-1} + cq^{-1/2}$, proposed in Sect. 3.2, for $a, b, c \geq 0$. Tables 1 and 2 show the relative error of the predicted execution time for Linpack and for BT, SP, and LU, respectively.

Table 1. Relative error of the predicted execution time for Linpack

	144	256	400	576	784	1024
Lomonosov, size = 16384	5.14%	6.20%	55.30%	21.92%	39.47%	
Lomonosov, size = 25600	1.10%	38.94%	7.65%	22.04%	143.61%	15.24%
Lomonosov-2, size = 25600	10.83%	10.40%	11.46%	18.05%	10.82%	9.67%
Lomonosov-2, size = 50000	9.95%	4.17%	0.75%	5.76%	9.86%	52.70%

Table 2. Relative error of the predicted execution time for BT, SP, and LU

	144	256	400	576	784	1024
BT Lomonosov-2, class A	10.47%	18.64%	31.50%	x	x	x
BT Lomonosov-2, class C	1.39%	6.01%	22.69%	7.70%	19.20	22.31%
SP Lomonosov-2, class C	12.21%	24.48%	26.07%	30.19%	9.13%	6.00%
LU Lomonosov-2, class C	19.60%	12.77%	50.91%	59.10%	36.29%	37.08%

An example of an approximating function for the Linpack execution time is shown in Fig. 2. Launches were performed on the Lomonosov-2 supercomputer with a fixed matrix size of 25 600.

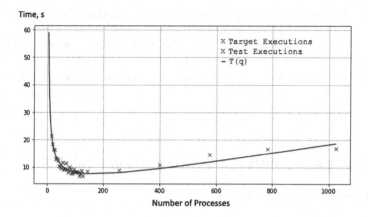

Fig. 2. Extrapolation function $T(q)$ for the Linpack execution time

It should be noted that the approximation by the function $T(q)$ on the configurations with $144, 256, 400, 576, 784,$ and 1024 processes for Linpack still produces a relative error of up to 15%. It imposes restrictions on the prediction accuracy that is achievable by using this kind of functions. At the same time, the prediction of the BT execution time with the function $T(q)$ is possible with a lower error. However, due to a stronger restriction on the number of test executions (the sizes of the configurations are squares of integers), the influence of the imbalance in prediction error, mentioned in Sect. 3.3, becomes even more perceptible. The obtained median relative error is higher for NPB benchmarks. The reason for this is the proximity of the real Linpack performance to the peak one, which can rarely be achieved for parallel applications in practice. In this case, the behavior of NPB benchmarks is more realistic. The abnormal error of 143.61% for Linpack may have been caused by the small matrix size, which can lead to cache effects.

4.2 Comparison of the Prediction Accuracy

The accuracy of the execution time predictions for the Linpack test on the Lomonosov and Lomonosov-2 supercomputers was compared with the predictions for the Linpack test described in [16]. The authors of [16] posed the interpolation problem for this case. In this section, we consider the corresponding extrapolation problem.

Table 3 shows the corresponding ranges of the median relative errors. The error of the predictions implemented in this section is similar to that given in [16] for runs with up to 512 processes. The prediction can be improved, for example, by opting for a method that enables a more precise consideration of the influence of large absolute errors for small relative errors, as it was described in Sect. 3.3, or by using a model with a large number of parameters.

Table 3. Median relative errors for the Linpack execution time prediction

Partial polynomial regression [16]	Neural networks [16]	Proposed approach
2.2–5.2% (up to 512 processes)	6.5–10.5% (up to 512 processes)	5.76–15.24 (up to 1024 processes)
		4.17–10.83% (up to 512 processes)

5   Comparison Method for Application Scalability

5.1   Concept of Scalability Metric

During the scalability study of different applications or several implementations of a single application, there arises the issue of finding a criterion for comparing their scalability. This criterion should be able to determine which application is more efficiently scaled and which values of the launch parameters produce the best results. Thus, it is necessary to introduce a metric that describes the scalability of an application in such a manner that it depends minimally on the application execution features and enables comparing applications with different launch parameters.

The metric described in [6] is indeed capable of reflecting scalability and does not require the intersection of the launch parameter values when comparing different applications. However, it turned out that the method used in [6] for calculating the metric is incorrect. This motivated this part of our paper.

5.2   Requirements on the Metric

The goal of studying the scalability of a parallel application is to find optimal launch conditions so that the efficiency of the application is maintained at a high level. Changes in the application scalability may occur for different reasons, for example, due to changes in the program launch parameters, the selected algorithm, or some hardware factors. The metric should help in solving the problem of finding both the optimal values of the application launch parameters and the optimal algorithm for implementing the program with minimal dependence on hardware factors.

Regarding dependence on hardware factors, we should notice that the definition formulated in [6] is not completely correct. Scalability is not only a feature of the "parallel program" but also of the computing system. It can significantly depend on the type of processor, organization and load of the communication environment, and many other hardware factors.

At best, minimal hardware dependence would allow comparing applications on different machines. However, another problem appears: execution of an application on different computers can sometimes be possible only with completely different ranges of launch parameters. The same problem may arise on identical machines when the access to the entire set of computing resources of the system is limited for some reason.

The proposed metric currently does not completely solve this problem since the application behavior is not the same in different ranges of launch parameters. In some ranges, the superlinear speedup effect may be observed due to a befitting location of data in the cache memory or due to a nonlinear complexity dependence between the application and the amount of data processed [25]. In other ranges, some dynamic characteristics of the application, for example, the efficiency or the execution time, are significantly influenced by the overhead of communication between processors. Thus, the execution time on a small number of processors usually decreases proportionally to their number, unlike what occurs when the application is run on a large number of processors. For this reason, the assumption made in [6] claiming that it is possible to introduce a metric that "allows comparing applications with launch parameter ranges that do not even intersect or have common values" is inappropriately optimistic.

Thus, the stability of metric values in case of computations on different machines or within different ranges is only a desirable requirement. The first condition that must be satisfied is the correctness of the metric. We call the metric *correct* if the number of points in the selected parameter range does not affect the value of the metric.

5.3   Definition of the Metric

When predicting scalability, the time was chosen as the dynamic characteristic that depends on the number of processors. Nevertheless, in the case of scalability of applications on different computing systems and with different input parameters, the values of the execution time are incomparable. Accordingly, preference was given to the parallelization efficiency: $E = \frac{S}{p}$. The main problem is to approximate empirical data with a continuous function. The function $T(q)$ introduced to predict the execution time is not adequate in this case since the dynamics of the efficiency and that of the execution time are different in the case of parallel applications. The main difference in their dynamics is that the efficiency does not increase in regions of low parallelism. In other regions, both quantities decrease, but the efficiency decreases slower.

In the light of the previous observations, the terms bq^{-1} and $cq^{-1/2}$ in $T(q)$ were replaced by others, given the high decay rate. The increasing function aq, $a \geq 0$, was left to compensate for decreasing functions on regions of poor parallelism. The final approximating function $E(q)$ looks as follows:

$$E(q) = aq - bq^{\frac{1}{2}} + c, \quad a, b, c \geq 0 \tag{3}$$

The coefficients of the function were selected as in the case of the function predicting the execution time. However, it would be wrong to introduce a metric based solely on the values of coefficients. Figure 3 confirms this.

The figure shows two graphs made with different steps in one set of process points. Both plots approximate points with a relative error less than 3.5%, which

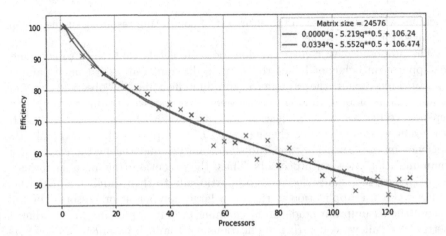

Fig. 3. The approximating function $E(q)$ for the Linpack benchmark on the Lomonosov supercomputer

means an almost perfect match in the current process range, but the coefficients of the linear term of the function $E(q)$ differ by several orders of magnitude. The values of the metric for these two graphs should be similar, so it was decided to use the integral of the function $E(q)$ as the metric value. Finally, the value of the metric $M(q)$ for an application executed on the process range $[q1; q2]$ was defined as

$$M(q, q_1, q_2) = \int_{q_1}^{q_2} \left(aq - bq^{\frac{1}{2}} + c\right) dq, \quad a, b, c \geq 0. \tag{4}$$

It should be noted that the value of the efficiency implicitly depends on the size of the matrix employed in the experiments. Therefore, the metric we defined also implicitly depends on the size of the matrix. Consequently, to make the results more precise when comparing parallel applications, the same input data size should be used, if possible.

6   Evaluation of the Metric

The verification of the proposed metric was conducted under similar conditions, described in Sect. 4, with the HPL and NPB BT benchmarks, but excluding some launch parameter values. In some tests, the process range was limited to 128 or 256 processes, and the matrix size ranged from 1024 to 51 200 for the HPL and from 63^3 to 408^3 for the NPB BT, depending on the purpose of the tests. For the HPL, the process grid size was chosen as close as possible to a square one. Due to the large execution time dispersion of the experiments, already noted in Sect. 4, about ten similar runs were made for each point of the parameter space, and the best result, i.e. that with the smallest execution time, was selected.

6.1 Verification of the Metric Correctness

The verification of the metric correctness is started by estimating the approximation error produced by the function $E(q)$, and only after obtaining acceptable values for the error, we start verifying the metric values themselves. The verification process can be described as follows. Firstly, for each matrix size, we calculated the coefficients of the approximating function in the entire process range $Q_{all} = Q(q_1, q_2, step)$, where q_1 and q_2 are, respectively, the minimum and maximum numbers of processes, and step is the step on the process grid. Next, we did the same in the subrange $Q_{part} = Q(q_1, q_2, k \cdot step)$, $k \in \mathbb{N}$, which was obtained from the first range by increasing the process grid step. Then, we calculated the relative error and the metric value in each case. Their values are given in Tables 4 and 5.

Table 4. Relative errors and metric values for HPL

Matrix size	Lomonosov		Lomonosov-2		
	14336	24576	14336	24576	51200
$Q(1, 128, 4)$	6154	8457	3933	5886	8208
	6.80%	3.35%	10.23%	9.70%	4.83%
$Q(1, 128, 8)$	6107	8379	3806	5807	8306
	5.81%	3.37%	9.40%	**10.42%**	4.61%
$Q(1, 128, 12)$	5994	8283	3816	5779	8124
	2.69%	2.62%	7.54%	7.58%	4.34%
$Q(1, 128, 16)$	6156	8438	3904	6006	8398
	5.45%	3.21%	6.44%	9.80%	3.22%
Metric dispersion	2.63%	2.05%	3.22%	**3.77%**	3.26%

Table 5. Relative errors and metric values for NPB BT (Lomonosov-2)

Matrix size	64^3	408^3
$Q(1, 121, step)$	9540	9731
	3.61%	3.06%
$Q(1, 121, 2step)$	9569	9482
	3.64%	3.62%
Metric dispersion	0.30%	2.55%

These tables show that the relative error of the approximation does not exceed, in the worst case, 10.5% for the HPL and 4% for NPB BT, for various computing systems and matrix sizes. Moreover, when the number of approximated points decreases, the error often drops. The metric values differ in less than 3.77% for each matrix size. We consider such errors acceptable and regard the metric as correct.

6.2 Comparison of Metric Values on Different Machines

Tables 4 and 5 point out that the metric values for the same launch parameters and the same application differ. This is one more proof that scalability is not only a property of the parallel application but also of the computing system which this application runs on. Based on the metric values, we can say that HPL scales better on the Lomonosov than it does on the Lomonosov-2.

Table 6 summarizes minimum and maximum values of the efficiency and average efficiency, for some cases of the process grid $Q(1, 128, 4)$. These data provide deeper insight into how metric values differ for different applications and computing systems.

Table 6. Values of the metric compared with the efficiency for HPL

HPL	Lomonosov		Lomonosov-2		
Matrix size	14336	24576	14336	24576	51200
Metric value	6154	8457	3933	5886	8208
Minimum efficiency	25%	46%	10%	23%	47%
Average efficiency	49%	67%	32%	47%	65%

6.3 Comparison of Different Applications

The comparison of different applications can be divided into three cases based on the process ranges: the ranges perfectly match each other, the ranges intersect, or they have no common points at all.

In the first case, when the ranges coincide, the comparison of applications is not complicated: we just need to compare the values of the metric on the current range. The biggest metric value characterizes the parallel application with better scalability. This fact follows from the method we used to introduce the metric.

In the second case, since the value of the metric is the integral of the positive function $E(q)$, it increases as the process range increases. Therefore, to compare applications on different process ranges, we should take their intersection and compare the scalability based on the values of the metric obtained in this intersection. Thus, this case can be reduced to the previous one.

Table 7 depicts an example of applications compared in different process ranges. This is an example taken from a real situation. In the case of a task of class D (a matrix of size 408^3), the number of processes can only start with

16 due to the particularities of NPB BT. The HPL test, on the contrary, has no restrictions. Launches on the Lomonosov-2 supercomputer were limited to 15 min. Consequently, the HPL test was executed on a number of processes equal or greater than 4, and NPB BT on 36. The intersection of their process ranges is [36; 256]. We calculated the efficiency by the formula $E_p = \frac{T_1}{pT_p}$. For this, we calculated the execution time on one process based on the minimum time for the number of processes, namely $T_1 = p_{min}T_{p_{min}}$. However, there is another problem: since the value of the metric implicitly depends on the size of the input data, it is necessary to determine which matrix size for the HPL is equivalent to the class D for BT, based on the number of operations.

No specific formulas for estimating the number of NPB BT test operations were found. Nonetheless, such characteristics as the number of operations per second and the execution time can be found in the output test file. Thus, determining the number of operations for a particular matrix size is not difficult. For the HPL benchmark, we determined the number of floating point operations by the formula $\frac{2n^3}{3} + 2n^2$, where n is the linear size of the matrix [26]. However, this is only a theoretical estimate of the number of operations that does not take into account the features of the compiler and the system, because of which the same application can be represented by a different number of instructions. Thus, to make the comparison correct, the HPL matrix size ($n = 44\,400$) corresponding to a class D task for NPB BT was determined from the information contained in the output file.

Table 7. Comparison of applications on different process ranges (Lomonosov-2)

	NPB BT, size = 408^3	HPL, size = $44\,400$
$Q(32, 256)$	20059	14088

As we can see, NPB BT has better scalability than HPL.

The third case is nontrivial even if the same application is compared since its behavior in different ranges can be completely different. The research of this case is still open.

7   Conclusions

The research described in this article was conducted in two directions. In the first part, we introduced a method for predicting scalability of parallel applications. This approach resorts to using the results of test executions on small configurations of computing systems to predict characteristics of the applications on larger configurations. In the second part, we formalized the requirements imposed upon the scalability metric of an application and formulated its desired properties. Next, we introduced a metric based on empirical data.

By means of computational experiments on the Lomonosov and Lomonosov-2 supercomputers using the HPL, NPB BT, SP, and LU benchmarks, we tested a method for predicting strong scalability of parallel applications on the basis of their execution times. The median relative error of the predictions ranged from 5.76 to 36.29%. We also analyzed the proposed metric and established its applicability.

We plan to achieve in further studies a greater universality of the metric. Firstly, this includes its extension to the multidimensional case, i.e. a metric that depends simultaneously on several launch parameters. Future studies should also comprise both a refinement of the method for predicting scalability and its generalization regarding wide and weak scaling.

Acknowledgements. The results were obtained with the financial support of the Russian Foundation for Basic Research (grant No. 18-29-03230). The research was carried out on the HPC equipment of the shared research facilities at Lomonosov Moscow State University.

References

1. Kazmina, K.P., Antonov, A.S.: Development of methods for predicting applications scalability on the configuration of supercomputers. In: Russian Supercomputing Days: Proceedings of the International Conference, pp. 858–869 (2018). https://doi.org/10.14489/vkit.2018.12.pp.045-056. (in Russian)
2. Linpack Benchmark. http://www.netlib.org/benchmark/linpackc.new
3. NAS Parallel Benchmarks. https://www.nas.nasa.gov/publications/npb.html
4. Sadovnichy, V., Tikhonravov, A., Voevodin, Vl., Opanasenko, V.: "Lomonosov": supercomputing at Moscow state university. In: Contemporary High Performance Computing: From Petascale toward Exascale (Chapman & Hall/CRC Computational Science), pp. 283–307. CRC Press, Boca Raton (2013)
5. Lomonosov-2 Supercomputer Characteristics. https://parallel.ru/cluster/lomonosov2.html
6. Antonov, A., Teplov, A.: Generalized approach to scalability analysis of parallel applications. In: Carretero, J., et al. (eds.) ICA3PP 2016. LNCS, vol. 10049, pp. 291–304. Springer, Cham (2016). https://doi.org/10.1007/978-3-319-49956-7_23
7. Yakovenko, P.N.: Analysis tools for parallel SPMD programs. In: Proceedings of the System Programming Institute RAS, vol. 3, no. 3, pp. 63–85 (2002). (in Russian)
8. Ivanov, D.E.: Scalable parallel genetic algorithm for building identifying sequences for modern multicore computing Systems. In: Control Systems and Machines, no. 1, pp. 25–32 (2011). (in Russian)
9. Gergel, V.P.: High performance computing for multi-core multiprocessor systems. Study Guide - Nizhny Novgorod. Publishing House of UNN them. N.I. Lobachevsky (2010). (in Russian)
10. Barnes, B.J., Rountree, B., Lowenthal, D.K., Reeves, J., de Supinski, B., Schulz, M.: A regression-based approach to scalability prediction. In: Proceedings of the ICS, pp. 368–377 (2008). https://doi.org/10.1145/1375527.1375580
11. Chatzopoulos, G., Dragojević, A., Guerraoui, R.: ESTIMA: Extrapolating ScalabiliTy of In-Memory Applications. ACM Trans. Parallel Comput. **4**, 2 (2017). https://doi.org/10.1145/2851141.2851159. Article 10

12. Ipek, E., de Supinski, B.R., Schulz, M., McKee, S.A.: An approach to performance prediction for parallel applications. In: Cunha, J.C., Medeiros, P.D. (eds.) Euro-Par 2005. LNCS, vol. 3648, pp. 196–205. Springer, Heidelberg (2005). https://doi.org/10.1007/11549468_24

13. Nadeem, F., Alghazzawi, D., Mashat, A., Fakeeh, K., Almalaise, A., Hagras, H.: Modeling and predicting execution time of scientific workflows in the Grid using radial basis function neural network. Cluster Comput. 20(3), 2805–2819 (2017). https://doi.org/10.1007/s10586-017-1018-x

14. Chen, T.-Y., Khalili, O., Campbell, R.L., Carrington, L., Tikir, M.M., Snavely, A.: Performance prediction and ranking of supercomputers. In: Advances in Computers, vol. 72, pp. 135–172 (2008). https://doi.org/10.1016/S0065-2458(08)00003-X

15. Escobar, R., Boppana, R.V.: Performance prediction of parallel applications based on small-scale executions. In: Proceedings of the HiPC, pp. 362–371 (2016). https://doi.org/10.1109/HiPC.2016.049

16. Lee, B.C., Brooks, D.M., de Supinski, B.R., Schulz, M., Singh, K., McKee, S.A.: Methods of inference and learning for performance modeling of parallel applications. In: Proceedings of the PPoPP, pp. 249–258 (2007). https://doi.org/10.1145/1229428.1229479

17. Marathe, A., Harris, R., Lowenthal, D.K., de Supinski, B.R., Rountree, B., Schulz, M.: Exploiting redundancy and application scalability for cost-effective, time-constrained execution of HPC applications on Amazon EC2. IEEE Trans. Parallel Distrib. Syst. 27(9), 2574–2588 (2016). https://doi.org/10.1109/TPDS.2015.2508457

18. Singh, K., et al.: Comparing scalability prediction strategies on an SMP of CMPs. In: D'Ambra, P., Guarracino, M., Talia, D. (eds.) Euro-Par 2010. LNCS, vol. 6271, pp. 143–155. Springer, Heidelberg (2010). https://doi.org/10.1007/978-3-642-15277-1_14

19. Rejitha, R.S., Benedict, S., Alex, S.A., Infanto, S.: Energy prediction of CUDA application instances using dynamic regression models. Computing 99(8), 765–790 (2017). https://doi.org/10.1007/s00607-016-0534-5

20. Escobar, R., Boppana, R.V.: Performance prediction of parallel scientific applications. HPDC Poster (2017)

21. Grobelny, E., Bueno, D., Troxel, I., George, A.D., Vetter, J.S.: FASE: a framework for scalable performance prediction of HPC systems and applications. Simulation 83(10), 721–745 (2007). https://doi.org/10.1007/978-3-642-25264-8_19

22. Zhai, J., Chen, W., Zheng, W., Li, K.: Performance prediction for large-scale parallel applications using representative replay. IEEE Trans. Comput. 65(7), 2184–2198 (2016). https://doi.org/10.1109/TC.2015.2479630

23. Levenberg, K.: A method for the solution of certain non-linear problems in least squares. Q. Appl. Math. 2(2), 164–168 (1944). https://doi.org/10.1090/qam/10666

24. Marquardt, D.: An algorithm for least-squares estimation of nonlinear parameters. SIAM J. Appl. Math. 11(2), 431–441 (1963). https://doi.org/10.1137/0111030

25. Ristov, S., Prodan, R., Gusev, M., Skala, K.: Superlinear speedup in HPC systems: why and when?. In: Federated Conference on Computer Science and Information Systems, vol. 8, pp. 889–898 (2016). https://doi.org/10.15439/2016F498

26. Linpack Benchmark Complexity. http://www.netlib.org/benchmark/performance.pdf

Parallel Numerical Algorithms

Using the AlgoWiki Open Encyclopedia of Parallel Algorithmic Features in HPC Education

Alexander Antonov$^{(\boxtimes)}$ and Vladimir Voevodin

Lomonosov Moscow State University, Moscow, Russia
{asa,voevodin}@parallel.ru

Abstract. AlgoWiki is an open encyclopedia of algorithm properties and their implementations on various hardware and software platforms. It can be used for various purposes, such as finding the optimal algorithm for addressing a particular problem, analyzing the information structure of an application, or comparing the efficiency of different implementations of the same algorithm. In this article, we consider several options for using AlgoWiki to illustrate various concepts and notions in the Body of Knowledge and Skills for parallel computation and supercomputer technologies. The approbation of these options is conducted as part of the courses "Supercomputers and Parallel Data Processing" and "Supercomputer Simulation and Technologies", taught in the Computational Mathematics and Cybernetics Department at Lomonosov Moscow State University.

Keywords: AlgoWiki · High-performance computing education · Structure of parallel algorithms · Information structure · Parallelism resource · Parallel programming · Supercomputers

1 Introduction

The demand for parallel computation is constantly growing. Technologies originally developed for supercomputers are becoming relevant for any computing devices, including mobile ones. This explains the growing need to train specialists in the field of parallel computation and supercomputer technologies. Software development in this field is quite active all around the world [1–5] and in Russia [6,7]. Interesting discussions on these issues are annually held within the framework of international seminars: EduHPC [8], EduPAR [9], Euro-EduPAR [10].

The Supercomputing Consortium of Russian Universities was established in Russia in 2008 [11]. The goal of the Consortium was to develop and implement a set of activities aimed at efficiently using the available higher-education capacities to develop and implement supercomputer technologies in Russian education, science, and industry. Since this moment, the Supercomputer Consortium has coordinated the development of educational programs for high-performance

L. Sokolinsky and M. Zymbler (Eds.): PCT 2019, CCIS 1063, pp. 77–90, 2019.
https://doi.org/10.1007/978-3-030-28163-2_6

computing in Russia. As of the end of 2018, the Supercomputing Consortium of Russian Universities features 48 permanent and 16 associate members.

From 2010 to 2012, the members of the Supercomputer Consortium played an active role in implementing a large-scale project for the President's Committee on Modernization and Technological Development of the Russian Economy called "Building a system to train highly skilled staff on supercomputing technologies and specialized software" ("Supercomputing Education"). The project involved dozens of Russian universities. The project featured a major body of work for developing a holistic supercomputer education system in Russia [12–14].

2   The Body of Knowledge and Skills

Part of the "Supercomputer Education" project was dedicated to building a Summary List (Body) of knowledge and skills (professional competencies) in the field of parallel computation and supercomputing technologies [15]. The Body describes the competencies in supercomputer education that students must possess once they graduate from the respective college or complete the corresponding training course, refresher course, or specialized training as part of special groups. The Body of Knowledge describes the structure and basis of the education process.

The main content of the Body of Knowledge is a description of the subject area of "Supercomputers and parallel computation," which clearly defines what should be taught and how the education process should be organized for each specific group of students. The structure of the Body of Knowledge follows the recommendations of such international professional communities as the Association for Computing Machinery (ACM) and the IEEE Computer Society [16, 17]. The Body of Knowledge is a set of five knowledge areas representing different aspects of the chosen specialization. The areas are divided into smaller units called sections which contain individual thematic modules within each respective area. Each section, in its turn, consists of a set of topics which represent the lowest level in the hierarchy for the given specialization.

The Body of Knowledge was the basis for a system for certifying training courses and programs on "Supercomputers and parallel computation" [18] and for the implementation of measures allowing to verify the aggregate content of training courses adopted in Russian universities for training, refresher training, and continuing education aimed at personnel working in the field of supercomputing technologies and parallel computation.

3   AlgoWiki, the Open Encyclopedia of Parallel Algorithmic Features

The AlgoWiki Open Encyclopedia of Parallel Algorithmic Features [19] has been developed since 2014 as an online project based on wiki-technologies. This project allows the entire computing community to participate in the description of algorithm properties. The project started with a universal structure for describing algorithm properties, which, in turn, enabled a common structure for explaining any computational algorithm [20,21]. When describing algorithms according to this structure, special emphasis is placed on their parallelism properties [22,23]. The description is divided into two parts. The first part describes the properties of an actual algorithm which are independent of the hardware and software platforms on which it will be implemented. The second part describes the properties of specific implementations of a given algorithm: both serial and parallel, intended for different hardware and software platforms [24].

Once the number of algorithms described [25] reached a certain limit, we needed to bring some order to this information. Since algorithms have no value in and of themselves but are needed to address specific tasks, the AlgoWiki Encyclopedia was complemented with a level of description related to the tasks being addressed. Each task can be addressed in various ways, so another level of description appeared related to the methods for addressing tasks. At the task and method levels, the descriptions can be prepared in an arbitrary manner; introducing a unified description structure for these levels can be the subject of further study for the project. Each method can be implemented via several algorithms, and each algorithm can have several implementations. Thus, it was natural to represent the subject area in a hierarchical form: a chain from task to method to algorithm to implementation [26].

Since the second part of an algorithm description gathers data, among other things, on execution results of the algorithm on various hardware and software platforms, we decided to use this data to conduct various comparisons. In this case, it is possible to perform comparisons using traditional computing platform ratings, similar to well-known supercomputer ratings (such as Top500, HPCG [27], Graph500 [28], and others), and to compare different implementations, algorithms, methods and tasks from the standpoint of their suitability for various hardware and software platforms [29]. The AlgoWiki Encyclopedia is working to implement "architectural profiles", which will allow subsets of tasks, methods, algorithms, and implementations to be singled out from the multitude of descriptions that best correspond to a specific type of software and hardware platforms.

4   The AlgoWiki Encyclopedia and the Body of Knowledge and Skills

When developing a training course on parallel computation and supercomputing technologies, it is important to take into account how training materials correspond to the Body of Knowledge and Skills. Many points in the Body of Knowledge must be illustrated with examples, and the AlgoWiki Open Encyclopedia of Parallel Algorithmic Features can be a great source of such examples. AlgoWiki furnishes plenty of useful information on all five areas of knowledge described by the Body of Knowledge.

Let us consider some examples of how the AlgoWiki Encyclopedia can be used to illustrate various concepts and notions from the Body of Knowledge and

Skills. However, it should be noted that the practical use of AlgoWiki is much wider and not limited to these few examples.

4.1 Knowledge Area 1. The Mathematical Basics of Parallel Computations

Section 1.1. Graph Models of Programs. AlgoWiki places special emphasis on the study of information graphs since these elements contain all of the information on an algorithm's information structure. An information graph is an oriented acyclic graph in which the vertices correspond to the operations of the algorithm, while the edges correspond to the actual information dependencies between them. Figure 1 shows an example of an information graph for an algorithm that multiplies a dense matrix by a vector, displaying the input and output data.

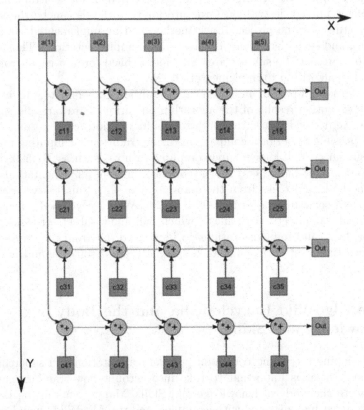

Fig. 1. Information graph for the multiplication of a dense matrix by a vector, displaying the input and output data

To study the parallelism resource of an algorithm, it may be necessary to conduct a study of its *parallel form*. Figure 1 is a representation of an information graph in which all of the vertices are partitioned into numbered subsets of levels, the initial point of each edge is located at a lower-numbered layer than the final point, and there may not be any edges between vertices of the same level. The parallel form can be visualized on an ordinary information graph by marking up the vertices belonging to different levels. The AlgoView interactive 3D visualization system for information graphs [30,31], created under the Algo-Wiki project, assists in visualizing the multilevel structure step by step. In this structure, the current step, previous steps, and subsequent steps of an algorithm are displayed in different colors. An example of the obtained image is given in Fig. 2.

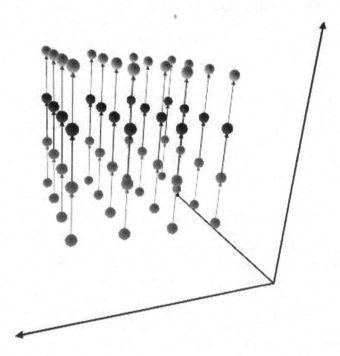

Fig. 2. Parallel form of an information graph for a dense-matrix multiplication algorithm

Section 1.2. Unlimited Parallelism Concept. The concept of unlimited parallelism stipulates that an algorithm is implemented on a parallel computing platform that does not impose any restrictions on its execution. There can be an infinite number of processors, they are all universal, working synchronously, having shared memory, and all information transfers are performed instantly and without conflicts [32].

Algorithm descriptions in AlgoWiki pay particular attention to the notion of parallel complexity, which is understood as the number of steps needed to execute the algorithm given an infinite number of processors (functional devices, computing nodes, cores, etc.), i.e. within the concept of unlimited parallelism. The parallel complexity of an algorithm is understood as the height of the canonical parallel form. Parallel complexity enables comparison of the parallelism resources of various algorithms. For example, the parallel complexity of an algorithm for summing vector elements by pairwise summation equals $\log_2 n$, the parallel complexity of a fast Fourier transform (Cooley–Tukey algorithm) for vectors with lengths equal to a power of two is $2 \log_2 n$, and the parallel complexity for dense-matrix multiplication is $2n^3$.

Section 1.3. Fine Information Structure of Programs.

The AlgoWiki Encyclopedia presents an extensive array of sample studies of parallel structure for various programs. They include all key *types of parallelism*:

- *finite parallelism*, determined by the information independence of certain fragments in the body of the program;
- *massive parallelism*, determined by the information independence of iterations within the program;
- *coordinate parallelism*, in which information-independent vertices lie in hyperplanes perpendicular to one of the coordinate axes;
- *skewed parallelism*, i.e. parallelism within the iteration space determined by the surfaces of unfolded levels.

Section 1.4. Equivalent Transformations of Programs.

Once the parallelism resource of a program fragment under investigation has been determined, the optimum usage option can be achieved by applying *equivalent transformations*, which are transformations of the code that preserve the exact result of the program execution. Since most parallelism resources are usually concentrated in loop structures, the most popular option for equivalent transformation is *elementary loop transformation*. Many examples of this transformation can be found in various algorithm implementations given in the AlgoWiki Encyclopedia. For example, one can study the efficiency of all possible loop interchanges in the main loop nest of the dense-matrix multiplication algorithm.

On the other hand, the available parallelism resources sometimes cannot be used if only equivalent transformations are applied. In certain occasions, *non-equivalent program transformations* can also be used. It should be noted, however, that the calculated results can vary in this case, for example, due to the accumulation of rounding errors in a different order of operation execution. A classic example of non-equivalent transformation for purposes of parallelization is summing arrays by pairwise summation, as per the information graph presented in Fig. 3.

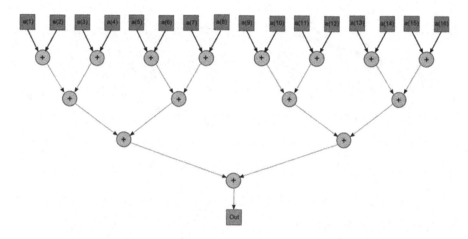

Fig. 3. Information graph for array summation by the pairwise summation algorithm

4.2   Knowledge Area 2. Parallel Computing Systems (Computing Basics)

Section 2.2. The Basics of Building Computing Systems. The AlgoWiki Encyclopedia does not directly include any information on building computer systems. However, for many topics related to using various types of devices (*scalar, vector, and pipeline functional units, accelerators, graphical processors, programmable logic processors (FPGA), specialized processors*), illustrations can be found in AlgoWiki in the form of algorithms and most suitable implementations for using them on a given type of device. The corresponding search can substantially simplify the "architectural profiles" mechanism currently being developed.

Traditional *memory access properties* are related to the computing system properties and are not directly explained in AlgoWiki either. However, a memory access property such as memory locality is determined by the properties of the algorithm and its implementation. This is why significant attention has to be paid to studying the locality of memory calls. Spatial and temporal localities are studied for all algorithms. Spatial locality is understood as the average distance between several consecutive memory calls. Temporal locality shows the average number of calls to the same memory address over the entire program execution time. Studying locality requires building memory access profiles for processing cores and key fragments. Figure 4 compares memory access profiles for six loop order options in the main computational core of the dense-matrix multiplication algorithm. Differences in the profiles determine different memory access properties, which in this case substantially affect the performance of the options under consideration. By analyzing the profiles, it is easy to explain why the fragments with the innermost loop of j show the highest performance in most modern architectures.

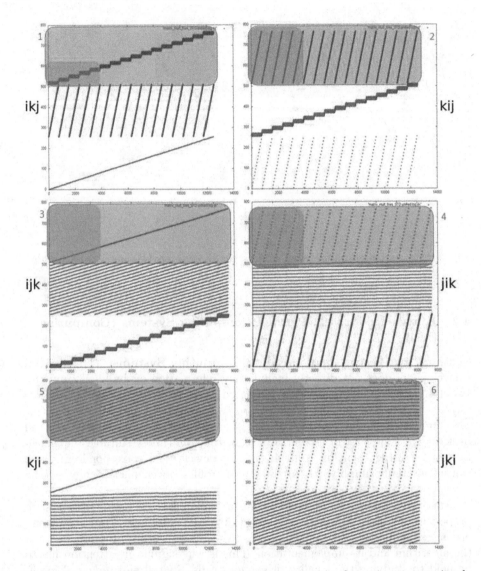

Fig. 4. Memory access profiles for six loop order options in the main computational core of the dense-matrix multiplication algorithm

Sections 2.4 ("Multi-CPU computing systems"), 2.5 ("Multi-CPU computing systems with shared memory"), 2.6 ("Multi-CPU computing systems with distributed memory"), and 2.7 ("Graphics Processing Units") describe different computing system properties, including those concepts that can be illustrated with AlgoWiki materials, such as *efficiency, performance, speedup,* and *scalability*. Illustrations for sections related to specific supercomputer architecture types can be found using the "architectural profiles" mechanism.

To illustrate the concept of *performance benchmarks*, the AlgoWiki Encyclopedia already describes the opportunity for using as a benchmark any implementation of any algorithm described. After obtaining data on the benchmark performance for a set of computing systems, it is possible to build ratings of these systems, either for a specific implementation or for an algorithm, a method, or even a task. Figure 5 shows an example rating for the Bellman–Ford algorithm [33–35], which computes the shortest possible path in a graph. The rating was built taking into account performance data from several implementations of this algorithm on the MSU Lomonosov and Lomonosov-2 supercomputers [36].

№	Problem	Algorithm	Implementation	Platform	Result (MTEPS)	CPU cores	Graph Type	Graph Size
1	Single Source Shortest Path	Bellman-Ford	RCC for GPU	Lomonosov-2	2129.0		SSCA-2	2^22
2	Single Source Shortest Path	Bellman-Ford	Graph500 MPI	Lomonosov	1611.0	8	SSCA-2	2^17
3	Single Source Shortest Path	Bellman-Ford	RCC for GPU	Lomonosov	1309.0		SSCA-2	2^20
4	Single Source Shortest Path	Bellman-Ford	RCC for GPU	Lomonosov	1300.0		SSCA-2	2^23
5	Single Source Shortest Path	Bellman-Ford	Ligra	Lomonosov-2	1187.0	14	RMAT	2^24
6	Single Source Shortest Path	Bellman-Ford	Ligra	Lomonosov-2	1100.0	14	RMAT	2^23
7	Single Source Shortest Path	Bellman-Ford	Ligra	Lomonosov-2	1075.0	14	RMAT	2^25
8	Single Source Shortest Path	Bellman-Ford	Ligra	Lomonosov-2	1035.0	14	RMAT	2^21
9	Single Source Shortest Path	Bellman-Ford	Ligra	Lomonosov-2	960.0	14	RMAT	2^22
10	Single Source Shortest Path	Bellman-Ford	Ligra	Lomonosov-2	874.0	14	RMAT	2^26
11	Single Source Shortest Path	Bellman-Ford	RCC for GPU	Lomonosov	687.0		RMAT	2^23
12	Single Source Shortest Path	Bellman-Ford	RCC for CPU	Lomonosov	609.169006	7	SSCA-2	2^19
13	Single Source Shortest Path	Bellman-Ford	RCC for CPU	Lomonosov	580.653015		SSCA-2	2^19
14	Single Source Shortest Path	Bellman-Ford	RCC for CPU	Lomonosov-2	564.0	14	RMAT	2^24
15	Single Source Shortest Path	Bellman-Ford	RCC for CPU	Lomonosov	546.664978	8	SSCA-2	2^19
16	Single Source Shortest Path	Bellman-Ford	RCC for GPU	Lomonosov	516.0		SSCA-2	2^25

Fig. 5. Ratings built for the Bellman–Ford algorithm

The *Linpack benchmark* algorithm [37] is also described in AlgoWiki, and this description can provide extensive material for studying the algorithm in detail if necessary.

Section 2.8. Computing Systems of Trans-PetaFLOPS and ExaFLOPS Performance. The *Top500* and *Top50* ratings of top-performance systems are based on Linpack benchmark results. However, the topic can be substantially complemented using the AlgoWiki Encyclopedia, not just by employing algorithm descriptions used by other known ratings (e.g., HPCG [38]), but also by applying the Top500 rating methodology to any other algorithm.

4.3   Knowledge Area 3. Parallel Programming Technologies (Basics of Software Engineering)

Section 3.1. General Parallel Application Development Principles. The *parallelization of serial programs* is one of the key objectives in the field of parallel computation. Parallelization must be based on a study of the parallelism resource and other properties of the algorithm and a consideration of the peculiarities of the target software and hardware platforms. In addition to a study of algorithm properties, the AlgoWiki Encyclopedia provides options for their serial and parallel implementations.

Section 3.3. Methods and Technologies for Parallel Application Development. When studying various *parallel programming technologies*, such as *MPI, OpenMP, CUDA*, and others, one can use as examples algorithms presented in the AlgoWiki Encyclopedia having implementations that resort to these technologies.

Section 3.4. Parallel Problem-Oriented Libraries and Program Sets. Many algorithms presented in *parallel problem-oriented libraries*, such as *BLAS, Lapack, Scalapack, FFTW, PETSc, MKL*, and others, are described in the AlgoWiki Encyclopedia. Studying these algorithms can help in selecting the right library functions.

4.4   Knowledge Area 4. Parallel Algorithms for Addressing Tasks

Section 4.1. General Parallel Algorithm Development Principles. The traditional *stages for addressing tasks in parallel computing systems* correspond well with the hierarchical representation of the subject area as a hierarchical form: a chain from task to method to algorithm to implementation, as in Algo-Wiki. In this case, the actual "Algorithm Classification" page can act as a good illustration of the concept being studied.

When studying *quality parameters of parallel programs*, such as *speedup, efficiency*, and *scalability*, one can use the data provided in the second part of the algorithm description in the AlgoWiki Encyclopedia. For example, the charts provided offer a good illustration of the scalability of algorithm implementations on specific computing systems. By way of illustration, Fig. 6 shows changes in the performance of the Linpack benchmark (HPL implementation) depending on the number of CPUs and the matrix size, obtained from running the test on the Lomonosov supercomputer.

For every algorithm described in AlgoWiki, an assessment of the *computational complexity of serial and parallel algorithms* is provided. The authors show *asymptotic evaluations* which help to understand the algorithm properties when working with large amounts of data.

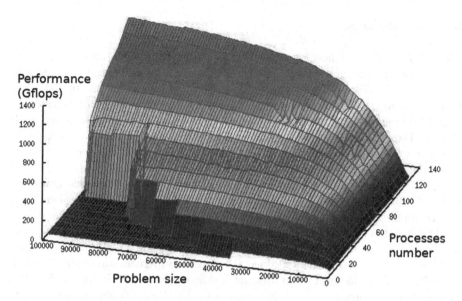

Fig. 6. Performance of the Linpack benchmark (HPL implementation) depending on the number of CPUs and matrix size

Section 4.2. Educational Algorithms for Parallel Programming. Almost all the algorithms in this section are already described in the AlgoWiki Encyclopedia. This can provide exhaustive data for studying the respective algorithms.

Algorithms from sections 4.3 ("Parallel algorithms for matrix computations"), 4.4 ("Parallel algorithms for data search and sorting"), 4.5 ("Parallel algorithms for graph computations"), 4.6 ("Parallel algorithms for solving partial differential equations"), 4.7 ("Parallel algorithms for solving optimization tasks"), 4.8 ("Monte Carlo parallel algorithms"), and 4.9 ("Parallel algorithms for other classes of computationally intensive tasks") are partially represented in the AlgoWiki Encyclopedia. Descriptions of algorithms not currently represented in AlgoWiki can easily be added by any member of the computing community.

4.5 Knowledge Area 5. Parallel Computations, Big Data Tasks, and Specific Subject Areas

The tasks listed in sections 5.1 ("Parallel methods for solving computationally complex tasks in Earth sciences"), 5.2 ("Parallel methods for solving computationally complex tasks in life sciences"), 5.3 ("Parallel methods for solving computationally complex tasks in engineering calculations"), 5.4 ("Parallel methods for solving computationally complex tasks in quantum chemistry"), and 5.5 ("Parallel methods for solving atomistic modeling tasks") can provide extensive material to add to the AlgoWiki Encyclopedia in the future.

5   Conclusions

The AlgoWiki Open Encyclopedia of Parallel Algorithmic Features is a project carried out at Lomonosov Moscow State University. In addition to classic use cases, the AlgoWiki Encyclopedia finds increasingly common use in the education process. Descriptions of methods, algorithms, and implementations from AlgoWiki are used with illustrative purpose in the courses "Supercomputers and Parallel Data Processing" and "Supercomputer Simulation and Technologies", which are taught in the Computational Mathematics and Cybernetics Department at Lomonosov Moscow State University.

In this article, we used AlgoWiki materials to illustrate various concepts of the Body of Knowledge and Skills in the field of parallel computation and supercomputer technologies, a building block in various educational courses.

The benefits of the AlgoWiki Encyclopedia are not limited to illustrative materials for various sections of numerous training courses. Throughout several years, fourth-year students taking the course "Supercomputers and Parallel Data Processing" in the Computational Mathematics and Cybernetics Faculty at Lomonosov Moscow State University have been given a practical assignment that includes algorithm descriptions in the AlgoWiki Encyclopedia. From 2016 to 2018, large-scale practical assignments were organized at the same faculty for second-year Master's degree students as part of the course "Supercomputer Modelling and Technologies" [39]. The assignment was modified slightly each year but always involved studying and describing a series of algorithm properties and implementations. Each practical assignment was given to about 200–250 Master's degree students, and the results were a valuable contribution towards the development of the AlgoWiki Encyclopedia.

The development of the AlgoWiki Open Encyclopedia of Parallel Algorithmic Features continues. In addition to expanding the description database, the authors are working on improving the functions implemented. Future plans include building a rating system for implementations, algorithms, methods, and tasks, as well as introducing "architectural profiles" of the Encyclopedia.

Acknowledgements. The results were obtained with the financial support of the Russian Foundation for Basic Research (grant N. 19-07-01030). The research was carried out on the HPC equipment of the shared research facilities at Lomonosov Moscow State University and was supported through the project RFMEFI62117X0011.

References

1. SIAM: Graduate Education for Computational Science and Engineering. SIAM Working Group on CSE Education (2014). http://www.siam.org/students/resources/report.php
2. Future Directions in CSE Education and Research. Report from a Workshop Sponsored by the Society for Industrial and Applied Mathematics (SIAM) and the European Exascale Software Initiative (EESI-2). http://wiki.siam.org/siag-cse/images/siag-cse/f/ff/CSE-report-draft-Mar2015.pdf

3. Prasad, S.K., et al.: NSF/IEEE-TCPP Curriculum Initiative on Parallel and Distributed Computing – Core Topics for Undergraduates, Version I. 55 p. (2012). http://www.cs.gsu.edu/~tcpp/curriculum

4. Qasem, A.: Modules for teaching parallel performance concepts. In: Prasad, S.K., Gupta, A., Rosenberg, A., Sussman, A., Weems, C. (eds.) Topics in Parallel and Distributed Computing, pp. 59–77. Springer, Cham (2018). https://doi.org/10.1007/978-3-319-93109-8_3

5. Ghafoor, S., Brown, D.W., Rogers, M.: Integrating parallel computing in introductory programming classes: an experience and lessons learned. In: Heras, D.B., Bougé, L. (eds.) Euro-Par 2017. LNCS, vol. 10659, pp. 216–226. Springer, Cham (2018). https://doi.org/10.1007/978-3-319-75178-8_18

6. Voevodin, V., Gergel, V.: Supercomputing education: the third pillar of HPC. Comput. Methods Softw. Dev. New Comput. Technol. **11**(2), 117–122 (2010)

7. Supercomputing Education in Russia. Supercomputing Consortium of the Russian Universities. Technical report (2012). http://hpc.msu.ru/files/HPC-Education-in-Russia.pdf

8. EduHPC-18: Workshop on Education for High-Performance Computing. https://grid.cs.gsu.edu/~tcpp/curriculum/?q=eduhpc18

9. 8th NSF/TCPP Workshop on Parallel and Distributed Computing Education (EduPar-18). https://grid.cs.gsu.edu/~tcpp/curriculum/?q=edupar18

10. 4th European Workshop on Parallel and Distributed Computing Education for Undergraduate Students (Euro-EDUPAR). http://www.euroedupar.tu-darmstadt.de/

11. HPC Consortium of Russian Universities. http://hpc-russia.ru

12. Voevodin, V., Gergel, V., Sokolinsky, L., Dyomkin, V., Popova, N., Bukhanovskiy, A.: Supercomputing education in Russia: results and perspectives. Vestnik Lobachevsky Univ. Nizhni Novgorod **1**(4), 203–209 (2012)

13. Antonov, A.S., et al.: Supercomputing education project: year 2012. Vestnik Lobachevsky Univ. Nizhni Novgorod **1**(1), 12–16 (2013)

14. Antonov, A.S., Voevodin, Vl.V., Gergel': A systematic approach to supercomputing education. Bull. South Ural State Univ. **2**(2), 5–17 (2013). Series "Computational Mathematics and Informatics"

15. The Body of Knowledge and Skills in Supercomputing Technologies. http://hpc-education.ru/?q=node/15

16. Computer Science Curricula (2013). http://ai.stanford.edu/users/sahami/CS2013

17. Prasad, S.K., Weems, C.C., Dougherty, J.P., Deb, D.: NSF/IEEE-TCPP curriculum initiative on parallel and distributed computing: status report. In: Proceedings of the 49th ACM Technical Symposium on Computer Science Education (SIGCSE 2018), pp. 134–135. ACM, New York (2018). https://doi.org/10.1145/3159450.3159632

18. Antonov, A.S., Voevodin, Vl.V., Odintsov, I.O.: Methodology for certification of training courses and programs in the area of studies "supercomputers and parallel computing". Vestnik Lobachevsky Univ. Nizhni Novgorod **2**(1), 13–18 (2014)

19. Open Encyclopedia of Parallel Algorithmic Features. http://algowiki-project.org

20. Antonov, A.S., Voevodin, Vad.V., Voevodin, Vl.V., Teplov, A.M., Frolov, A.V.: First version of an open encyclopedia of algorithmic features. Vestnik UGATU **19**(2(68)), 150–159 (2015). Series "Management, Computer Engineering and Informatics"

21. Voevodin, V., Antonov, A., Dongarra, J.: AlgoWiki: an open encyclopedia of parallel algorithmic features. Supercomput. Front. Innov. **2**(1), 4–18 (2015)

22. Voevodin, V., Antonov, A., Dongarra, J.: Why is it hard to describe properties of algorithms? Proc. Comput. Sci. **101**, 4–7 (2016). https://doi.org/10.1016/j.procs.2016.11.002
23. Voevodin, Vl., Antonov, A., Voevodin, Vad.: What do we need to know about parallel algorithms and their efficient implementation? In: Prasad, S.K., Gupta, A., Rosenberg, A., Sussman, A., Weems, C. (eds.) Topics in Parallel and Distributed Computing, pp. 23–58. Springer, Cham (2018). https://doi.org/10.1007/978-3-319-93109-8_2
24. Antonov, A., Voevodin, Vad., Voevodin, Vl., Teplov, A.: A study of the dynamic characteristics of software implementation as an essential part for a universal description of algorithm properties. In: 24th Euromicro International Conference on Parallel, Distributed, and Network-Based Processing Proceedings, 17–19 February, pp. 359–363 (2016)
25. Antonov, A., et al.: Parallel processing model for Cholesky decomposition algorithm in AlgoWiki project. Supercomput. Front. Innov. **3**(3), 61–70 (2016). https://doi.org/10.14529/jsfi160307
26. Antonov, A., Frolov, A., Konshin, I., Voevodin, Vl.: Hierarchical domain representation in the AlgoWiki encyclopedia: from problems to implementations. In: Sokolinsky, L., Zymbler, M. (eds.) PCT 2018. CCIS, vol. 910, pp. 3–15. Springer, Cham (2018). https://doi.org/10.1007/978-3-319-99673-8_1
27. Top500 Supercomputer Sites. https://www.top500.org
28. Graph 500. https://graph500.org
29. Antonov, A., Dongarra, J., Voevodin, Vl.: AlgoWiki project as an extension of the Top500 methodology. Supercomput. Front. Innov. **5**(1), 4–10 (2018). https://doi.org/10.14529/jsfi180101
30. Antonov, A.S., Volkov, N.I.: An AlgoView web-visualization system for the AlgoWiki project. Commun. Comput. Inf. Sci. **753**, 3–13 (2017). https://doi.org/10.1007/978-3-319-67035-5_1
31. Antonov, A., Volkov, N.: Interactive 3D representation as a method of investigating information graph features. In: Voevodin, V., Sobolev, S. (eds.) RuSCDays 2018. CCIS, vol. 965, pp. 587–598. Springer, Cham (2019). https://doi.org/10.1007/978-3-030-05807-4_50
32. Voevodin, V., Voevodin, Vl.: Parallel Computing, p. 608. BHV-Petersburg, St. Petersburg (2002)
33. Bellman, R.: On a routing problem. Q. Appl. Math. **16**, 87–90 (1958)
34. Ford, L.R.: Network Flow Theory. Rand.org, RAND Corporation (1958)
35. Moore, E.F.: The shortest path through a maze. In: International Symposium on the Theory of Switching, pp. 285–292 (1959)
36. Sadovnichy, V., Tikhonravov, A., Voevodin, V., Opanasenko, V.: Lomonosov: supercomputing at Moscow state university. In: Contemporary High Performance Computing: From Petascale Toward Exascale. Chapman & Hall/CRC Computational Science, Boca Raton, pp. 283–307 (2013)
37. Dongarra, J.J., Luszczek, P., Petitet, A.: The LINPACK benchmark: past, present and future. Concurr. Comput.: Pract. Exp. **15**(9), 803–820 (2003)
38. Dongarra, J., Heroux, M.A., Luszczek, P.: HPCG benchmark: a new metric for ranking high performance computing systems. Technical report. Electrical Engineering and Computer Science Department, Knoxville, Tennessee, UT-EECS-15-736 (2015)
39. Antonov, A., Popova, N., Voevodin, Vl.: Computational science and HPC education for graduate students: paving the way to exascale. J. Parallel Distrib. Comput. **118P1**, 157–165 (2018). https://doi.org/10.1016/j.jpdc.2018.02.023

On an Integrated Computational Environment for Numerical Algebra

Valery Il'in$^{(\boxtimes)}$

Institute of Computational Mathematics and Mathematical Geophysics SB RAS,
Novosibirsk State University, Novosibirsk, Russia
`ilin@sscc.ru`

Abstract. This paper describes the conception, general architecture, data structure, and main components of an Integrated Computational Environment (ICE) for the high-performance solution of a wide class of numerical algebraic problems on heterogeneous supercomputers with distributed and hierarchical shared memory. The tasks considered include systems of linear algebraic equations (SLAEs), various eigenvalue problems, and transformations of algebraic objects with large sparse matrices. These tasks arise in various approximations of multidimensional initial boundary value problems on unstructured grids. A quite large variety of types of matrices, featuring diverse structural, spectral, and other properties are allowed; there can also be a wide diversity of algorithms for computational algebra. There are relevant issues associated with scalable parallelism through hybrid programming on heterogeneous multiprocessor systems, MPI-processes, multithread computing, and vectorization of operations, including those without formal constraints on the number of degrees of freedom and on the number of computing units. The numerical methods and technologies are implemented in the KRYLOV library, which provides the integrated subsystem of the ICE. There are various technical requirements imposed upon the software: extendibility of the set of problems and algorithms, adaptation to the evolution of supercomputer architecture, ability to reuse external products, and coordinated participation of development groups taking part in the project. The end goal of these requirements is to provide a product featuring a long life cycle, high performance, and general acceptance among end users of diverse professional backgrounds.

Keywords: Computational algebra · Sparse matrix ·
Iterative method · High performance · Krylov subspaces ·
Scalable parallelism · Heterogeneous supercomputer ·
Domain decomposition

1 Introduction

Solving systems of linear algebraic equations (SLAEs) is by far the main task in computational algebra, which is, in turn, the key stage when solving problems of mathematical modeling of processes and phenomena. After discretization, approximation, and linearization are carried out, all the diversity of initial

© Springer Nature Switzerland AG 2019
L. Sokolinsky and M. Zymbler (Eds.): PCT 2019, CCIS 1063, pp. 91–106, 2019.
https://doi.org/10.1007/978-3-030-28163-2_7

statements (e.g., classical and generalized, differential and/or integral, stationary and evolutionary, linear and nonlinear) leads to the necessity of solving SLAEs. This stage is a "bottleneck" in large-scale computational experiments where the number of resources required by applications grows, as a rule, nonlinearly with respect to the number of degrees of freedom and accounts for the main part in estimates of the computational complexity. Among other tasks of numerical algebra, we can mention eigenvalue problems, systems of nonlinear algebraic equations (SNLAEs), matrix equations, and various matrix and/or vector transformations.

Naturally, the main object of our attention is the solution of "large" and "difficult" systems of equations on modern heterogeneous multiprocessor computing systems (MPS) with distributed and hierarchical shared memory. A SLAE with a number of unknowns N of the order of 10^8 to 10^{11} is considered to be large, and a SLAE in which the matrices have condition numbers between 10^{12} and 10^{15} is considered to be difficult. We henceforth assume that arithmetic operations are performed with numbers in standard 64-bit double-precision floating-point format, which gives a relative error of about 10^{-15}. It is important to emphasize that with the advent of supercomputers capable of post-petaFLOPS performances and more, the task of solving SLAEs does not lose its relevance. Indeed, as a matter of fact, "appetite comes with eating" and, as supercomputer performances increase, both the order of current implementations of algebraic systems and their condition numbers also grow. One can confidently predict that, in the near future, double precision will not be sufficient for stable calculations. For example, the problem of using "smart" arithmetic operations with variable numbers of digits, including representations with numbers of bits both greater than 100 and less than 50, will become unavoidable.

SLAEs of considerable interest are those obtained after the original multidimensional boundary value problems for differential equations or their generalized variational statements are approximated by finite-difference methods, finite volume methods, finite element methods, or discontinuous Galerkin methods (FDM, FVM, FEM, or DGM, respectively) on unstructured grids. The matrices generated in this manner have two important features. First, they are sparse and banded, i.e. the total number of nonzero entries is relatively small (NNZ $\ll N^2$) and, moreover, all of them are located in a band of width $M \ll N$ around the main diagonal. Secondly, their portraits have an irregular structure, i.e. the numbers of non-diagonal nonzero elements in the matrix lines can be specified only by enumeration. This leads to the need of saving the matrices in sparse compressed formats (for example, Compressed Sparse Row (CSR)) which rely on storing in memory only the nonzero entries and their numbers. This fact implies a significant specificity for the software implementations of the algorithms and slows down the process of access to the values of the matrix entries.

An important feature of SLAEs that significantly affects algorithms created for their solution is the block structure of the matrix. It is determined by the specific properties of the initial boundary value problem, the grid and approximation methods used, and by the ordering of the vector entries. The most complicated matrices, in the structural sense, are those arising from grid approximations of multidimensional interdisciplinary problems described by systems of functional equations with many unknown vector functions (for example, density distributions of simulated substances present temperature, velocity, electromagnetic fields, etc.).

The main means for solving large systems are iterative methods, thanks to their less strict requirements on the amount of memory used and the number of computational operations needed. The most effective and common are the preconditioned algorithms in the Krylov subspaces. During decades, these algorithms have been used in active research. From the large amount of literature devoted to them, we will restrict our references in this paper to the monographs [1, 2].

A special place in these issues corresponds to the problem of scalable parallelization of algorithms on MPS. Here, the main tool is the domain decomposition method (DDM), which has become one of the current divisions of computational algebra and to which special monographs, conferences, and Internet sites have been devoted (see [3, 4]). The primary principle of the DDM is the partitioning of the original problem for a complex calculation domain into interconnected auxiliary problems for subdomains. Then, each subproblem can be synchronously solved approximately by a processor of a supercomputer. As for the parallelization technologies, the scaling is attained using MPI-processes, multithreaded computing, and operation vectorization. In a sense, multigrid approaches appeared as an alternative to the DDM (see [5] and the literature cited there). Those approaches give asymptotically optimal (from the theoretical standpoint) estimations of the computation volume which are inadequate for parallelization on MPS.

The study of iterative methods and their practical applications has recently been a matter of research in two major directions. The first of them is the development of efficient methods for preconditioning the initial SLAEs with the aim of improving their conditioning. To this end, numerous algorithms for approximate matrix vectorization have been deeply involved (see the reviews [1, 2]). The second direction is the development of iterative processes assuming they are considered in Krylov subspaces. In this case, there are various approaches associated with a number of methods: deflation, aggregation, coarse-grid correction, low-rank matrix approximations, etc. [6, 7].

Another important fact is that a quite large number of algebraic methods have already been implemented and are widely available over computer networks. A rather complete list of such methods is given in [8]. In this respect, it is worth mentioning the BLAS and SPARSE BLAS libraries of vector-matrix operations, containing, in particular, high-performance implementations on supercomputers of different types. The use of such standard functions in applications significantly increases, at large, the efficiency of the designed software.

The objective of this study is to develop the concept, architecture, data structure, and main components of an Integrated Computing Environment (ICE) for the high-performance solution of a wide class of SLAEs. We focus on the creation of a product that can be effectively used in applied software systems and is capable of motivating a high demand among end users of diverse professional backgrounds. The considered mathematical software is a part of the KRYLOV library [9], which is a subsystem of the Basic System Modeling (BSM) [10], designed to support all major technological stages of mathematical modeling.

The paper is organized as follows. In Sect. 2, we classify the types of SLAEs considered in the paper. Sections 3 and 4 offer a brief presentation of methods for solving algebraic systems and existing software for their implementation. The next section describes the general structure and main components of the KRYLOV library. In the Conclusions section, we discuss plans for a further research on the development of an integrated tool environment for solving problems of computational algebra. For the sake of brevity, we often omit the references to terms and concepts from computational algebra that can be easily found on the Internet.

2   Classification of the Considered Tasks

From a formal point of view, we consider a trivial mathematical problem involving a SLAE:

$$Au = f, \quad A = \{a_{l,m}\} \in \mathcal{R}^{N,N}, \quad u = \{u_l\}, \quad f = \{f_l\} \in \mathcal{R}^N, \tag{1}$$

where, for simplicity, the matrix A of order N is assumed to be real, square, and positive definite in the sense of the condition

$$(Av, v) \geq \delta \|v\|^2, \quad \delta > 0, \quad (v, w) = \sum_{i=1}^{N} v_i w_i, \quad \|v\|^2 = (v, v). \tag{2}$$

However, symmetry is not required. Most algorithms considered below are applicable to more general SLAEs: complex (Hermitian or non-Hermitian), degenerate, or non-definite. A particular attention is given to sparse systems arising from the approximation of multidimensional boundary value problems by various methods, such as finite differences, finite volume, finite elements, and the discontinuous Galerkin algorithms on unstructured grids (see [8] and the literature cited therein). This type of grid equations is, for simplicity, written as

$$a_{i,i}u_i + \sum_{j \in \omega_i} a_{i,j}u_j = f_i, \quad i \in \Omega^h, \tag{3}$$

where ω_i denotes the set of off-diagonal nonzero entries placed on the i-th row of the matrix. On the other hand, sometimes matrices are presented in a block form,

$$A_{q,q}u_q + \sum_{r \in \Omega q} A_{q,r}u_r = f_q, \quad q = 1, \ldots, P, \tag{4}$$

$$A_{q,r} \in \mathcal{R}^{N_q, N_r}, \quad f_q \in \mathcal{R}^{N_q}, \quad N_1 + \ldots + N_P = N,$$

where Ω_q is the set of numbers of nonzero matrix computing blocks located in the q-th block row.

In the case of grid algebraic equations, the block representation of a SLAE can be visually associated with the geometric decomposition of the grid computation domain Ω to non-intersecting subdomains Ω_q:

$$\Omega = \Omega_1 + \ldots + \Omega_P, \quad \text{where} \quad \Omega_q \cap \Omega_r = 0 \quad \text{for} \quad q \neq r. \tag{5}$$

Relations (5) can also be interpreted as an algebraic decomposition if Ω_q is simply considered as a subspace of N_q induces corresponding to the u_q or f_q subvector.

The transformation of matrices to the block form is associated with renumbering vector components and matrix rows, which can generate a large number of computational methods and technologies. For example, based on the initial algebraic decomposition of a grid computational domain of the form of (5) without intersections, it is possible to form a decomposition with parametrized intersections for different numbers of common grid layers. Moreover, for large matrices, it is natural to consider their memory representation in a distributed way, both in the sense of various physical processors and at the logical level of MPI processes. Similar problems of renumbering and shaping various block structures also arise in efficient multigrid approaches to solving SLAEs (see the review in [5]).

The structural properties of the matrices under consideration strongly depend on the features of the initial boundary value problems and on the methods applied for their approximation. Typical block characteristics and matrix portraits can be classified according to the type of a system of equations for the corresponding applied problems: heat and mass transfer, hydro-gas dynamics, stress-strain state, multiphase filtration, electromagnetism, and others. It is also important to emphasize that, from theoretical and practical points of view, we are not interested in solving a specific SLAE on a fixed grid with a characteristic step h but a series of algebraic systems of the same type on a sequence of condensed (possibly nested) grids.

During last decades, approximations of higher orders have been made more exact, reaching an error in the numerical solution of the order of $O(h^\gamma)$, where $\gamma > 1$. As the order grows (up to $\gamma = 2, 4, 6, \ldots$), the number NNZ of nonzero matrix entries and the computational complexity of the algorithms increase, but the amount of memory required to ensure a given accuracy of the result significantly decreases. The last factor leads to a reduction in communications, and it is well known that communications not only slow down the overall computational process but also constitute the most energy-intensive operations.

From a technological point of view, it is also important to classify matrices according to the methods used f or storing them. It should be noted that, in addition to the universal CSR format, there are other common matrix representations and software converters from one format to another (see, for example, the Intel MKL library [11]). Let us mention, in particular, "small-block" formats, which are built similarly to CSR but store matrices of a fixed small size instead of numeric entries of the matrix. It is also natural that more efficient formats can be created for some matrices of simple structure. For example, for a symmetric matrix, it suffices to store only the main diagonal and one of its triangular parts (above or below the main diagonal). In the case of a Toeplitz, a quasi-Toeplitz, or a band matrix, the problem of storing its nonzero entries is absolutely trivial. Finally, let us note a special type called the block-structured matrices, characterized by different storage methods for different blocks. Also, representations can be highly efficient when using quasi-structured grids [12], in which the computational domain is divided into subdomains with different types of grids, including regular and uniform ones.

We should also note such tasks as the solution of SLAEs at each time step using implicit approximations for the numerical integration of multidimensional initial boundary value problems. In [7], for example, it was shown that a special choice of initial approximations using the method of least squares significantly reduces the computational complexity.

3   Preconditioned Methods in Krylov Subspaces

In this section, we give a brief overview of modern iterative algorithms for solving SLAEs, their parallelization, and software implementations in public libraries.

3.1   Multi-conditional Methods for Semi-conjugate Directions

Preconditioned iterative methods in Krylov subspaces are currently the main approaches to solving large sparse SLAEs. By their computational complexity and resource-intensity, and by the methods of implementation, these methods are divided into two groups: for symmetric and for asymmetric algebraic systems. The first group includes the efficient conjugate gradient (CG) and the conjugate residual (CR) methods, based on short (two-term) recursions. To solve asymmetric systems, there are algorithms based on the biorthogonalization of the computed vectors, also using short recursions. Examples from this group are the BiCG and the BiCR methods, as well as their stabilized versions: BiCGStab and BiCRStab. These algorithms are less developed from the theoretical and practical standpoints. The most popular and reliable are the generalized minimal residual methods (GMRES or FGMRES), which have theoretically an optimal rate of convergence of iterations and high practical reliability (robustness), but are based on computations of long vector recursions.

We consider methods of semi-conjugate directions that are equivalent in the rate of convergence of iterations and are a direct generalization of the CG and CR algorithms for asymmetric SLAEs. We give a formal presentation of these iterative processes in a "multi-conditional" version with the possibility of using several preconditioning matrices at each step; the number of such matrices and their form may vary. These methods belong to the class of block methods since they use several directional vectors. Therefore, it is natural to consider the Krylov subspaces generated by these vectors as blocks (see [13] and the references cited therein):

$$r^0 = f - Au^0, \quad n = 0, \ldots, \quad u^{n+1} = u^n + P_n \bar{a}_n,$$
$$r^{n+1} = r^n - AP_n \bar{a}_n = r^q - AP_q \bar{a}_q - \cdots - AP_n \bar{a}_n, \quad 0 \le q \le n, \qquad (6)$$
$$P_n = (p_q^n, \ldots, p_{M_n}^n) \in \mathcal{R}^{N, M_n}, \quad \bar{a}_n = (a_{n,1}, \ldots, a_{n,1})^\top \in \mathcal{R}^{M_n}.$$

Here, $\bar{a}_n$ are the coefficient vectors and P_n is a matrix in which each of its M_n columns is a direction vector p_l^n associated with the corresponding precondition matrix $B_{n,l}$, $l = 1, \ldots, M_n$, whose form is not determined yet. Relation (6) has a remarkable property, namely if the directional vectors satisfy the orthogonality conditions

$$(Ap_k^n, A^\gamma p_{k'}^{n'}) = \rho_{n,k}^{(\gamma)} \delta_{n,n'}^{k,k'}, \quad \rho_{n,k}^{(\gamma)} = (Ap_k^n, A^\gamma p_{k'}^{n'}),$$
$$\gamma = 0, 1, \quad n' = 0, 1, \ldots, n-1, \quad k, k' = 1, 2, \ldots, M_n, \qquad (7)$$

then the following relations are valid for the residual functionals:

$$\Phi_n^{(\gamma)}(r^{n+1}) \equiv (r^{n+1}, A^{\gamma-1} r^{n+1}) = (r^q, A^{\gamma-1} r^q) - \frac{\sum\limits_{k=q}^{n} \sum\limits_{l=1}^{M_n} (r^q, A^\gamma P_l^k)^2}{\rho_{k,l}^{(\gamma)}}, \qquad (8)$$

unless the coefficients are defined by the formulas

$$\alpha_{n,l} = \frac{\sigma_{n,l}}{\rho_{n,n}^{(\gamma)}}, \quad \sigma_{n,l} = (r^0, A^\gamma P_l^n), \qquad (9)$$

in which case functional (8) attains a minimum in the block subspace

$$\mathcal{H}_n = \mathrm{Span}\{r^0, Ap_1^0, \ldots, Ap_{M_0}^0, \ldots, Ap_1^n, \ldots, Ap_{M_n}^n\}, \quad M = 1 + M_0 + \cdots + M_n, \quad (10)$$

if $\gamma = 1$. For a symmetric matrix A, the minimum of $\Phi_n^{(\gamma)}$ is attained for any value of γ.

The orthogonality conditions are satisfied, in particular, if the directional vectors are determined with the help of certain preconditioning non-degenerate matrices $B_{n,l}$ from the following recurrence relations:

$$p_l^0 = B_{0,l}^{-1} r^0, \quad p_l^{n+1} = B_{n+1,l}^{-1} r^{n+1} - \sum_{k=0}^{n} \sum_{l=1}^{M_k} \beta_{n,k,l}^{(\gamma)} p_l^k, \quad n = 0, 1, \ldots;$$

$$B_{n,l} \in \mathcal{R}^{N,N}, \quad i = 1, \ldots, M_n; \quad \gamma = 0, 1,$$

$$\bar{\beta}_{n,k}^{(\gamma)} = \{\beta_{n,k,l}^\gamma\} = \left(\beta_{n,k,1}^{(\gamma)} \ldots \beta_{n,k,M_n}^{(\gamma)}\right)^T \gamma \in \mathcal{R}^{M_n}. \qquad (11)$$

$$\beta_{n,k,l}^{(\gamma)} = -\frac{\left(A^\gamma p_l^k, AB_{n+1,l}^{-1} r^{n+1}\right)}{\rho_{n,l}^\gamma}, \quad n = 0, 1, \ldots; \quad k = 0, \ldots, n; \quad l = 1, \ldots, M_n.$$

In this case, formula (11) defines the multi-preconditioning of the Krylov subspace. In the case when the matrices A and $B_{n,l}$ in formulas (6)–(11) are symmetric, we obtain preconditioned methods for conjugate gradients and conjugate residuals (MPCG and MPCR) for $\gamma = 0, 1$, respectively, whereas $\beta_{n,k,l}^{(\gamma)} = 0$ for $k < n$, and the direction vectors are computed from two-term recursions.

If $A \neq A^{\top}$, then, apparently, the most appropriate is the multi-preconditioned semi-conjugate residual method (MPSCR). In this case, when solving ill-conditioned asymmetric SLAEs associated with a large number of iterations due to the resource-intensity of long recursions (mainly due to an increase in the amount of memory used), they have to be shortened by force, which is done by either introducing the so-called restart procedure or limiting the number of orthogonalized directional vectors (or matrices with multi-preconditioning), or by applying both approaches simultaneously. In all these cases, the rate of convergence of the iterations drops, sometimes very noticeably, which is the inevitable price of saving memory.

To overcome this degrading effect, we will consider the application of the method of least squares (MLS) [6], restricting ourselves to using restarts in a "pure form". Assume, for simplicity, that the restarts are periodically repeated through the same number m of iterations. This means that at each iteration of number $n_t = mt$, $t = 0, 1, \ldots$, the residual vector is calculated not from the recurrence relations (6), but from the original equation, i.e.

$$r^{n_t} = f - A u^{n_t}. \tag{12}$$

Then, the recursion is used again in the usual way. More precisely, such an iterative process is conveniently described by the two-index notation of the corresponding numbers of consecutive approximate vectors

$$u^{n_t} = u^{t,0}, \quad u^n = u^{t,k}, \quad k = n - n_t \quad \text{for} \quad n \in [n_t, n_{t+1}].$$

Let us assume that we already know the values of the "restart" approximations $u^{n_0}, u^{n_1}, \ldots, n_0 = 0$. For the correction of the last iterative approximation, we will write a linear combination of the vectors

$$\hat{u}^{n_t} = u^{n_t} + b_1 v_1 + \ldots + b_t v_t = u^{n_t} + V_t \bar{b}, \quad \bar{b} = (b_1, \ldots, b_t)^{\top},$$
$$V_t = \{v_k = u^{n_k} - u^{n_{k-1}}, \; k = 1, \ldots, t\} \in \mathcal{R}^{N,t}, \tag{13}$$

with coefficients b_n that will be found from the generalized solution of the overdetermined system obtained after multiplying Eq. (13) by the original matrix A:

$$W_t \bar{b} = r^{n_t} = f - A u^{n_t}, \quad W_t = A V_t. \tag{14}$$

There are several ways to solve SLAE (14): using SVD or QR decompositions for the matrix W_t, finding the generalized inverse matrix W_t^+ using the Greville formulas, calculating the normal solution by the method of least squares (using the left Gauss transformation, i.e. multiplying the common parts of Eq. (14) on the left by $W_t^{\top}$), or applying a "lightweight" (by condition number) transformation by reducing to the square system

$$V_t^{\top} A V_t \bar{b} = V_t^{\top} r^{n_t}. \tag{15}$$

In all these cases, after finding the vector $\bar{b}$ and carrying out the correction of the iterative approximation by formula (15), the next "restart" begins with the calculation of the residual

$$r^{n_t+1} = f - Au^{n_t+1}, \quad u^{n_t+1} = \hat{u}^{n_t}. \tag{16}$$

3.2 Parallel Domain Decomposition Methods

Historically, the DDM came into existence in the 19th century as an alternating Schwartz method for the theoretical study of geometrically complex boundary value problems by reducing them to simpler ones. If we turn from the differential equations to the grid ones, then, in algebraic terms, the initial decomposition of a domain into subdomains corresponds to the use of the Seidel block method. In due course, this method was transformed into the additive Jacobi–Schwartz block method, which is a natural way to parallelize the computational process on an MPS. Currently, both geometric and algebraic interpretations of the DDM exist on equal terms, mutually generalizing each of the approaches.

Referring to the block representation of SLAE (4), we can write down the iterative Jacobi–Schwartz method in the following form:

$$B_{q,q}u_q^{n+1} \equiv (A_{q,q}+\theta D_q)u_q^{n+1} = f_r+\theta D_q u_q^n - \sum_{r\in\Omega_q} A_{q,r}u_r^n, \quad r = 1,\ldots,P, \tag{17}$$

where $\theta \in [0,1]$ is the iterative parameter and D_q is the diagonal matrix defined by the relation

$$D_q e = \sum_{r\in\Omega_q} A_{q,r}e, \quad e = (1,\ldots,)^\top \in \mathcal{R}^{Nq}.$$

Let us turn to the geometric interpretation of this algorithm and emphasize that the solution of each q-th Eq. (12) is an independent solution of the auxiliary boundary value problem in the subdomain Ω_q. In this case, the formally introduced values θ and D_q correspond to the use of interface conditions between the contacting subdomains (see [13] for more details).

The iterative process (12) can be represented in the form:

$$u^{n+1} = u^n + B_1^{-1}(f - Au^n), \quad u^n = \{u_q^n\}, \quad B_1 = \text{block} - \text{diag}\{B_{q,q}\}, \tag{18}$$

where B_1 is the preconditioning matrix of this version of the DDM, which can be used in formulas (7), thereby generating preconditioned methods in the Krylov subspaces.

Let us now consider the possibility of accelerating the described block iterative Krylov-type methods based on the deflationary approach (see review in [14]). In this case, in addition to the conventional variational and/or orthogonal properties of computational successive approximations, supplementary conditions of orthogonality are imposed on them to the specially introduced m-dimensional fixed deflation subspace associated with a rectangular matrix

$$V = (v_1 \ldots v_m) \in \mathcal{R}^{N,m}.$$

For the sake of brevity, we will discuss the use of the deflation method as applied to the conjugate residual algorithm for solving a SLAE with a symmetric matrix A. First, there is an approach to optimizing, in a certain sense, the vector of the initial iterative approximation u^0. Let an arbitrary vector u^{-1} be given. Then, we define the vectors

$$u^0 = u^{-1} + Vc, \quad r^0 = r^{-1} - AVc. \tag{19}$$

The vector of unknown coefficients $c = (c_1, \ldots, c_m)^T$ in (19) is determined (assuming formally that $r^0 = 0$) from the solution of the overdetermined system

$$Wc = AVc = r^{-1}. \tag{20}$$

Applying the method of least squares (MLS) to (20), we find the normal solution

$$c = (W^T W)^+ W^T r^{-1},$$

providing a minimum norm of the residual vector:

$$\begin{aligned}
r^0 &= T_0 r^{-1} \quad T_0 = I - W(W^T W)^{-1} W^T, \\
(r^0, r^0) &= (r^{-1}, r^{-1}) - (W(W^T W)^{-1} W^T r^{-1}, r^{-1}) \\
&= (W^T W z, z) - ((W^T W)^{-1} z, z), \quad z = W^T r^{-1},
\end{aligned} \tag{21}$$

where the matrix T_0 is a symmetric (orthogonal) projector with the following properties:

$$T_0 = T_0^T = T_0^2, \quad W^T T_0 = T_0 W = 0,$$

that is, the space $\mathrm{Span}(W)$ belongs to the kernel $\mathcal{N}(T_0)$. Further, from the initial direction vector in the form

$$p^0 = r^0 - V(W^T W)^{-1} W^T A r^0 = B r^0, \quad B = I - V(W^T W)^{-1} W^T A, \tag{22}$$

we obtain the deflation orthogonality conditions

$$W^T r^0 = 0, \quad W^T A p^0 = 0, \tag{23}$$

for the vectors r^0 and p^0. Moreover, the introduced matrix B has the following easily verifiable orthogonal properties:

$$W^T A B = 0, \quad BV = 0. \tag{24}$$

Further iterations of the derived Deflated Conjugate Residual deflation algorithm are performed according to the "standard" formulas of the method of conjugate residuals with a preconditioning matrix, whose role in this case is played by B from (22):

$$\begin{aligned}
u^{n+1} &= u^n + \alpha_n p^n, \quad \alpha_n = \sigma_n / \rho_n, \quad \rho_n = (A p^n, A p^n), \\
r^{n+1} &= r^n - \alpha_n A p^n, \quad \sigma_n = (A B r^n, r^n), \\
p^{n+1} &= B r^{n+1} + \beta_n p^n, \quad \beta_n = \sigma_{n+1} / \sigma_n.
\end{aligned} \tag{25}$$

In this case, at each iteration, the minimum of the residual norm $\|r^n\|$ in the preconditioned Krylov subspace

$$\mathcal{K}_n(A,\ r^0, B) = \mathrm{Span}\big(r^0, ABr^0, \ldots, (AB)^{n-1}r^0\big),$$

and the computed vectors satisfy the orthogonality conditions

$$(ABr^k, r^n) = \sigma_n \delta_{k,n}, \quad (Ap^k, Ap^n) = \rho_n \delta_{k,n}, \ \ k = 0, 1, \ldots, n-1,$$
$$W^T r^n = 0, \quad W^T Ap^n = 0, \ \ n = 0, 1, \ldots. \tag{26}$$

Note that the introduced preconditioning matrix B is degenerate since $BW = 0$. Nevertheless, relations (24) and (26) ensure the orthogonality of all residual vectors to the kernel $\mathcal{N}(\bar{A}) = \mathrm{Span}(W)$ of the matrix $\bar{A} = AB$, thereby ensuring the convergence of iterative process (21)–(25).

The considered deflationary approach is quite universal and was studied in different ways under the names of aggregation methods, coarse-grid correction, and low-rank approximations of matrices (see [13]). Deflationary preconditioning matrices of the form B_z from (17)–(20), in particular, can be effectively used in conjunction with B_z in the multi-preconditioned semi-conjugate residuals (MP-SCR). In this case, we will in fact make use of the coarse-grid correction to accelerate the DDM, and each column of the matrix V from (14)–(17) corresponds to certain subdomains (see [13,14]).

On the whole, it can be said that there are a lot of versions of the DDM in Krylov subspaces, as well as a lot of adjoining multigrid methods, but here we do not dwell on them. For example, according to the methodology of quasi-structured grids in individual subdomains, auxiliary special-type SLAEs can be solved either by fast Fourier transform algorithms, or by optimal implicit alternating direction methods, or by modern adaptive incomplete factorization techniques, or by low-rank approximations of HSS matrices (hierarchical semi-separable approximations; see [15]), and so on.

4   On the Technology of Programming Parallel Algorithms

The final performance of computer-aided implementations of the SLAE solution methods obviously depends on two main factors: the mathematical efficiency of the application of the algorithms (which we already discussed in the previous section) and the adaptability of their mapping onto the supercomputer architecture, which determines the quality of the parallelization of a computing process of heterogeneous type with heterogeneous arithmetic devices (universal CPUs) or specialized accelerators of the type of GPGPU or Intel Phi, functioning with distributed or hierarchical shared memory.

In the framework of two-level methods for decomposing functioning domains in Krylov subspaces, scalable parallelization is generally achieved by means of hybrid programming. At the top level, the Jacobi–Schwartz block matrices for the subdomains with interprocessor exchanges are implemented in a distributed way by organizing the MPI processes. At a lower level, the simultaneous solution of auxiliary SLAEs in the subdomains is performed on multi-core processors with shared memory, by using multithread computing. Additional acceleration can be accomplished using vectorization of operations in the AVX-type command system.

To ensure a high performance, several technological aspects need to be considered. First, to avoid idle processors that implement the solution of various auxiliary SLAEs, it is necessary to construct a balanced partition of the domain into subdomains, something that is not generally an easy task. Strictly speaking, all SLAEs in subdomains must be solved at the same time; this is highly problematic to attain on heterogeneous devices. For example, the equality of the dimensions of the algebraic systems does not ensure their being solved synchronously, even in case of identical calculators. The second point is associated with a decrease in communication losses. One of the possibilities of the task is associated with matching of data transfer and arithmetic operations in time, and the other, with the formation of information buffers and special management of communication operations. In general, it should be noted that the study and design of optimal computational schemes, including parallel ones, is mainly an experimental work whose success largely depends on the quality of planning and on the equipment used.

When developing a new generation of mathematical software, one cannot but bear in mind that there is already a huge number of publicly available libraries that implement computational algebra methods, including high quality ones and those adapted for modern supercomputer platforms, in which a considerable intellectual potential is incorporated.

The world scientific community engaged in numerical methods for linear algebra has historically turned out to be well organized, has its own regular journals and conferences, and has actively participated in contact groups throughout many years. What is even more important, these activities are not only committed with the development and theoretical analysis of new methods but also with the development of software and experimental research into the efficiency and performance of computer implementations on modern platforms.

The greatest success has been achieved in computational algebra problems involving relatively small matrices, both dense and sparse, corresponding to the BLAS and SPARSEBLAS libraries which have become standard and are widely used (for example, in the Intel MKL library, which features one of the most effective "direct solvers": PARDISO). Taking into account the current state of affairs, it can be said that the fundamental issues concerning software problems are basically close to each other, although research is steadily conducted to improve both algorithms and their implementations on newly emerging computer architectures.

Regarding the development and testing of methods for large problems with sparse matrices, the research in this case is carried out mostly in an academic style since it is associated with general high-tech problems of mathematical modeling, including the construction of grids, approximations of initial equations, and others, that are related to indivisible stages of a single computing process. In particular, the construction of subdomains for large tasks must be started at the stage of generation of grids and distributed among processors, since the global matrix may not fit in the memory of a single computing node.

At the same time, libraries for solving large sparse SLAEs are available in fairly large numbers. Among the best known, we can mention SPARSE KIT, PETSc, HYPRE (see an informative review in [8]), and LParSol [16]. During the last decades, however, a noticeable trend in the development of software for mathematical modeling has been the creation of integrated computing environments focused on coordinating the participation of various groups in the design of products with a long life cycle, showing a steady development of functionality and adaptation to the evolution of computer architectures. To a certain extent, DUNE [17], INMOST [18], and BSM [10], which include libraries of algebraic solvers, can serve as examples of such developments.

5   Technical Requirements for the Creation of an ICE

An overview of the classes of tasks and methods for solving them suffices to see that the creation of mathematical software that claims to be an integrated computing environment for a broad range of problems is a big and complex process that requires significant investments and the involvement of highly qualified specialists from various fields. Obviously, such a knowledge-intensive and resource-demanding project should be planned on a strict methodological basis which can be defined through the following list of technical requirements for the content of research and development (R & D):

- A wide range of functionality with a regular support on modern scientific and technological levels. In our particular case, this implies a flexible expansion of the composition of solved SLAEs by means of different spectral, variational, structural, and other properties, and also with methods used and adapted to the specific features of the algebraic systems and information methods for their presentation and architectural specifications used by MPS. In particular, redundant functionality should generally provide for the selection of the best algorithm from the library to solve a specific problem to achieve highly contradictory features as of efficiency and universality.
- High performance of the code, with scalable parallelization of computations using hybrid programming tools, and without formal restrictions either on the number of degrees of freedom of the implemented tasks and algorithms or on the number of used computing nodes, cores, and other devices in heterogeneous supercomputers with distributed and hierarchical shared memory.
- Adaptability to the evolution of computer platforms and the multi-versatility of the functional core of the library, provided by modern component technologies and through the coordination of the external and internal working interfaces of the software-tool kernel.
- Universal and convertible data structures that are consistent with existing common formats and have the ability to reuse external software products that represent a high intellectual potential.

- Multilingual and cross-platform software functional content, ability to benefit from different styles of teamwork and openness for coordinated participation in the project of various groups of developers.
- Availability of user and internal interfaces that are both intelligent and friendly and focus on wide applications and active demand in diverse production areas by specialists of various professional backgrounds. In particular, it is necessary to distinguish between the support of effective activities of mathematicians-developers of new algorithms, information technologists (including experts in parallelization), and the end user, who needs only a minimum of information about the "backstage" activity of an ICE.

These architectural principles focus on the long life cycle of the product developed, as well as on the high productivity of the programming work of the participants in such a high-tech project. At the same time, the focus on super-computing involves not only the speed of the algorithms but also the intelligence of the tool environment and skilled work with big data to avoid communication losses.

Naturally, the system infrastructure of the ICE must include the following library tools, necessary for the active development, maintenance, and efficient operation of a production software product:

- means for automated testing and comparative analysis of the efficiency of algorithms on characteristic SLAEs from methodological and practical problems, including the international matrix collections MATRIX MARKET, FLORIDA, BOEING, and others, as well as matrix generators for standard applications;
- user documents with descriptions of source data and features of the use of library algorithms, including examples (EXAMPLEs) of running solvers and recommendations on their choice for specific types of SLAEs, as well as archives of computational scenarios with the calculation results;
- configuration management tools and support for multi-version of the functional core of a library, providing a flexible expansion of the composition of computing modules and their adaptation to computer platforms;
- means of forming interfaces with developers, end users, and external software products, as well as the integration of the formed library modules with application software packages;
- regularly updated knowledge base on methods and technologies of computational algebra, containing up-to-date information with cognitive analysis and target search for scientific publications, software implementations, and sets of test examples.

Some prototypes of such developments, mainly of the information type, are the integration project "Tree of Mathematics" (www.mathtree.ru), supervised by the Siberian Branch of the Russian Academy of Sciences, and the ALGO WIKI project (www.parallel.ru), developed by Moscow State University.

It should be borne in mind that at the stages of development, verification, validation, and testing of computational library modules, as well as in the course of trials or production operations, program codes are in different forms and require the corresponding methods of support. Different stages of product availability can be defined depending on the organization of the technological process, from a trial or an experimental version (alpha version) to a final product with high-quality performance indicators. An extensive range of features in the considered products, from the level of requirements of the modern scientific world to a high technological level, determines the hierarchy of qualifications of the project participants: from experts of academic background to specialists in supercomputing technologies.

It should be specially noted the importance of creating educational versions of an ICE, accompanied by cognitive techniques and relevant materials, aimed at students with university lecturers and professionals in advanced computer skills training courses.

6   Conclusions

The main result of this study is that we managed to give a concrete expression to conceptual propositions concerning the creation of mathematical software of new generation for the high-performance solution of a wide class of computational algebra problems. The software was constructed in the framework of an integrated computing environment focused on a long life cycle and steady development of a functional core and is able to adapt to the evolution of computer architectures. It has been designed to support the coordinated participation of various groups of developers and the active reuse of external software products. It intends to respond to the demand of end users from various professional areas. The system content of such a project is designed to significantly increase the productivity of programmers through intelligent automation tools for building parallel algorithms and their mapping onto a supercomputer architecture. In an industrial language, the considered ICE should ensure the transition to the production of the means of production in applied software systems as a basis for predictional mathematical modeling.

Acknowledgements. The work was supported by the Russian Foundation for Basic Research (grants No. 16-29-15122 and No. 18-01-00295).

References

1. Saad, Y.: Iterative Methods for Sparse Linear Systems, 2nd edn. SIAM (2003)
2. Il'in, V.P.: Finite Element Methods and Technologies. ICM&MG SBRAS, Novosibirsk (2007)
3. Dolean, V., Jolivet, P., Nataf, F.: An Introduction to Domain Decomposition Methods: Algorithms, Theory and Parallel Implementation. SIAM, Philadelphia (2015)
4. Domain Decomposition Methods. http://ddm.org
5. Shapira, Y.: Matrix-Based Multigrid: Theory and Application. Springer, Heidelberg (2008). https://doi.org/10.1007/978-0-387-49765-5

6. Il'in, V.P.: Two-level least squares methods in Krylov subspaces. J. Math. Sci. **232**, 892–902 (2018)
7. Il'in, V.: High-performance computation of initial boundary value problems. In: Sokolinsky, L., Zymbler, M. (eds.) PCT 2018. CCIS, vol. 910, pp. 186–199. Springer, Cham (2018). https://doi.org/10.1007/978-3-319-99673-8_14
8. Dongarra, J.: List of freely available software for linear algebra on the web (2006). http://netlib.org/utk/people/JackDongarra/la-sw.html
9. Butyugin, D.S., Gurieva, Y.L., Il'in, V.P., Perevozkin, D.V., Putuchov, A.V., Skopin, I.N.: Functionality and technologies of the algebraic solvers in the library Krylov. Herald SUSU Ser. Comput. Math. Inf. **2**(3), 92–103 (2013). (in Russian)
10. Il'in, V.P., Gladkih, V.S.: Basic system of modelling (BSM): the conception, architecture and methodology. In: Proceedings of the International Conference Modern Problems of Mathematical Modelling, Image Processing and Parallel Computing (MPMMIP & PC-2017), pp. 151–158. DSTU Publ. Rostov-Don. (2017). (in Russian)
11. Intel Math Kernel Library. Reference Manual. http://software.intel.com/sites/products/documentation/hpc/composerxe/enus/mklxe/mk-manual-win-mac/index.html
12. Il'in, V.P.: Mathematical Modeling. Part 1: Continuous and Discrete Models. SBRAS Publ., Novosibirsk (2017). (in Russian)
13. Il'in, V.P.: Multi-preconditioned domain decomposition methods in the Krylov subspaces. In: Dimov, I., Faragó, I., Vulkov, L. (eds.) NAA 2016. LNCS, vol. 10187, pp. 95–106. Springer, Cham (2017). https://doi.org/10.1007/978-3-319-57099-0_9
14. Gurieva, Y.L., Il'in, V.P.: On parallel computational technologies of augmented domain decomposition methods. In: Malyshkin, V. (ed.) PaCT 2015. LNCS, vol. 9251, pp. 35–46. Springer, Cham (2015). https://doi.org/10.1007/978-3-319-21909-7_4
15. Solovyev, S.: Multifrontal hierarchically solver for 3D discretized elliptic equations. In: Dimov, I., Faragó, I., Vulkov, L. (eds.) FDM 2014. LNCS, vol. 9045, pp. 371–378. Springer, Cham (2015). https://doi.org/10.1007/978-3-319-20239-6_41
16. Aleinicov, A.Y., Barabanov, R.A., Bartenev, Y.G.: An application of parallel solvers for SLAEs in the applied packages for engineering. In: Proceedings of the International Conference "Supercomputing and Mathematical Model", pp. 102–110. Unicef Publ. (2015). (in Russian)

Comparative Analysis of Parallel Computational Schemes for Solving Time-Consuming Decision-Making Problems

Victor Gergel[(⊠)] and Evgeny Kozinov

Lobachevsky State University of Nizhni Novgorod, Nizhni Novgorod, Russia
gergel@unn.ru, evgeny.kozinov@itmm.unn.ru

Abstract. We consider in this paper an efficient approach to the parallel solution of complex multicriterial optimization problems using heterogeneous computing systems. The complexity of these problems can be very high since the criteria that are to be optimized can be multiextremal and the computation of criteria values can be time-consuming. In the framework of the proposed approach, the multicriterial optimization problem is reduced to the solution of a series of global optimization problems by means of the convolution of the partial criteria with different sets of parameters. To solve the series of global optimization problems, we apply an efficient information-statistical method of global search. Parallel computations are implemented through the simultaneous solution of several global optimization problems. We present in this paper a comparative analysis of various methods for parallel computations and the results of numerical experiments.

Keywords: Decision making · Multicriteria global optimization · Parallel calculations · Dimensionality reduction · Search of information · Numerical experiments

1 Introduction

Decision making is inherent to almost all fields of human activity. A variety of mathematical formulations have been proposed for the formal description of decision-making problems. These problems relate to many classes of optimization approaches, such as unconstrained optimization, nonlinear programming, global optimization, etc. Multicriterial optimization (MCO) problem statements are widely used in the most complex situations of decision making. An opportunity to set several criteria is a distinctive property of MCO problems which allows for a more precise formulation of the requirements to the optimality of the chosen decisions. MCO is currently a field of intensive studies; see, for example, the monographs [1–6] and reviews of scientific and practical results in [7–10].

The possibility of contradictions between the partial criteria of efficiency, which makes it impossible to achieve the optimal (the best) values for all partial

© Springer Nature Switzerland AG 2019
L. Sokolinsky and M. Zymbler (Eds.): PCT 2019, CCIS 1063, pp. 107–121, 2019.
https://doi.org/10.1007/978-3-030-28163-2_8

criteria simultaneously, is an important feature of multicriterial optimization problems. As a result, the determination of some compromised (efficient, non-dominated) decisions, in which the achieved values with respect to some partial criteria satisfy the given requirements for the necessary level of efficiency, is usually taken as solution of an MCO problem.

The present work is devoted to solving MCO problems which are used for the description of decision-making problems in the design of complex objects and systems. In such applications, the partial criteria can take a complex multiextremal form, and therefore computing the values of the criteria and constraints may require a large number of computations.

Besides, an opportunity for correcting the MCO problem statements when changing the perceptions of the requirements to the optimality of the chosen decisions is allowed within the framework of the considered approach. Thus, in the case of redundancy of the partial criteria set, an opportunity to transform some criteria into constraints is allowed. Or, otherwise, if the set of feasible variants is insufficient, some constraints can be transformed into criteria, and so on. Such an opportunity to change the MCO problem statements is a source of additional computational costs of the search for optimal decisions.

In the present paper, we present the results of the investigations we have performed on generalization of decision-making problem statements [11,12] and on the development of highly efficient global optimization methods that use all the search information obtained in the course of computations [13–17]. Parallel algorithms developed earlier are presented in [14–17]. The proposed algorithms are able to efficiently use hundreds and thousands of computing cores [18]. Within the scope of this paper, a comparative analysis of the implemented parallel algorithms on modern computing systems is carried out. We also compare implementations of algorithms for shared memory, distributed memory, and heterogeneous computing systems. As a result of a comparative analysis, an optimal running configuration is selected.

The structure of the paper is as follows. In Sect. 2, we formulate the multicriterial optimization problem. In Sect. 3, we consider the reduction of multicriterial problems to scalar optimization ones by means of minimax convolution of the partial criteria and dimensionality reduction through the use of Peano space-filling curves. In Sect. 4, we provide arguments to substantiate the possibility of enhancing the efficiency of computations by reusing search information. In Sect. 5, we describe the general organization scheme of parallel computations allowing the maximum use of the computing potential of modern supercomputer systems. In Sect. 6, we present the results of numerical experiments confirming the viability of the proposed approach. Finally, in the Conclusions, we discuss the results obtained and suggest possible directions for future research.

2   Statement of the Multiple Multicriterial Optimization Problem

We propose the following generalized multilevel model for the formal description of the process of search for efficient variants in complex decision-making problems.

1. At the highest level, a decision-making problem (DMP) is defined within the framework of the proposed model. The optimal values of the parameters should be determined for this problem according to the available requirements to optimality. In the most general form, a DMP can be defined by a vector function of characteristics,

$$w(y) = \big(w_1(y), w_2(y), \ldots, w_M(y)\big), \ y \in D, \tag{1}$$

where $y = (y_1, y_2, \ldots, y_N)$ is the vector of constructive parameters and $D \in R^N$ is the domain of feasible values, which usually is an N-dimensional hyperinterval,

$$D = \{y \in R^N : a_i \le y_i \le b_i, \ 1 \le i \le N\}, \tag{2}$$

for two given vectors a and b.

The values of the characteristics $w(y)$ are supposed to be nonnegative. Moreover, when these values decrease, the efficiency of the chosen variants increases. It is also assumed that the characteristics $w_j(y)$, $1 \le j \le M$, can be multiextremal and computing their values can require large numbers of computations. In addition, within the framework of the approach under consideration in this paper, we assume that the characteristics $w_j(y)$, $1 \le j \le M$, satisfy the Lipschitz condition

$$|w_j(y_1) - w_j(y_2)| \le L_j \|y_1 - y_2\|, \ 1 \le j \le M, \tag{3}$$

where L_j is the Lipschitz constant corresponding to the characteristic $w_j(y)$, $1 \le j \le M$, and $\| \cdot \|$ denotes the Euclidean norm in R^N. It is important to note that the fulfillment of the Lipschitz condition corresponds to practical applications. Indeed, small variations of the parameters $y \in D$ lead, as a rule, to limited variations of the corresponding values of the characteristics $w_j(y)$, $1 \le j \le M$.

In general, the DMP model is defined by the following set of elements:

$$S = \langle y, w(y), D, a, b \rangle. \tag{4}$$

This model is supposed to remain the same and does not change throughout the course of computations.

2. The requirements to the optimality of the chosen variants $y \in D$ of the DMP can be defined in the following way. First of all, we should find characteristics $w_j(y)$, $1 \le j \le M$, for which the achievement of the minimum possible values is necessary. By defining a set of indices $F = \{i_1, i_2, \ldots, i_s\}$ of such characteristics, we define the vector criterion of efficiency

$$f(y) = \big(w_{i_1}(y), w_{i_2}(y), \ldots, w_{i_s}(y)\big). \tag{5}$$

Sufficient levels of efficiency for characteristics $w_j(y)$, $1 \leq j \leq M$, not included in the vector criterion of efficiency defined by a set of indices $G = \{j_1, j_2, \ldots, j_m\}$ should be defined by a tolerance vector $q = (q_1, q_2, \ldots, q_m)$. The availability of tolerances enables one to define a vector function of constraints

$$g(y) = (g_1(y), g_2(y), \ldots, g_m(y)), \ g_l(y) = w_{j_l}(y) - q_l, j_l \in G, \ 1 \leq l \leq m. \quad (6)$$

The constraints $g(y)$ define the feasible search domain

$$Q = \{y \in D : g(y) \leq 0\}. \quad (7)$$

With the criteria of efficiency and constraints formulated in this way, we can pose the multicriterial optimization (MCO) problem

$$f(y) \to \min, \ y \in Q, \quad (8)$$

defined on the basis of the model S from (4) using the elements listed above:

$$P = \langle S, F, G, q \rangle. \quad (9)$$

The scheme proposed covers many existing optimization problem statements. In the case $s = 1$, $m = 0$, the general statement becomes a global optimization problem. For $s = 1$ and $m > 0$, the general statement defines a nonlinear programming problem. When $s > 1$ and $m > 0$, it becomes a constrained multicriterial problem.

3. The use of multicriterial optimization problem statements reduces the complexity of the formal description of decision-making problems since it becomes possible to define several partial criteria instead of defining a unified "global" criterion of efficiency, as is necessary in MCO problems. No doubt, such an approach is a simpler task for the decision maker determining the requirements to the optimality of the decisions made. It is worth noting also that the multicriterial definition of the efficiency criteria corresponds to the practice of decision making in various fields of applications.

The scheme of the above-considered MCO problem statement was first proposed in [11] and was widely used in the solution of many applied decision-making problems. At the same time, the results of the practical application of this approach have demonstrated that the formulation of a single MCO problem statement may be difficult when altering the ideas of the necessary requirements to optimality. Thus, in the case of redundancy of the partial criteria set, the transformation of some criteria into constraints might be appropriate. Or, vice versa, if a small feasible domain Q is obtained, the loosening of the tolerances q or transforming some constraints into criteria may be required. As a result, an opportunity for simultaneous formulation of several MCO problems,

$$\mathbb{P}_t = \{P_1, P_2, \ldots, P_t\}, \quad (10)$$

will be allowed within the framework of the extended model of optimal choice as a further development of scheme (1)–(9). It is worth noting that the set of problems $\mathbb{P}$ can change in the course of computations as a consequence of the addition of new problems or the elimination of already existing ones:

$$\mathbb{P}' = \mathbb{P} +/- \mathbb{P}. \tag{11}$$

The solution of the problems from set $\mathbb{P}$ can be performed sequentially or simultaneously in time-sharing mode or in parallel if several computing devices are available. Undoubtedly, the simultaneous method of solving the problems is preferable since the results obtained in the course of computations allow to adjust promptly the current set of problems $\mathbb{P}$. In the case of parallel computations, the total time of problem solving can be reduced substantially.

In general, the model (1)–(11), proposed for the search of optimal decisions, defines a new class of optimization problems, namely multiple multicriterial global optimization (MMGO) problems.

3   Successive Reduction of Multicriterial Global Optimization Problems

As it has been mentioned above, the partial criteria of efficiency in MCO problems are usually contradictory. This feature of MCO problems implies that the attainment of optimal (the best) values with respect to all partial criteria simultaneously cannot be ensured. In such a situation, particular compromised (effective, non-dominated) decisions, in which the achieved values with respect to some specific partial criteria cannot be improved without simultaneous worsening with respect to some other criteria, are usually understood as solutions to an MCO problem. In the limiting case, it may be required to find all effective (Pareto-optimal) decisions $PD(Q) \subset Q$.

Due to the high relevance of this topic, a large number of methods have been developed and widely used for the solution of MCO problems (see, for example, [2–10]). Some of these algorithms ensure the attainment of the numerical estimates of the whole Pareto set $PD(Q)$ [3, 27–29]. Along with the usefulness of this approach (all effective decisions of the MCO problem are found), the use of these methods appears to be difficult due to the high computational complexity of finding the estimates of the Pareto set. Besides, the estimate of the whole set $PD(Q)$ might be redundant in cases when it would suffice to find several particular effective decisions to solve the MCO problem (this happens quite often in practice). As a result, the more extensively used approach to the solution of MCO problems is based on the scalarization of the vector criterion into some general scalar criterion of efficiency, the optimization of which can be performed with the use of already existing optimization methods. Among these, we can mention the weighted sum method, the compromise programming method, the weighted min-max method, goal programming, and many other algorithms (see, for example, [2–6, 22]).

The general approach used in the present work consists in the reduction of the solution of an MMGO problem to the solution of a series of single-criterion global optimization problems[1]:

$$\min \varphi(y) = F(\lambda, y), \ g(y) \leq 0, \ y \in D, \tag{12}$$

where $g(y)$ is the vector function of constraints from (6); $F(\lambda, y)$ is the minimized convolution of the partial criteria $f_i(y(x))$, $1 \leq i \leq s$, of MCO problem (5),

$$F(\lambda, y) = \max (\lambda_i f_i(y), 1 \leq i \leq s), \tag{13}$$

using the vector of weighting coefficients

$$\lambda = (\lambda_1, \lambda_2, \ldots, \lambda_s) \in \Lambda \subset R^s : \sum_{i=1}^{s} \lambda_i = 1, \ \lambda_i \geq 0, \ 1 \leq i \leq s. \tag{14}$$

The approach we have developed includes one more step of conversion of the problems $F(\lambda, y)$ from (12) being solved, namely to perform a reduction of dimensionality using Peano space-filling curves (evolvents) $y(x)$ which provide an unambiguous mapping of the interval $[0, 1]$ onto an N-dimensional hypercube D [19,21]. As a result of this reduction, the multidimensional global optimization problem (12) is reduced to the one-dimensional problem

$$F(\lambda, y(x^*)) = \min \{F(\lambda, y(x)) : g(y(x)) \leq 0, x \in [0, 1]\}. \tag{15}$$

The considered dimensionality reduction scheme superimposes the multidimensional problem with the Lipschitzian minimized function $F(\lambda, y)$ to a one-dimensional problem, in which the reduced function $F(\lambda, y(x))$ satisfies a uniform Hölder condition, i.e.

$$\left| F(\lambda, y'(x)) - F(\lambda, y''(x)) \right| \leq H \|x' - x''\|, \ x', x'' \in [0, 1], \tag{16}$$

where the constant H is defined by the relation $H = 2L\sqrt{N+3}$, N is the dimensionality of the optimization problem from (1), and L is the Lipschitz constant for the function $F(\lambda, y)$ from (12). The dimensionality reduction makes it possible to apply (after the necessary generalization) many well-known highly efficient one-dimensional global optimization algorithms for the solution of problems (12) (see, for instance, [19,21,30–36]).

To obtain several effective decisions (or to estimate the whole Pareto domain), problem (12) should be solved several times for the corresponding sets of values of the elements of the weighting vector $\lambda \in \Lambda$. In this case, the set of MCO problems $\mathbb{P}$ from (10) which are required for solving the initial decision-making problem is transformed into a wider set of scalar optimization problems (12)

$$\mathbb{F}_T = \{F_i(\lambda_i, y) : \lambda_i \in \Lambda, 1 \leq i \leq T\}, \tag{17}$$

where it is possible that several problems (12) with different values of the coefficients $\lambda \in \Lambda$ correspond to each problem $P \in \mathbb{P}$.

---

[1] On account of the initial assumptions on possible multiextremality of the characteristics $w_j(y)$, $1 \leq j \leq M$, from (1).

4 Improving the Efficiency of Computations by Reusing Search Information

As we already mentioned, the solution of MCO problems P from (10) and the corresponding global optimization problems F from (15) may require a large number of computations even when solving a single particular problem. In the case when a series of problems from the sets $\mathbb{P}$ and $\mathbb{F}$ is to be solved, the required quantity of computations can be much larger. And, therefore, overcoming the computational complexity of decision-making problems of the considered class is a necessary condition for the practical application of the proposed approach. An intensive use of the whole search information obtained in the course of computations can be a viable approach to the solution of this problem.

The numerical solution of global optimization problems is usually reduced to a sequential computation of the values of the characteristics $w(y)$ at the points y^i, $0 \leq i \leq k$, of the search domain D [19,23–26]. The data obtained as a result of the computations can be represented as a matrix of search information (MSI):

$$\Omega_k = \{(y^i, w^i = w(y^i))^T : 1 \leq i \leq k\}. \tag{18}$$

By scalarization of vector criterion (12) and application of dimensionality reduction (15), Ω_k in (18) can be transformed into the matrix of the search state (MSS),

$$A_k = \{(x_i, z_i, g_i, l_i)^T : 1 \leq i \leq k\}, \tag{19}$$

where

- x_i, $1 \leq i \leq k$, are the reduced points of the executed global search iterations at which the criteria values have been computed;
- z_i, g_i, $1 \leq i \leq k$, are the values of the scalar criterion and of the constraints at points x_i, $1 \leq i \leq k$, for the current optimization problem $F(\lambda, y(x))$ from (15) which is being solved, i.e.

$$z_i = \varphi(x_i) = F(\lambda, y(x_i)), \quad g_i = g(y(x_i)), \quad 1 \leq i \leq k; \tag{20}$$

- l_i, $1 \leq i \leq k$, are the indices of the global search iterations at which the points x_i, $1 \leq i \leq k$, were generated; these indices are used to store the correspondence between the reduced points and the multidimensional ones at the executed iterations, i.e.

$$y^j = y(x_i), \quad j = l_i, \quad 1 \leq i \leq k. \tag{21}$$

The matrix of the search state A_k contains search information transformed into the current scalar reduced problem (15) which is being solved. Moreover, the search information in the MSS is arranged according to the values of the coordinates of the points x_i, $1 \leq i \leq k$, for a more efficient execution of the global search algorithms; the arranged representation of the points is reflected by the use of a lower index:

$$x_1 \leq x_2 \leq \cdots \leq x_k. \tag{22}$$

The representation of the search information obtained in the course of computations in the form of matrices Ω_k and A_k provides the basis for the development of efficient optimization search procedures. The availability of such information allows to perform an adaptive choice of the points for the execution of the global search iterations taking into account the results of all computations completed earlier,

$$y^{k+1} = Y(\Omega_k), \ k = 1, 2, \ldots \tag{23}$$

where Y is the rule for computing the points y^i, $0 \leq i \leq k$, of the applied optimization algorithm. Such an adaptive choice of the points of executed search iterations can accelerate the determination of the effective decisions. In the case of global optimization problems, the accumulation of the entire search information and the use of rules of type (23) is indeed mandatory. Any reduction of the search information in the determination of the points to be executed would result in the execution of excessive global search iterations.

It should be noted that the availability of the MSS Ω_k from (18) makes it possible to transform the results of all preceding computations z_i, $1 \leq i \leq k$, stored in the MSS A_k, into the values of the current optimization problem $F(\lambda, y(x))$ from (15). These recalculations are based on the values of the characteristics w^i, $1 \leq i \leq k$, which were computed earlier and stored in Ω_k without any repeated time-consuming computation of the values of $w(y)$ from (1). The updated values z_i, $1 \leq i \leq k$, can be used both for any new parameters of the MCO problem statement P from (9) and for any new values of the convolution coefficients λ from (14), i.e.

$$w_i \xrightarrow{\lambda, P} (z_i, g_i), \ 1 \leq i \leq k, \ \forall \lambda \in \Lambda, \ P \in \mathbb{P}. \tag{24}$$

Therefore, all the search information can be used to continue the computations in full. An efficient use of this information becomes an important requirement in the selection of the methods used to solve the problems $F(\lambda, y(x))$ from (15). The reuse of search information can provide a continuous decrease in the number of computations required to solve each subsequent optimization problem, up to the execution of just a few iterations only to find the next effective decision.

5   Parallel Computations in Multiple Multicriterial Global Optimization Problems

The sequential solution of the set of problems $\mathbb{F}$ from (17) to find several globally optimal decisions increases substantially the number of computations required to solve the decision-making problems. The parallel solution of the problems $F(\lambda, y(x))$ from the set $\mathbb{F}$ is a promising and simple method of accelerating the computations. For sufficiently large numbers of computer nodes (processors), the time required for the solution of the whole set of problems $\mathbb{F}$ is determined by the computation time for the problem $F(\lambda, y(x))$, whose solution takes the longest time. This approach can be implemented quite easily. However, it does not solve the problem of computational complexity when it is necessary to expand the set of problems $\mathbb{F}$ throughout the course of the computations.

The method considered above for the organization of parallel computations can be further developed by using the information compatibility of the problems of set $\mathbb{F}$. Indeed, as we noted in Sect. 4, the values of the characteristics w^i, $1 \leq i \leq k$, computed when solving any problem $F_1(\lambda, y(x))$, can be transformed into the values of any other problem $F_2(\lambda, y(x))$. This result provides the basis for the following general scheme of parallel computations for the simultaneous solution of the problems of set $\mathbb{F}$.

1. The Distribution of the Problems. Before starting the computations, the problems of the set $\mathbb{F}$ from (17) must be distributed among the computing nodes of a multiprocessor system. This distribution can be quite varied: for solving one problem from the set $\mathbb{F}$, various numbers of computing cores ($q, q \geq 1$) and computing nodes ($p, p \geq 1$) can be allocated. An opportunity to use several computing nodes $p > 1$ for the solution of the same problem is provided by applying multiple evolvents $y_i(x)$, $1 \leq i \leq p$ (see [13,14]). Provided that we are not planning to employ the processor cores of a computing node in full when solving one problem from the set $\mathbb{F}$, we could use the same node to solve several optimization problems, depending on the number of processors available at the computer node and the number of cores in each of them.

2. The Choice of the Optimization Algorithms. The proposed scheme of parallel computations is a general one: various methods of multiextremal optimization can be applied for solving the problems on each computing node (see, for example, [19,21,30–36]). The main condition imposed upon the selection of the algorithms is that the methods must use the search information Ω_k and A_k to increase the efficiency of global search. As it was already noted above, in Sect. 3, the global optimization algorithms should be generalized to solve the reduced one-dimensional global optimization problems of type $F(\lambda, y(x))$ from set $\mathbb{F}$. The possibility to apply the proposed approach was substantiated using as example efficient global search algorithms developed within the framework of the information-statistical theory of multiextremal optimization (see [12–21]).

3. The Execution of Global Search Iterations. The execution of global search iterations for each problem $F(\lambda, y(x))$ from the set $\mathbb{F}$ is performed in parallel. The execution of every iteration on each computing node includes the following operations:

1. The choice of q, $q \geq 1$, points x^i, $1 \leq i \leq q$, of the next global search iteration is performed according to the optimization method rule given in (23). The points are chosen taking into account the available search information Ω_k and A_k. The number q, $q \geq 1$, of generated points is determined by the number of computing cores employed to solve the problem $F(\lambda, y(x))$.

2. For each chosen one-dimensional point $x^i \in [0,1]$, $1 \leq i \leq q$, of the current global search iteration, the multidimensional image $y^i \in D$, $1 \leq i \leq q$, is determined by the mapping $y(x)$. Then, each computed image $y^i \in D$, $1 \leq$

$i \leq q$, is sent to all employed computing nodes, so as to prevent that the same points of the domain D are chosen more than once when solving the problems of the set $\mathbb{F}$.

3. The values of the characteristics $w(y)$ from (1) are computed at all points $y^i \in D$, $1 \leq i \leq q$. For each point $y^i \in D$, $1 \leq i \leq q$, these computations are performed in parallel using different computing cores. The computed values of the characteristics $w(y)$ are sent to all employed computing nodes to include the data obtained into the search information Ω_k and A_k.

4. Updating Search Information. Before starting the next global search iteration, the availability of data sent from other computing units (processors or cores) is checked; the received data should be included in the search information.

According to the suggested computational scheme of parallel computations, each computing unit contains an identical copy of the matrix of search information Ω_k from (18); the matrices A_k from (19) contain different sets of global search points x_i, $1 \leq i \leq k$, because of the use of different evolvents $y(x)$ for the dimensionality reduction and the different values of the scalar criterion and the constraints z_i, g_i, $1 \leq i \leq k$, corresponding to the problems $F(\lambda, y(x))$ from the set $\mathbb{F}$.

Within the framework of such computational scheme, it becomes possible to expand the set of problems $\mathbb{F}$ being solved at any moment of the computations. Indeed, to solve a new problem $F(\lambda, y(x))$, it is enough to allocate an additional computing unit, copy the set Ω_k, and create the corresponding matrix A_k.

It is clear that the increase in global search efficiency due to the use of the sets of search information Ω_k and A_k depends on the optimization algorithms employed. An analysis of the elements of the above-considered scheme of parallel computations applied to information-statistical global optimization algorithms can be found in [12–17]. In Sect. 6, we present the results of the numerical experiments carried out for a complete evaluation of the efficiency of the scheme we used to choose the best parameters for using the computing resources of high-performance supercomputer systems.

6   Results of Numerical Experiments

The numerical experiments were carried out on the "Lobachevsky" supercomputer, at the University of Nizhni Novgorod (operation system: CentOS 6.4, management system: SLURM). Each supercomputer node is equipped with two Intel Sandy Bridge E5-2660 processors (2.2 GHz, 64 Gb RAM). Each processor has eight cores (making a total of 16 CPU cores available on each computer node). We used the Intel C++ 17.0 compiler to generate the executable program code.

The numerical experiments, conducted according to the general scheme of parallel computations considered in Sect. 5, made use of efficient algorithms of multicriterial multiextremal optimization [12–17] developed within the framework of the information-statistical theory of global optimization [19]. The efficiency of these algorithms has been proven by many numerical experiments.

In the present work, we present the results of a series of numerical experiments performed to choose the best parameters of parallel computations (the number of computing cores and computer nodes employed, the number of multicriterial optimization subproblems to be solved simultaneously, the number of evolvents used for the dimensionality reduction, etc.). Each experiment included the solution of 30 MCO problems, each one six-dimensional and five-criterial, i.e. $N = 6$, $s = 5$. Multiextremal functions defined by the GKLS generator (see [37]) were used as criteria. This generator produces multiextremal optimization problems having properties given a priori, such as the number of local minima, sizes of their attractors, the global minimum point, the value of the objective function at this point, etc.

The values of the parameters used in the numerical experiments were as follows. The computation stop condition in the solution of the MCO problems was set for a predefined accuracy $\varepsilon = 0.05$. The estimates of the Pareto domain $PDA(f, D)$ were constructed by solving 100 subproblems of the family $\mathbb{F}$ from (17) for various values of the coefficients $\lambda \in \Lambda$. In the experiments, 10 computing nodes of the supercomputer were used, each having two 8-core processors (which gives a total of 160 computing cores used for computations in each experiment).

At the beginning of the series of experiments, the parallel computations were performed using only a single supercomputer node (2 processors, 16 computing cores using shared memory). The results of these experiments are contained in Table 1.

Table 1. Evaluation of the efficiency of a single supercomputer node for the solution of 30 six-dimensional five-criterial MCO problems

Cores	Search information	Iterations	S1	S2
1	Not used	26 813 722.7	1.0	
1	Used	9 103 069.6	2.9	1.0
8	Used	1 291 720.0	20.8	7.0
16	Used	609 169.5	44.0	14.9

The number of computing cores employed in the experiments is given in the *Cores* column. The *Search information* column contains information on reuse of the search information when solving the series of subproblems of the family $\mathbb{F}$ from (17) for different values of the coefficients $\lambda \in \Lambda$. The *Iterations* column presents the average number of global search iterations executed in the solution of a single subproblem of the MCO problem. The *S1* column shows the overall speedup achieved in parallel computations for the corresponding number of computing cores. Finally, the *S2* column shows the speedup of computations with respect to the serial optimization algorithm, in which the search information is used.

As follows from the results in Table 1, even when using a single computing core, one can achieve a computation speedup by almost a factor of three by reusing the search information obtained during the optimization. The overall speedup achieved in parallel computations is 44 times using the 16 cores of a single supercomputing node.

Ten supercomputer nodes (i.e. a total of 160 computing cores) were used in all the experiments. The parameters of the experiments were the following: the number of subproblems of the family $\mathbb{F}$ from (17) being solved simultaneously, the number of computing nodes employed for solving a single MCO subproblem, and the number of test problems (produced by the GKLS generator) that are solved simultaneously (a total of 30 six-dimensional five-criterial test MCO problems were solved in each experiment). The results of the numerical experiments are summarized in Table 2.

Table 2. Evaluation of efficiency of ten supercomputing nodes for the solution of a series of 30 six-dimensional five-criterial MCO problems

Nodes	NumMCO	SubProblems	SubNodes	Iterations	S1	Scaling
1	1	1	1	609 169.5	44.0	1.0
10	1	10	1	75 045.8	357.3	8.1
10	1	5	2	77 643.6	345.3	7.8
10	1	1	10	108 700.0	246.7	5.6
10	2	5	1	76 819.0	349.1	7.9
10	2	1	5	81 132.8	330.5	7.5
10	**5**	**2**	**1**	**66 618.6**	**402.5**	**9.1**
10	5	1	2	77 995.0	343.8	7.8
10	10	1	1	73 999.7	362.3	8.2

The number of employed supercomputing nodes is given in the *Nodes* column. The *NumMCO* column contains the number of test MCO problems (produced by the GKLS generator) solved simultaneously. The *SubProblems* column lists the number of subproblems of the family $\mathbb{F}$ from (17) solved simultaneously. The *SubNodes* column gives the number of computing nodes employed for solving one MCO subproblem. The *Iterations* column indicates the average number of global search iterations executed for the solution of one single subproblem of the MCO problem. The *S1* column indicates the overall speedup achieved in parallel computations for the corresponding number of supercomputing nodes. Finally, the *Scaling* column shows the speedup of parallel computations with respect to the results obtained when using a single computing node.

As follows from the results presented in Table 2, the best speed up in the considered series of experiments was achieved for the simultaneous solution of five MCO problems of the test class when two subproblems of the family $\mathbb{F}$ from (17) were solved in parallel for each problem, using a single supercomputing node for each subproblem (the parameters of the corresponding experiments are shown in bold in Table 2). In this case, the speedup factor achieved in parallel computations exceeded 400 (it should be noted that we used in the experiment 10 computing nodes with a total of 160 cores).

7   Conclusions

In the present paper, we proposed an efficient computational scheme of parallel computations for solving complex multicriterial optimization problems with non-convex constraints in which the optimization criteria can be multiextremal and the computation of criteria values can require a large quantity of computations. The approach proposed is based on the reduction of the multicriterial problems to nonlinear programming problems by means of minimax convolution of the partial criteria, dimensionality reduction through the use of Peano space-filling curves, and application of efficient information statistical and global optimization methods which implement a novel index scheme of accounting for the constraints instead of the penalty functions, which are commonly used.

The developed general computational scheme includes concurrent parallel computational methods for computing systems with shared memory, distributed parallel computations using multiple mappings for dimensionality reduction, and multilevel nesting of parallel computations for high-performance computing systems. In general, the proposed computational scheme can provide for an efficient application of high-performance computing systems with large numbers of cores/processors for the solution of complex global optimization problems. Besides, this general scheme can be used for the organization of parallel computations for a wide range of algorithms to solve time-consuming problems of decision making (in particular, for information-statistical global optimization algorithms).

We carried out a series of experiments to determine the optimal parameters of parallel computations. The results of our research showed that the suggested approach leads to a significant reduction of the computational costs of the solution of multicriterial optimization problems with non-convex constraints.

Finally, this research has shown the viability of the approach we have developed, but it requires further investigation. First of all, numerical experiments on the solution of multicriterial optimization problems should be conducted for larger numbers of partial criteria of efficiency and for larger dimensions of the optimization problems.

Acknowledgements. This research was supported by the Russian Science Foundation (project No. 16-11-10150 "Novel efficient methods and software tools for time-consuming decision making problems using supercomputers of superior performance").

References

1. Parnell, G.S., Driscoll, P.J., Henderson, D.L. (eds.): Decision Making in Systems Engineering and Management, 2nd edn. Wiley, Hoboken (2011)
2. Miettinen, K.: Nonlinear Multiobjective Optimization. Springer, Heidelberg (1999)
3. Marler, R.T., Arora, J.S.: Multi-Objective Optimization: Concepts and Methods for Engineering. VDM Verlag (2009)
4. Ehrgott, M.: Multicriteria Optimization, 2nd edn. Springer, Heidelberg (2010)
5. Collette, Y., Siarry, P.: Multiobjective Optimization: Principles and Case Studies (Decision Engineering). Springer, Heidelberg (2011)
6. Pardalos, P.M., Žilinskas, A., Žilinskas, J.: Non-Convex Multi-Objective Optimization. Springer, Heidelberg (2017)
7. Figueira, J., Greco, S., Ehrgott, M. (eds.): Multiple Criteria Decision Analysis: State of the Art Surveys. Springer, New York (2005)
8. Hillermeier, C., Jahn, J.: Multiobjective optimization: survey of methods and industrial applications. Surv. Math. Ind. **11**, 1–42 (2005)
9. Zavadskas, E.K., Turskis, Z., Kildiene, S.: State of art surveys of overviews on MCDM/MADM methods. Technol. Econ. Dev. Econ. **20**, 165–179 (2014)
10. Cho, J.-H., Wang, Y., Chen, I.-R., Chan, K.S., Swami, A.: A survey on modeling and optimizing multi-objective systems. IEEE Commun. Surv. Tutor. **19**(3), 1867–1901 (2017)
11. Strongin, R.G., Gergel, V.P., Markin, D.L.: Multicriterion multiextreme optimization with nonlinear constraints. In: Lewandovski, A., Volkovich, V. (eds.) LNEMS, vol. 351, pp. 120–127. Springer, Laxenburg (1988)
12. Gergel, V.P., Kozinov, E.A.: Accelerating multicriterial optimization by the intensive exploitation of accumulated search data. AIP Conf. Proc. **1776**, 090003 (2016). https://doi.org/10.1063/1.4965367
13. Gergel, V.P., Kozinov, E.A.: Efficient multicriterial optimization based on intensive reuse of search information. J. Glob. Optim. **71**(1), 73–90 (2018). https://doi.org/10.1007/s10898-018-0624-3
14. Gergel, V.P., Kozinov, E.A.: Accelerating parallel multicriterial optimization methods based on intensive using of search information. Procedia Comput. Sci. **108**, 1463–1472 (2017)
15. Gergel, V., Kozinov, E.: Parallel computing for time-consuming multicriterial optimization problems. In: Malyshkin, V. (ed.) PaCT 2017. LNCS, vol. 10421, pp. 446–458. Springer, Cham (2017). https://doi.org/10.1007/978-3-319-62932-2_43
16. Gergel, V., Kozinov, E.: Efficient methods of multicriterial optimization based on the intensive use of search information. In: Kalyagin, V., Nikolaev, A., Pardalos, P., Prokopyev, O. (eds.) Models, Algorithms, and Technologies for Network Analysis. PROMS, vol. 197, pp. 27–45. Springer, Cham (2017). https://doi.org/10.1007/978-3-319-56829-4_3
17. Gergel, V., Kozinov, E.: An approach for parallel solving the multicriterial optimization problems with non-convex constraints. In: Voevodin, V., Sobolev, S. (eds.) RuSCDays 2017. CCIS, vol. 793, pp. 121–135. Springer, Cham (2017). https://doi.org/10.1007/978-3-319-71255-0_10
18. Gergel, V., Kozinov, E.: GPU-based parallel computations in multicriterial optimization. In: Voevodin, V., Sobolev, S. (eds.) RuSCDays 2018. CCIS, vol. 965, pp. 88–100. Springer, Cham (2019). https://doi.org/10.1007/978-3-030-05807-4_8
19. Strongin, R., Sergeyev, Y.: Global Optimization with Non-convex Constraints. Sequential and Parallel Algorithms. Kluwer Academic Publishers, Dordrecht (2nd ed. 2013, 3rd ed. 2014)

20. Strongin, R., Gergel, V., Grishagin, V., Barkalov, K.: Parallel Computations for Global Optimization Problems. Moscow State University Press, Moscow (2013). (in Russian)
21. Sergeyev, Y.D., Strongin, R.G., Lera, D.: Introduction to Global Optimization Exploiting Space-Filling Curves. Springer, Heidelberg (2013)
22. Eichfelder, G.: Scalarizations for adaptively solving multi-objective optimization problems. Comput. Optim. Appl. **44**, 249–273 (2009)
23. Floudas, C.A., Pardalos, M.P.: Recent Advances in Global Optimization. Princeton University Press, Princeton (2016)
24. Locatelli, M., Schoen, F.: Global Optimization: Theory, Algorithms, and Applications. SIAM (2013)
25. Arora, R.K.: Optimization: Algorithms and Applications. CRC Press, Boca Raton (2015)
26. Bazaraa, M.S., Sherali, H.D., Shetty, C.M.: Nonlinear Programming: Theory and Algorithms. Wiley, Hoboken (2006)
27. Evtushenko, Y.G., Posypkin, M.A.: A deterministic algorithm for global multi-objective optimization. Optim. Methods Softw. **29**(5), 1005–1019 (2014)
28. Žilinskas, A., Žilinskas, J.: Adaptation of a one-step worst-case optimal univariate algorithm of bi-objective Lipschitz optimization to multidimensional problems. Commun. Nonlinear Sci. Numer. Simul. **21**, 89–98 (2015)
29. Markin, D.L., Strongin, R.G.: Uniform estimates for the set of weakly effective points in multi-extremum multicriterion optimization problems. Comput. Math. Math. Phys. **33**(2), 171–179 (1993)
30. Piyavskij, S.A.: An algorithm for finding the absolute extremum of a function. Comput. Math. Math. Phys. **12**, 57–67 (1972)
31. Shubert, B.O.: A sequential method seeking the global maximum of a function. SIAM J. Numer. Anal. **9**, 379–388 (1972)
32. Breiman, L., Cutler, A.: A deterministic algorithm for global optimization. Math. Program. **58**(1–3), 179–199 (1993)
33. Törn, A., Žilinskas, A.: Global Optimization. LNCS, vol. 350. Springer, Heidelberg (1989). https://doi.org/10.1007/3-540-50871-6
34. Evtushenko, Y.G.: Numerical Optimization Techniques. Translations Series in Mathematics and Engineering. Springer, Berlin (1985)
35. Sergeyev, Y.D.: Global one-dimensional optimization using smooth auxiliary functions. Math. Program. **81**, 127–146 (1998)
36. Gergel, V.P.: A method of using derivatives in the minimization of multiextremum functions. Comput. Math. Math. Phys. **36**, 729–742 (1996)
37. Gaviano, M., Kvasov, D.E., Lera, D., Sergeyev, Y.D.: Software for generation of classes of test functions with known local and global minima for global optimization. ACM Trans. Math. Softw. **29**(4), 469–480 (2003)

On the Convergence Acceleration and Parallel Implementation of Continuation in Disconnected Bifurcation Diagrams for Large-Scale Problems

Nikolay M. Evstigneev$^{(\boxtimes)}$

Federal Research Center "Computer Science and Control",
Institute for System Analysis, Russian Academy of Science, Moscow, Russia
evstigneevnm@yandex.ru

Abstract. The automated construction of disconnected bifurcation diagrams for large-scale steady-state problems is a difficult task. Standard continuation methods are not suited for Krylov-type solvers when dealing with disconnected solutions and, to allow branch switching, they require that the vectors that span the linear operator kernel at bifurcation points be known. An alternative approach to Krylov-type solvers was suggested by Farrell, Beentjes, and Birkisson in 2016, based on the idea of solution deflation. In this paper, we modify the aforementioned method by changing the algorithm to accelerate convergence and rearranging it to be more suited for parallel architectures together with the pseudo-arclength continuation method. We provide a detailed explanation of the corresponding serial and parallel algorithms. Furthermore, we demonstrate the efficiency and correctness of the algorithms by constructing bifurcation diagrams for the 1D Bratu problem and the stationary Kuramoto–Sivashinsky equation considered in 1D and 2D.

Keywords: Newton's methods · Continuation · Solution deflation ·
Automated bifurcation analysis · Disconnected bifurcations ·
Branch switching · Krylov methods

1 Introduction

We consider the problem of constructing the bifurcation diagram for a large-scale steady-state problem. A bifurcation diagram is the grouping of the parameter space and phase space representation induced by the topological equivalence of the system phase space [1]. Each group is presented by a curve in parameter–phase space representation, called a diagram space. Classical bifurcation diagrams for maps and low-dimensional systems are available at Wikipedia. The

This work was supported by the Russian Foundation for Basic Research (grant No. 18-29-10008 mk).

L. Sokolinsky and M. Zymbler (Eds.): PCT 2019, CCIS 1063, pp. 122–138, 2019.
https://doi.org/10.1007/978-3-030-28163-2_9

task of constructing a bifurcation diagram for each curve on the diagram can be formulated as the *homotopy* (or *continuation*) problem, for we are interested in the continuous morphism of the system solutions as a parameter continuously changes [2]. For example, in fluid dynamics problems, a natural homotopy parameter is the Reynolds number. A solution found for low values of the Reynolds number (a task that is computationally cheaper) can be continuously extended to high values of the Reynolds number with less computational efforts. This kind of approach can be found in many papers (we recommend [3] as a very good review paper). The same method was used by the author in [4] to find steady-state solutions to the 3D Kolmogorov problem. There exists another method, called piecewise linear continuation, which allows one to trace paths of solutions to nonlinear equations by building $n - 1$-dimensional simplexes, but it is not well suited for large-scale systems.

The problem of bifurcation diagram construction for a single parameter $\lambda \in \mathbb{R}$ can be formulated as follows. Consider $\mathbf{F} \colon \mathbb{R}^n \times \mathbb{R} \to \mathbb{R}^n$,

$$\mathbf{F}(\mathbf{x}, \lambda) = \mathbf{0}, \tag{1}$$

where $\mathbf{F}$ is a continuous nonlinear function and $\mathbf{x}$ is a solution vector. Equation (1) defines a one dimensional curve in $\mathbb{R}^{n+1}$ which can be continued for given values $\lambda_{\min} \leq \lambda \leq \lambda_{\max}$ of the parameter. Then, the linear stability of problem (1) should be analyzed at selected points in the given parameter interval. A point $(\mathbf{x}^*, \lambda^*)$ in which the linear stability of the system changes (i.e. the Jacobi matrix $\mathbf{F}(\mathbf{x}^*, \lambda^*)_{\mathbf{x}}$ has eigenvalues with zero real parts) is called a bifurcation point. Then, another solution curve emanated from the bifurcation point can be constructed in the same manner. To change a branch, one must perform a continuation from the detected bifurcation point in the direction of the eigenvectors that span the unstable manifold. This is an example of supercritical bifurcation detection and branch switching. More complicated examples of bifurcations are subcritical bifurcations, bifurcations in symmetry-broken systems, and multi-stability. The detection of subcritical bifurcations requires inverse branch continuation. Bifurcation diagrams for symmetry-broken systems are discussed in detail in [5, Chapter IV]. The diagrams for such systems are disconnected, which means that there exist such curves in the diagram space that have no interaction for any considered parameter values. Therefore, the continuation techniques cannot be applied directly and some additional work is required. The multi-stable case is the most difficult since the linear stability of the main branch can be maintained (i.e. the system is linear stable for all values of the parameters), but the basis of attraction tends to measure zero and the system switches to other solutions as a result of finite perturbations. Examples of such behavior in fluid dynamics are Poiseuille and plane Couette flow problems.

Automated construction of bifurcation diagrams is described in a number of papers. The majority of these methods are aimed at low-order maps or ODE systems. Examples of such methods are AUTO [6], CONTENT [7], MATCONT [8], and DSTOOL [9]. The continuation of time-periodic problems for partial differential equations (PDEs) is possible through PDECONT [10], which performs a timestepper-based continuation. Some methods can also perform the analysis of PDEs under conditions of low discretization order thanks to the use of direct methods (LU and QR algorithms).

The question of using an automated construction of bifurcation diagrams for large-scale problems with Newton–Krylov-type solvers is a challenge. Only a few methods of this kind have been suggested, e.g., in [11]. Parallel computations are mostly used in the Newton–Krylov part, whereas branch movement is performed serially. This approach is used in other papers (see, e.g., [12]). A parallel version of diagram construction was proposed in [13], where a parallel movement along a branch is suggested. The recommended approach results in an aggressive optimization of the step length at the cost of redundancy in the concurrent computation. A similar approach is proposed in a recent conference paper [14]. Note that only a few papers use inherently parallel construction of bifurcation diagrams.

The goal of the present paper is to develop a method that executes bifurcation diagram construction in parallel using new ideas associated with solution deflation for bifurcation switching in large-scale systems. We use parallel computations in both diagram construction and large-scale system solution. The paper is laid out as follows. First, we offer an overview of bifurcation diagram construction methods and formulate their main idea. Then, we present a modified deflation approach for convergence to new solutions together with pseudo-arclength continuation. Next, we outline the parallel implementation of the suggested algorithm. Finally, we provide some examples and draw conclusions based on the research results.

2   Overview of Methods

All methods for bifurcation diagram construction consist mainly of the items discussed above: continuation problem for a specific branch, detection of bifurcation points, and branch switching. Let us give a short description of each item.

2.1   Continuation Problem

The homotopy or continuation of problem (1) is usually carried out by using the predictor-corrector method [15]. Two methods are applied.

One is the natural continuation [15, 16], which can be formulated as follows. Let $(\mathbf{x}^0, \lambda^0)$ be a solution that satisfies (1), e.g., some known analytical solution or a trivial solution. This is the initial point. At step k, we have a solution $(\mathbf{x}^k, \lambda^k)$ and introduce a parameter increment $\Delta\lambda^k$. Then, we solve the system as follows:

1. Predictor step:

$$\mathbf{x}^{k+1,0} = \mathbf{x}^k + \Delta\lambda^k \left(\mathbf{F}_{\mathbf{x}^k}\right)^{-1} \left(\mathbf{F}_{\lambda^k}\right). \tag{2}$$

2. Corrector step (Newton's method with $\lambda^{k+1} = \lambda^k + \Delta\lambda^k$):

$$\begin{cases} \mathbf{F}_{\mathbf{x}^{k+1,l}}\delta\mathbf{x} = -\mathbf{F}(\mathbf{x}^{k,l}, \lambda^{k+1}), \\ \mathbf{x}^{k+1,l+1} = \mathbf{x}^{k+1,l} + \delta\mathbf{x}, \end{cases} \tag{3}$$

until $\|\mathbf{F}(\mathbf{x}^{k+1,l+1}, \lambda^{k+1})\| \leq \varepsilon$ with $0 < \varepsilon \ll 1$. The final solution at point $k+1$ is $\mathbf{x}^{k+1} := \mathbf{x}^{k+1,l+1}$ for the parameter value λ^{k+1}.

This algorithm fails at bifurcation points in which the Jacobi matrices $\mathbf{F}_{\mathbf{x}^k}$ and/or $\mathbf{F}_{\mathbf{x}^{k+1,l}}$ are singular (e.g., saddle-node bifurcations, pitchfork bifurcations). In this case, another algorithm is used which provides an augmented linear system that is not singular at these points.

The pseudo-arclength continuation method uses an additional parameter s which introduces dependencies for the pair $(\mathbf{x}(s), \lambda(s))$ with the following gauge:

$$(d\mathbf{x}/ds, d\mathbf{x}/ds) + (d\lambda/ds)^2 = 1, \tag{4}$$

where $(\cdot, \cdot)$ is an inner product. Equation (4) ensures that the arc trajectory is pulled in the tangent direction to the branch. The method is formulated as follows. Let $(\mathbf{x}^0, \lambda^0)$ be a solution that satisfies (1), e.g., some known analytical solution or a trivial solution. This is an initial point with the additional assumption that the Jacobi matrix is not singular at this point. At the k step, we have a solution $(\mathbf{x}^k(s), \lambda^k(s))$ and introduce an arc increment Δs^k. From the parametrization and (1), we can obtain the following equation for the branch slope:

$$\mathbf{F}_{\mathbf{x}}d\mathbf{x}/ds + \mathbf{F}_{\lambda}d\lambda/ds = 0. \tag{5}$$

If this is the initial step ($k = 0$), then we obtain the slopes by introducing the vector $\mathbf{z} = \frac{d\mathbf{x}^k}{ds} / \frac{d\lambda^k}{ds}$, as follows:

$$\mathbf{F}_{\mathbf{x}^0}\mathbf{z} = -\mathbf{F}_{\lambda^0}, \quad d\lambda^0/ds = \pm\left((\mathbf{z}, \mathbf{z}) + 1\right)^{-1/2}, \quad d\mathbf{x}^0/ds = (d\lambda^0/ds)\mathbf{z}, \tag{6}$$

where the choice of the sign (plus or minus) depends on the direction of the continuation. At each step, we solve the augmented system as follows:

1. Predictor step:

$$\begin{cases} \mathbf{x}^{k+1,0} = \mathbf{x}^k + \frac{d\mathbf{x}^k}{ds}\Delta s^k, \\ \lambda^{k+1,0} = \lambda^k + \frac{d\lambda^k}{ds}\Delta s^k. \end{cases} \tag{7}$$

2. The corrector step imposes orthogonality between the vector increment and the correcting trajectory. We use Newton's method:

$$\begin{cases} \begin{pmatrix} \mathbf{F}_{\mathbf{x}^{k+1,l}} & \mathbf{F}_{\lambda^{k+1,l}} \\ \left(\mathbf{x}^{k+1,l} - \mathbf{x}^k\right)^T & \left(\lambda^{k+1,l} - \lambda^k\right) \end{pmatrix} \begin{pmatrix} \delta\mathbf{x} \\ \delta\lambda \end{pmatrix} = \begin{pmatrix} -\mathbf{F}(\mathbf{x}^{k+1,l}, \lambda^{k+1,l}) \\ 0 \end{pmatrix}, \\ \begin{cases} \mathbf{x}^{k+1,l+1} = \mathbf{x}^{k+1,l} + \delta\mathbf{x}, \\ \lambda^{k+1,l+1} = \lambda^{k+1,l} + \delta\lambda, \end{cases} \end{cases} \tag{8}$$

until $\|\mathbf{F}(\mathbf{x}^{k+1,l+1}, \lambda^{k+1,l+1})\| \leq \varepsilon$, with $0 < \varepsilon \ll 1$. The final solution at point $k+1$ is $\mathbf{x}^{k+1} := \mathbf{x}^{k+1,l+1}$, $\lambda^{k+1} := \lambda^{k+1,l+1}$ for the given value of Δs^k.

3. In the tangent estimation step, we solve the following linear system from (4) and (5):

$$\begin{pmatrix} \mathbf{F}_{\mathbf{x}^{k+1}} & \mathbf{F}_{\lambda^{k+1}} \\ (\frac{d\mathbf{x}^k}{ds})^T & \frac{d\lambda^k}{ds} \end{pmatrix} \begin{pmatrix} \frac{d\mathbf{x}^{k+1}}{ds} \\ \frac{d\lambda^{k+1}}{ds} \end{pmatrix} = \begin{pmatrix} \mathbf{0} \\ 1 \end{pmatrix}. \tag{9}$$

This method allows one to pass bifurcation points while continuing the solution. Indeed, the matrix of the augmented linear system in (8) is non-singular even when the Jacobian of the original system is singular with kernel dimension equal to one. The value of Δs^k can be adjusted to increase the convergence rate.

While the solution of the augmented system for small-scale problems is a simple task, it becomes a challenge in large-scale problems, where direct matrix linear algebra cannot be applied. It is very difficult to solve the rank-one change of linear systems (8), (9) even when one can efficiently obtain the solution of the original linear system in (3) by using iterative methods. A preconditioner matrix for the Jacobian is no longer applicable. Therefore, the use of the pseudo-arclength continuation becomes a major issue in large-scale computations.

2.2 Bifurcation Detection and Branch Switching Problem

The detection of bifurcations along the trajectory path requires one to introduce some indicative functions that change parameters while moving on the branch trajectory. A common function of this sort, used in almost all continuation software, is either the sign of the Jacobi matrix determinant (available for small-scale systems with LU decomposition) or the sign of $\mathrm{Re}(\lambda_j)$, where λ_j are the system eigenvalues with largest real parts, obtained by solving the exact or approximate spectral problem for the Jacobi matrix. In large-scale problems, the Lanczos or Arnoldi variants of eigenvalue solvers are used. For a parallel solution of the largest real eigenvalues problem, one can apply either the Parpack or the author's implementation of the implicitly restarted Arnoldi method (IRA) [17] for multiple graphical processing units (GPUs). Practically, there are no other ways to detect bifurcations. Almost all branch switching techniques rely on the idea of continuation in the direction of the null-space basis vectors at bifurcation points [16].

All these approaches to bifurcation detection require substantial computational efforts.

2.3 Deflated Continuation

An alternative approach, suggested in [18], is based on solution deflation for nonlinear partial differential equations [19]. The idea of this approach stems from the deflation method for polynomials [20]. The theory of the multi-convergent Newton method is also developed in the above-mentioned papers. Assume that $(\mathbf{x}^{(1)}, \lambda^*)$ satisfies (1) with nonsingular Fréchet derivative $\mathbf{F}'(\mathbf{x}^{(1)}, \lambda^*)$. Let us assume that there exists another pair $(\mathbf{x}^{(2)}, \lambda^*)$ that also satisfies (1) but is unknown. Construct the modified problem

$$\mathbf{G}(\mathbf{x}, \lambda) := M(\mathbf{x}, \mathbf{x}^{(1)}) \mathbf{F}(\mathbf{x}, \lambda) = \mathbf{0}, \tag{10}$$

where M is a deflation operator of the residual $\mathbf{F}$. This deflated residual satisfies two properties:

1. Preservation of solutions of (1), i.e. $\mathbf{G}(\mathbf{x}) = 0$ if and only if $\mathbf{F}(\mathbf{x}) = 0$, $\forall \mathbf{x} \neq \mathbf{x}^{(1)}$.
2. Newton's method applied to $\mathbf{G}$ for some initial guess $\mathbf{x} \neq \mathbf{x}^{(1)}$ does not converge to the original solution $\mathbf{x}^{(1)}$, i.e. $\lim\limits_{\mathbf{x} \to \mathbf{x}^{(1)}} \inf \|\mathbf{G}(\mathbf{x})\| > 0$.

If Newton's method converges to a new solution $\mathbf{x}^{(2)}$, then it can be deflated as well and the process is repeated until Newton's method fails to converge. We may assume that all solutions for λ^* are found. The continuation process, suggested in [18], is built around this deflation strategy. This method can be applied to large-scale systems using Newton–Krylov-type methods. The Jacobian of deflated operator (10) is a rank-one perturbation of the original operator, so it becomes dense even if the original matrix is sparse. However, it was shown in [19] that this problem can be solved for large-scale systems. If a preconditioner $\mathbf{P}_F$ is available for the original undeflated problem, then a new preconditioner $\mathbf{P}_G$ for the deflated problem can be constructed by using only the matrix-vector product operator and the action of the original preconditioner by the Sherman–Morrison formula. However, this method has some drawbacks. First, it is impossible to detect a bifurcation point during the continuation process. Second, this method cannot pass saddle-node points. Third, this method is undecidable, which means that if Newton's method fails to converge, this does not imply that there are no other solutions.

We may formulate now our approach to the problem. All calculations presented here were performed using an 8-core (16 in hyper-threading) Intel Xeon E5-2640V2 Ivy Bridge and up to six K40 Nvidia GPUs installed in one chassis. The hardware is in the author's possession.

3   Deflated Pseudo-arclength Continuation

Note that the continuation and deflation methods are, practically, data independent. Therefore, continuation and deflation can be executed in parallel. The deflation of solutions provides a distinguished root and continuation will start from this root, adding further solutions at certain points that will be deflated. In this paper, we outline our approach which uses combinations of all abovementioned techniques and is aimed at parallel execution of bifurcation diagram construction for large-scale problems solved by the Newton–Krylov method. Our approach uses parallelism for both diagram construction and system solution.

We start our method by updating the corrector for the pseudo-arclength continuation method (8). Let us rewrite the corrector step in an equivalent form:

$$\begin{cases} \mathbf{F}_{\mathbf{x}^{k+1,l}}\delta\mathbf{x} + \mathbf{F}_{\lambda^{k+1,l}}\delta\lambda = -\mathbf{F}(\mathbf{x}^{k+1,l}, \lambda^{k+1,l}), \\ \left(\mathbf{x}^{k+1,l} - \mathbf{x}^k\right)^T \delta\mathbf{x} + (\lambda^{k+1,l} - \lambda^k)\delta\lambda = 0. \end{cases} \tag{11}$$

At this point, we should note that this method needs correction near turning points. More precisely, we choose Δs^k in such a manner that $\alpha = (\lambda^{k+1,l} - \lambda^k)$, $|\alpha| \geq \varepsilon > 0$. This ensures that a turning point is passed without difficulties and that the condition number of the augmented-system matrix is determined by the condition number of $\mathbf{F}_{\mathbf{x}^{k+1,l}}$. In practice, we set $\varepsilon = 1.0 \times 10^{-6}$. Then, $\delta\lambda = -\left(\mathbf{x}^{k+1,l} - \mathbf{x}^k\right)^T \delta\mathbf{x}/\alpha$, and the linear system for step (11) of Newton's method can be rewritten as

$$\left(\mathbf{F}_{\mathbf{x}^{k+1,l}} - 1/\alpha\mathbf{F}_{\lambda^{k+1,l}}\left(\mathbf{x}^{k+1,l} - \mathbf{x}^k\right)^T\right)\delta\mathbf{x} = -\mathbf{F}(\mathbf{x}^{k+1,l}, \lambda^{k+1,l}),$$
$$\delta\lambda = 1/\alpha\left(\mathbf{x}^{k+1,l} - \mathbf{x}^k\right)^T \delta\mathbf{x}. \tag{12}$$

Let us assume that we are using the Newton–Krylov numerical method for the original problem (1) and we already have a well tuned left preconditioner $\mathbf{P}_F^{-1}$ to efficiently solve system (3). In more detail, the linear system of the natural continuation corrector method (3) is used as follows:

$$\mathbf{P}_F^{-1}\mathbf{F}_{\mathbf{x}^{k+1,l}}\delta\mathbf{x} = -\mathbf{P}_F^{-1}\mathbf{F}(\mathbf{x}^{k,l}, \lambda^{k+1}). \tag{13}$$

Now we want to construct the preconditioner for the extended system (12):

$$\mathbf{R}_F^{-1}\left(\mathbf{F}_{\mathbf{x}^{k+1,l}} - 1/\alpha\mathbf{F}_{\lambda^{k+1,l}}\left(\mathbf{x}^{k+1,l} - \mathbf{x}^k\right)^T\right)\delta\mathbf{x} = -\mathbf{R}_F^{-1}\mathbf{F}(\mathbf{x}^{k+1,l}, \lambda^{k+1,l}), \tag{14}$$

where

$$\mathbf{R}_F^{-1} := \left(\mathbf{P}_F - 1/\alpha\mathbf{F}_{\lambda^{k+1,l}}\left(\mathbf{x}^{k+1,l} - \mathbf{x}^k\right)^T\right)^{-1}. \tag{15}$$

The new preconditioner is a full matrix even if $\mathbf{P}_F$ is sparse. The only way to work with $\mathbf{R}_F$ efficiently is to use $\mathbf{P}_F$, and the correction must be applied in a matrix-free way, i.e. only the matrix-vector application is provided. This can be achieved by using the Sherman–Morrison formula [21]. Denote $-1/\alpha\mathbf{F}_{\lambda^{k+1,l}} = \mathbf{b}$ and $\left(\mathbf{x}^{k+1,l} - \mathbf{x}^k\right) = \mathbf{c}$. Then, we can rewrite (15) as

$$\left(\mathbf{P}_F + \mathbf{b}\mathbf{c}^T\right)^{-1} = \mathbf{P}_F^{-1} - \frac{\mathbf{P}_F^{-1}\mathbf{b}\mathbf{c}^T\mathbf{P}_F^{-1}}{1 + \mathbf{c}^T\mathbf{P}_F^{-1}\mathbf{b}}. \tag{16}$$

An additional constraint on (17) is that $1 + \mathbf{c}^T\mathbf{P}_F^{-1}\mathbf{b} = 1 - \left(\mathbf{x}^{k+1,l} - \mathbf{x}^k\right)^T$ $\mathbf{P}_F^{-1}\mathbf{F}_{\lambda^{k+1,l}}/\alpha \neq 0$, which can be verified during calculations. Finally, we obtain

$$\mathbf{R}_F^{-1} := \mathbf{P}_F^{-1} + \frac{\mathbf{P}_F^{-1}\mathbf{F}_{\lambda^{k+1,l}}\left(\mathbf{x}^{k+1,l} - \mathbf{x}^k\right)^T\mathbf{P}_F^{-1}}{\alpha + \left(\mathbf{x}^{k+1,l} - \mathbf{x}^k\right)^T\mathbf{P}_F^{-1}\mathbf{F}_{\lambda^{k+1,l}}}. \tag{17}$$

If we use the original preconditioner in a matrix-vector application form with the residual vector $\mathbf{r}$, then the computation of (17) is also carried out in a matrix-free way. First, we calculate $\mathbf{p} = \mathbf{P}_F^{-1}\mathbf{F}_{\lambda^{k+1,l}}$, followed by the application of the original preconditioner to the residual vector: $\mathbf{q} = \mathbf{P}_F^{-1}\mathbf{r}$. Next, we compute the inner product $\left(\mathbf{x}^{k+1,l} - \mathbf{x}^k\right)^T \mathbf{q}$ and multiply the result by the vector $\mathbf{p}$. The vector $\mathbf{p}$ is temporally stored and used in the denominator and in the left inverse. Finally, we obtain

$$\mathbf{R}_F^{-1}\mathbf{r} = \mathbf{q} + \frac{\left(\left(\mathbf{x}^{k+1,} - \mathbf{x}^k\right)^T \mathbf{q}\right)\mathbf{p}}{\alpha - \left(\mathbf{x}^{k+1,l} - \mathbf{x}^k\right)^T \mathbf{p}}. \tag{18}$$

Thus, the new preconditioner is twice more expensive than the original one, with two more calls of vector dot products. If the original preconditioner is used in a matrix explicit sparse form (e.g., SGS or ILU), then the new preconditioner is thrice more expensive but can be applied in the same way. The same approach with preconditioners is used to solve linear system (9) for tangent vectors.

Now we turn our attention to the deflation of solutions. The general idea is to use pseudo-arclength continuation for some time and then apply deflation at selected parameter points, thereby performing branch switching for connected and disconnected branches. Firstly, we give more details on an alternative approach to using deflation operators. Let us assume that for a given parameter value λ we have K solutions $\mathbf{x}_1, \mathbf{x}_2, \ldots, \mathbf{x}_K$ that we need to deflate. In this case, we suggest the following modified deflated problem (see (10)):

$$\mathbf{G} := \frac{1}{K}\sum_{j=1}^{K} M(\mathbf{x}, \mathbf{x}_j)\mathbf{F}(\mathbf{x}, \lambda) = \frac{1}{K}\operatorname{diag}\left(\sum_{j=1}^{K}\left(\|\mathbf{x} - \mathbf{x}_j\|_2^{-p} + \sigma\right)\right)\mathbf{F}(\mathbf{x}, \lambda) = \mathbf{0}, \tag{19}$$

where $\operatorname{diag}()$ is a diagonal matrix. The Jacobi matrix $\mathbf{G}_{\mathbf{x}}$ can be written explicitly as

$$\mathbf{G}_{\mathbf{x}} := \frac{1}{K}\sum_{j=1}^{K} M(\mathbf{x}, \mathbf{x}_j)\mathbf{F}(\mathbf{x}, \lambda)_{\mathbf{x}} - \mathbf{F}(\mathbf{x}, \lambda)\frac{p}{K}\sum_{j=1}^{K}\|\mathbf{x} - \mathbf{x}_j\|_2^{(-p-2)}(\mathbf{x} - \mathbf{x}_j)^T. \tag{20}$$

Here we use the equality $(\|f(x)\|_2)' = \|f(x)\|_2^{-1}f_x^T f(x)$. Again, a preconditioner can be applied by using the Sherman–Morrison formula. This deflation operator is different from the original one given in [19] since it relies on summation instead of multiplication. Nevertheless, the conclusion of the theorem in [19] remains true, and (19) leads to a decrease in number of operations. In practice, we notice that deflation operator (19) works faster than the original one. Finally, we use the set of known solutions S and Newton's method to solve the deflation problem in a maximum of N steps:

$$\begin{cases} \mathbf{G}_{\mathbf{x}^l}(S)\,\delta\mathbf{x} = -\mathbf{G}(S, \mathbf{x}^l), \\ \mathbf{x}^{l+1} = \mathbf{x}^l + \delta\mathbf{x}, \end{cases} \quad l = 1, 2, \ldots. \tag{21}$$

The method stops when either it converges, i.e. $\|\mathbf{G}(S, \mathbf{x}^{l+1})\|_2 \le \varepsilon \ll 1$, or $l + 1 \ge N$.

Algorithm 1. Pseudo-arclength continuation

1: **function** PSEUDO_ARCLENGTH_CONTINUATION($\mathbf{F}$, $\mathbf{F_x}$, $\mathbf{F_\lambda}$, $\mathbf{P}_F$, λ, $\mathbf{x}$, $\lambda_{\min}$, $\lambda_{\max}$, Δs, $\Delta s_{\min}$, $\Delta s_{\max}$, N_{steps}, ε, $d\lambda$, dF, $direction$)
2:     $k = 0$; $\Delta s^0 \leftarrow \Delta s$; $\lambda \leftarrow \lambda_{\min}$; $s \leftarrow 0$, $S \leftarrow (\mathbf{x}, \lambda, \text{dim\_reduction}(\mathbf{x}, s)$;
3:     **while** ($\lambda \leq \lambda_{\max}$) or ($k < N_{steps}$) **do**
4:         $status \leftarrow$ false;
5:         **while** ($\Delta s_{\min} < \Delta s^k < \Delta s_{\max}$) or ($status ==$ false) **do**
6:             SOLVE((7), $status$, $direction$); SOLVE((8),ε, $status$); SOLVE((9),$status$).
7:             **if** ($status ==$ false) **then**
8:                 $\Delta s^k \leftarrow$ Update(Δs^k);
9:         **if** ($status ==$ false) **then**
10:             **return** ($status$, S, k);
11:         $s \leftarrow s + \Delta s^k$;
12:         $dimF = \text{dim\_reduction}(\mathbf{x})$
13:         $S \leftarrow S \cup (\mathbf{x}, \lambda, dimF, s)$;
14:         $k \leftarrow k + 1$;
15:         $check\_coincide \leftarrow$ CHECK_REDUCED_COLLISION$((S, d\lambda, dF), \varepsilon)$;
16:         **if** ($check\_coincide ==$ true) **then**
17:             **if** (INTERPOLATE_SOLUTION_CURVE($\mathbf{F}$, $\mathbf{F_x}$, $\mathbf{P}_F$, S, $\mathbf{x}$, λ, N_{Newton}, ε)==true) **then**
18:                 **return** ($status$, S, k);
        return ($status$, S, k);

4   Parallel Deflated Pseudo-arclength Continuation

We provide the main algorithms for deflated pseudo-arclength continuation. Each call to the numerical method is referenced to appropriate equations. Note that the execution of Newton–Krylov methods can be carried out either in serial or in parallel.

The method consists of two main blocks: pseudo-arclength continuation and solution deflation. The variant of pseudo-arclength continuation implemented in Algorithm 1 takes an initial point, the necessary operators, matrices (or operators), and convergence control parameters. It basically solves systems (7), (8), and (9) and modifies Δs^k to ensure a faster convergence by means of the function call Update(Δs^k). This function either increases or decreases the arc step length depending on the residual norm or the number of iterations of linear solvers. This function also verifies whether the trajectory coincides with the one previously calculated. This is done by means of the dimension reduction function dim_reduction($\mathbf{x}$), which can be some vector semi-norm, and a call to the solution interpolation algorithm. The *Check_reduced_collision* function can check whether some last steps of the continuation process are in the ε-tube of other points in the current solution path. This algorithm requires two scalars, $d\lambda$ and dF, which are set by the user and are usually functions of the parameter range and solution vector norms. The algorithm returns a set containing solutions and parameters along the arc together with arc points and a reduced dimension vector in the set S, a flag of either success or failure, and the number

Algorithm 2. Deflation

1: **function** DEFLATION($\mathbf{F}$, $\mathbf{F_x}$, $\mathbf{P}_F$, $\mathbf{x}$, Q, λ, N_{Newton}, ε)
2:     $l = 0$; $\mathbf{G} \leftarrow$ USE(19)(($\mathbf{F}$, Q)); $\mathbf{G_x} \leftarrow$ USE(20)(($\mathbf{F_x}$, Q)); $\mathbf{P}_G \leftarrow$ USE(17)($\mathbf{P}_F$);
3:     **repeat**
4:         SOLVE SINGLE NEWTON STEP((21));
5:         $l \leftarrow l + 1$; $residual \leftarrow \|\mathbf{G}(Q, \mathbf{x}^{l+1})\|_2$;
6:     **until** $(l < N_{Newton})$ or $(residual > \varepsilon)$
7:     **if** $residual \leq \varepsilon$ **then**
8:         **return** $(l, \mathbf{x}^{l+1}, residual)$;
9:     **else**
10:         **return** $(l, \text{null\_ptr}, residual)$;

---

of steps. Below, deflation Algorithm 2 is considered. It accepts all operators and matrices (or matrix-vector operators), a set Q of all solutions and parameters to be deflated, a parameter value, an initial guess vector, and convergence control parameters. It either returns a true flag with the solution vector and number of iterations, or a false flag with an empty vector.

The resulting algorithm is based on the MASTER-SLAVES format. The algorithm accepts the data required to perform the continuation, with two additional parameters *proc_solve*, *total_proc*. The first one is the number of threads used to solve the problem by Newton–Krylov methods. The second is the total number of threads. The algorithm returns a bifurcation diagram (by the master process), as well as continued paths of parameters and solutions (by slave processes). The master process algorithm uses a single thread and is only employed for control and storage of light data (the bifurcation diagram itself without solutions). A slave process can be in either of seven modes, namely 'D' for deflating, '$C_{p/m}$' for continuation on positive/negative direction, 'R' for data receiving, 'S' for data sending, 'F' for idle, and 'T' for termination. Firstly, the master process initializes the statuses of all slaves and sets them to 'D', except for one that is used to perform the continuation and is set to, say, 'C_p'. Next, it monitors the statuses of the slaves. If a slave reaches some knot, it signals to the master. If the knot has continuation status, then the slave process is set to idle. If a knot has deflating status, then deflation is paused, the deflating process in the knot receives an additional solution, which is added to the deflation list, and then, it resumes the deflation. A process with the '$C_{p/m}$' property proceeds with the continuation in the same direction. If a slave process signals with 'F', then the master process checks whether there are vacant knots available. If there are any, then the master process sets the deflation process. If there are no vacant knots available, the master process checks whether it is required to perform the continuation in the opposite direction. If yes, then the process is executed. Otherwise, the status of the slave remains 'F'. If all slaves have status 'F', then the master signals 'T' to all processes. A slave process can change its status from 'D' to either '$C_{p/m}$' (if a new root is found) or 'F' (if nothing is found). A slave process continues from status 'S' in the previously given direction, whereas from status 'R', it continues the deflation. Each slave process sends light bifurcation diagram data to the master process but keeps all other heavy data locally. Data exchange occurs only when a new solution is added to the deflation list or when a new process starts the deflation in a previously vacant knot.

The code for the parallel deflation process is written in C++ and CUDA C for the solution of large-scale problems on GPUs.

5   Examples

Here we present some application examples and timings for the algorithms. In all computations, we assume that $p = 2$ and $\sigma = 1$ in (19) and (20).

5.1   1D Bratu Problem

This problem is a well-known benchmark for continuation testing [22]. It is called the Liuoville–Bratu–Gelfand problem [23] and is formulated for a function $u(x)$ as

$$u_{xx} + \lambda\,e^u = 0, \; x \in [0, 1], \; u(0) = u(1) = 0. \tag{22}$$

The problem has the trivial point solution $(\mathbf{0}, 0)$. Moreover, it has a single saddle-node bifurcation at $\lambda^* \sim 3.514$. For the discretization of the problem, we used the Bubnov–Galerkin method with 100 Gegenbauer polynomials. All matrices were solved by direct methods with a single thread. We compared three variants: the simple pseudo-arclength continuation by Algorithm 1, and the parallel algorithm with two, three, and four threads. As set of deflation parameter points (knots), we took the set $D = \{0.8, 3\}$. So, the last comparison for four threads has two spare threads. The results of the two-threaded algorithm are portrayed in Fig. 1(left). The crosses denote the solutions that were deflated when the continuation process reached a knot. The circles show the beginning of the continuation process from a knot, i.e when the deflation process in a knot finds a solution. The ratio of the wall time for the serial continuation version to that for the parallel one is illustrated in Fig. 1(right). The optimal time for this particular setup was achieved when there was one extra thread available for the number of knots in the set D. We were still unable to achieve the ideal acceleration in this problem. The results show a good agreement with benchmark data (e.g., from [22]). The value of the parameter at the saddle-node bifurcation is $\lambda_C = 3.51391235$, which gives an error of 0.0054%.

5.2   1D Kuramoto–Sivashinsky Equation

Let us consider a well-known system of PDEs with quadratic nonlinearity, called the Kuramoto–Sivashinsky equation. Excellent exact results for this system are available for benchmarking (see [24]). The system for $u(x, t)$ with Dirichlet boundary conditions is defined as

$$u_t + 4u_{xxxx} + \lambda\,(u_{xx} + 2uu_x) = 0, \; u(0, t) = u(\pi, t) = 0, \; x \in [0, \pi]. \tag{23}$$

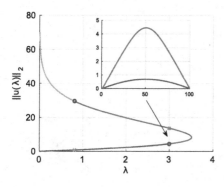
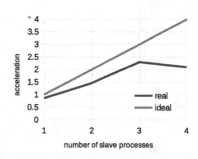

Fig. 1. Bifurcation diagrams of the 1D Bratu problem and two solutions (left). The circles represent the beginning of the parallel continuation process, while the crosses represent the deflated solutions. Wall-time comparison with the serial version (right)

It has a trivial solution $u(x,t) = \mathbf{0}$ which exists for all λ. We considered only stationary solutions of (23), which means that $u_t = 0$. We solved the system by employing a pseudo-spectral method with DST(sin transfer) featuring 511 degrees of freedom. Further, we dealt with the nonlinearity by using zero-padding (3/2-rule). The Jacobian is a full matrix, so only a matrix-vector application was considered. The system was solved taking Newton–Krylov methods as a basis for all algorithms, with a preconditioner constructed by means of a low-order finite difference approximation of the problem, which was solved by the multigrid method, again as a matrix-free preconditioner. This allowed us to verify the concept of matrix-free operations during the continuation and deflation stages. The same preconditioner strategy was successfully used in [4] for a Navier–Stokes 3D system with large spectral discretization (up to $512 \times 256 \times 256$).

We applied the parallel Algorithm 1 to the considered problem. The results are shown in Fig. 2. For the diagram space, we used the L_2 vector norm in Fourier space (see [24]). As set of deflation parameter points, we chose the set $D = \{5, 17, 60, 75, 107, 148\}$. The solutions that were found by the deflation process at these points are marked with circles in Fig. 2. The parallel version of the continuation algorithm used up to seven CPU slave threads, and each thread performed the Newton–Krylov steps serially.

The obtained diagram is shown in Fig. 2 for $\lambda \leq 80$; it fully coincides with exact results given in [24]. In the whole parameter range considered, we detected 19 pitchfork bifurcations, 8 transcritical bifurcations, 12 saddle-node bifurcations, and 4 Hopf bifurcations. To detect bifurcation points and their type, we resorted to a CPU version of the implicitly restarted Arnoldi method as an eigensolver. We also detected two disconnected branches in the interval $92.8 \leq \lambda \leq 133.7$; their paths are shown with dashed curves in Fig. 2. The solutions for these branches in physical space are shown with different colors in the top left graph for $\lambda = 107$. We will not dwell on details touching nonlinear dynamics, so we only provide this diagram as a benchmark, an illustration, and a wall time measurement problem.

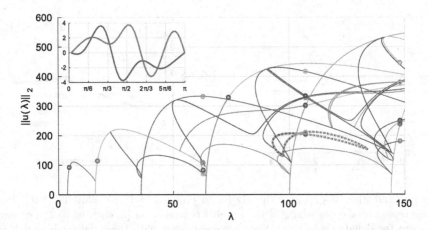

Fig. 2. Bifurcation diagrams of the solutions to the stationary 1D Kuramoto–Sivashinsky (KS) equation for $0 \leq \lambda \leq 150$. The disconnected solutions are shown in the top left graph; their paths are indicated with dashed curves. The circles represent the initial solutions found by deflation.

We compared the serial and parallel algorithms with regards to wall-time execution. It was found that the ratio of the wall time for the successful convergence of the deflation step to that for a single pseudo-arclength continuation step is about 50 to 100. This allowed us to exploit the parallel diagram construction much more efficiently than in the previous subsection. We discovered that the optimal number of knots must be, again, one less than the number of threads. With an optimal balance of threads between continuation and deflation, we obtained an almost perfectly linear acceleration (see Fig. 3(left)). The initial drop in efficiency is related to the growing load on the master process.

5.3 2D Kuramoto–Sivashinsky Equation

This is a 2D variant of the previous problem, considered for a smooth vector function $u \colon [0, 2\pi]^2 \to \mathbb{R}$ with periodic boundary conditions:

$$\lambda \left(2uu_x + 2uu_y + \triangle u\right) + 4\triangle^2 u = 0, \ u \in \mathbb{T}^2 : \mathbf{x} \in [0, 2\pi]^2. \tag{24}$$

We used a pseudo-spectral method for the discretization. The system was solved with the cuFFT library on Nvidia GPUs. We considered 4 177 936 degrees of freedom (taking reality conditions into account), which is equal to 2048×2048 Fourier harmonics. The preconditioner strategy is similar to the one applied in the previous subsection: we used a geometric multigrid method for solving a low-order system which served as a preconditioner for the main Krylov BICGStab(L) solver. We noted that the suggested approach with preconditioners works efficiently for large linear systems. Moreover, we observed a maximum 30% iteration increase in the linear solver.

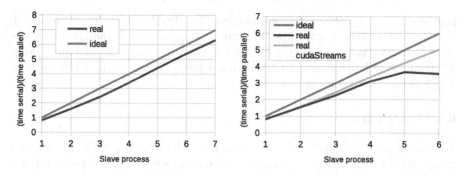

Fig. 3. Acceleration of the parallel algorithms for 1D KS equations (left) and 2D KS equations (right)

For each slave process, we used a single GPU, so the maximum total number of slave processes was equal to six. We considered the following set of parameter knots: $D = \{1, 5, \ldots, 45\}$. The acceleration results are given in Fig. 3(right). Our first version of the code used a single CUDA stream for computations and MPI communications in slaves. We can see an obvious degradation of acceleration, especially for five and six GPUs. The updated version of the code used separate CUDA streams for computations and communications, resulting in much better scaling, which is shown in Fig. 3.

Although there are some papers regarding 2D KS equations, there is only one that deals with bifurcation analysis (see [25]) but it is focused mainly on nonstationary solutions. The rich and complex nonlinear dynamics and bifurcation analysis of stationary problem (24) will be considered in future papers. We conducted an analysis of stationary solutions only for benchmarking and timing

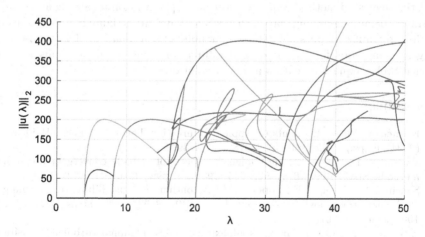

Fig. 4. Bifurcation diagrams of the solutions to the stationary 2D Kuramoto–Sivashinsky (KS) equation for $0 \leq \lambda \leq 50$ in L_2 norm

purposes. The bifurcation diagram obtained for $0 \leq \lambda \leq 50$ is shown in Fig. 4. We can see how complex the system becomes compared with the 1D case in the same parameter range (see Fig. 2). We detected at least 27 pitchfork bifurcations, 32 saddle-node bifurcations, 12 transcritical bifurcations, 10 Hopf bifurcations, and 7 disconnected solutions.

6   Conclusions

This paper is a milestone in the year-long research dedicated to the automation and parallel construction of bifurcation diagrams with disconnected solutions. We offered a review of the methods used for bifurcation diagram construction and employed some suitable techniques that have been developed in recent years, including the deflation of solutions of PDEs [18, 19]. We successfully applied some modifications of these methods to Newton–Krylov-type solvers. Moreover, we constructed disconnected bifurcation diagrams by using the pseudo-arclength continuation method. We described in detail parallel algorithms for the auto- mated construction of bifurcation diagrams and gave some examples of the application of these algorithms to test problems. The bifurcation diagrams for 2D Kuramoto–Sivashinsky equations were presented in this paper for the first time.

We can conclude that the suggested method is capable of successfully con- structing in parallel bifurcation diagrams for large-scale systems. An adequate balance of workload is required since we use a master-slave formulation as well as optimization of data transfer. This can be a bottleneck considering that we need to send solutions from one process to another to update a list of deflated solutions. Due to space restrictions, we have omitted in this paper information related to successful results achieved by using the suggested preconditioning strategy. This information will be provided in future papers. It should be noted that the suggested method requires additional optimization regarding automatic workload balancing among slave processes and automatic launch of eigensolvers in the proximity of possible bifurcation points. Our main goal is to use the suggested methods for the numerical bifurcation analysis of problems for the Navier–Stokes equations with stationary solutions.

References

1. Kuznetsov, Yu.A.: Elements of Applied Bifurcation Theory. Springer, Heidelberg (2004). https://doi.org/10.1007/978-1-4757-3978-7
2. Watson, L.T.: Numerical linear algebra aspects of globally convergent homotopy methods. SIAM Rev. **28**(4), 529–545 (1986). https://doi.org/10.1137/1028157
3. Sanchez, J., Marques, F., Lopez, J.M.: A continuation and bifurcation technique for Navier-Stokes flows. J. Comput. Phys. **180**, 78–98 (2002). https://doi.org/10.1006/jcph.2002.7072
4. Evstigneev, N., Magnitskii, N.: Nonlinear dynamics of laminar-turbulent transition in generalized 3D Kolmogorov problem for incompressible viscous fluid at symmet- ric solution subset. J. Appl. Nonlinear Dyn. **6**, 345–353 (2017). https://doi.org/10.5890/JAND.2017.09.003

5. Golubitsky, M., Schaeffer, D.: Singularities and Groups in Bifurcation Theory: Volume I. Applied Mathematical Sciences, vol. 51. Springer, Heidelberg (1985). https://doi.org/10.1007/978-1-4612-5034-0
6. Wang, X.J., Doedel, E.J.: AUTO94P: an experimental parallel version of AUTO, Technical report, Center for Research on Parallel Computing, California Institute of Technology, Pasadena CA 91125. CRPC-95-3 (1995)
7. Kuznetsov, Yu.A., Levitin, V.V.: CONTENT: a multiplatform environment for continuation and bifurcation analysis of dynamical systems. Centrum voor Wiskunde en Informatica, Kruislaan 413, 1098 SJ Amsterdam, The Netherlands (1997)
8. Dhooge, A., Govaerts, W., Kuznetsov, Yu.: MATCONT: a Matlab package for numerical bifurcation analysis of ODEs. ACM Trans. Math. Softw. **29**, 141–164 (2003). https://doi.org/10.1145/980175.980184
9. Back, A., Guckenheimer, J., Myers, M.R., Wicklin, F.J., Worfolk, P.A.: DsTool: computer assisted exploration of dynamical systems. Not. Am. Math. Soc. **39**(4), 303–309 (1992)
10. Lust, K., Roose, D., Spence, A., Champneys, A.R.: An adaptive Newton-Picard algorithm with subspace iteration for computing periodic solutions. SIAM J. Sci. Comput. **19**(4), 1188–1209 (1998). https://doi.org/10.1137/S1064827594277673
11. Böhmer, K., Mei, Z., Schwarzer, A., Sebastian, R.: Path-following of large bifurcation problems with iterative methods. In: Doedel, E., Tuckerman, L.S. (eds.) Numerical Methods for Bifurcation Problems and Large-Scale Dynamical Systems. IMA, vol. 119, pp. 37–65. Springer, New York (2000). https://doi.org/10.1007/978-1-4612-1208-9_2
12. Govaerts, W.J.F.: Numerical Methods for Bifurcations of Dynamic Equilibria. SIAM, Philadelphia (2000). https://doi.org/10.1137/1.9780898719543
13. Aruliah, D.A., van Veen, L., Dubitski, A.: PAMPAC: a parallel adaptive method for pseudo-arclength continuation. ACM Trans. Math. Softw. **42**(1), Article 8 (2016). https://doi.org/10.1145/2714570
14. Marszalek, W., Sadecki, J.: 2D bifurcations and chaos in nonlinear circuits: a parallel computational approach. In: 2018 15th International Conference on Synthesis, Modeling, Analysis and Simulation Methods and Applications to Circuit Design (SMACD), Prague, pp. 1–300 (2018). https://doi.org/10.1109/SMACD.2018.8434908
15. Abbott, J.P.: Numerical continuation methods for nonlinear equations and bifurcation problems. Ph.D. thesis, Australian National University (1977). https://doi.org/10.1017/S0004972700010546
16. Seydel, R.: Practical Bifurcation and Stability Analysis. Interdisciplinary Applied Mathematics, vol. 5. Springer, Heidelberg (2010). https://doi.org/10.1007/978-1-4419-1740-9
17. Evstigneev, N.M.: Implementation of implicitly restarted arnoldi method on Multi-GPU architecture with application to fluid dynamics problems. In: Sokolinsky, L., Zymbler, M. (eds.) PCT 2017. CCIS, vol. 753, pp. 301–316. Springer, Cham (2017). https://doi.org/10.1007/978-3-319-67035-5_22
18. Farrell, P.E., Beentjes, H.L.C., Birkisson, A.: The computation of disconnected bifurcation diagrams. arXiv:1603.00809 (2016)
19. Farrell, P.E., Birkisson, A., Funke, S.W.: Deflation techniques for finding distinct solutions of nonlinear partial differential equations. SIAM J. Sci. Comput. **37**, A2026–A2045 (2015). https://doi.org/10.1137/140984798
20. Wilkinson, J.H.: Rounding Errors in Algebraic Processes. Notes on Applied Science, vol. 32. H.M.S.O. (1963)

21. Sherman, J., Morrison, W.J.: Adjustment of an inverse matrix corresponding to a change in one element of a given matrix. Ann. Math. Stat. **21**(1), 124–127 (1950). https://doi.org/10.1214/aoms/1177729893
22. Doedel, E., Keller, H.B., Kernevez, J.P.: Numerical analysis and control of bifurcation problems. II. Bifurcation in infinite dimensions. IJBC **1**, 745–772 (1991). https://doi.org/10.1142/S0218127491000555
23. Gelfand, I.M.: Some problems in the theory of quasi-linear equations. Am. Math. Soc. Transl. Ser. **2**(29), 295–381 (1963)
24. Arioli, G., Koch, H.: Computer-assisted methods for the study of stationary solutions in dissipative systems, applied to the Kuramoto-Sivashinski equation. Arch. Ration. Mech. Anal. **197**(3), 1033–1051 (2010). https://doi.org/10.1007/s00205-010-0309-7
25. Kalogirou, A., Keaveny, E.E., Papageorgiou, D.T.: An in-depth numerical study of the two-dimensional Kuramoto-Sivashinsky equation. Proc. Math. Phys. Eng. Sci. **471**(2179), 20140932 (2015). https://doi.org/10.1098/rspa.2014.0932

LRnLA Lattice Boltzmann Method: A Performance Comparison of Implementations on GPU and CPU

Vadim Levchenko[1]($\boxtimes$)(iD), Andrey Zakirov[1](iD), and Anastasia Perepelkina[1,2](iD)

[1] Keldysh Institute of Applied Mathematics, Moscow, Russia
lev@keldysh.ru,mogmi@narod.ru
[2] Kintech Lab Ltd., Moscow, Russia
zakirov@kintechlab.ru

Abstract. We present an implementation of the Lattice Boltzmann Method (LBM) with Locally Recursive non-Locally Asynchronous (LRnLA) algorithms on GPU and CPU. The algorithm is based on the recursive subdivision of the domain of the dD1T space-time simulation and loosens the memory-bound limit for numerical schemes with local dependencies. We show that LRnLA algorithm allows to overcome the main memory bandwidth limitations in both CPU and GPU implementations. For CPU, we find the data layout that provides alignment for the full use of AVX2/AVX512 vectorization. For GPU, we devise a procedure for pairwise CUDA-block synchronization applied to the implementation of the LRnLA algorithm, which previously worked only on CPU. The performance on GPU is higher, as it is usual in modern implementations. However, the performance gap in our implementation is smaller, thanks to a more efficient CPU version. Through a detailed comparison, we show possible future applications for both the CPU and the GPU implementations of the lattice Boltzmann method in the complex setting.

Keywords: LRnLA · LBM · Temporal blocking · Time skewing · GPU · Vectorization

1 Introduction

The Lattice Boltzmann Method (LBM) is a method of Computational Fluid Dynamics (CFD). Inside its application range, it benefits from the simplicity of the numerical procedure and the ease of the straightforward implementation. For this reason, it is also a good benchmark for performance studies of algorithms and computer hardware [8]. However, in an attempt to maximize performance, implementations may become increasingly complicated.

We previously implemented the LBM with Locally Recursive non-Locally Asynchronous (LRnLA) algorithms on CPU [9]. The LRnLA approach [5] offers an opportunity to overcome the memory throughput limitations; the achieved performance confirmed the efficiency of the approach.

© Springer Nature Switzerland AG 2019
L. Sokolinsky and M. Zymbler (Eds.): PCT 2019, CCIS 1063, pp. 139–151, 2019.
https://doi.org/10.1007/978-3-030-28163-2_10

Our next challenge is to surpass the performance of modern LBM GPU codes by using the LRnLA method.

2 The Numerical Method

Among its numerous variations, the specific lattice Boltzmann method [12] is defined by:

- A **set of discrete speeds** c_i, $i = 1, 2, \ldots, Q$, which are the links between the cells of the numerical grid. Each cell has a number of Discrete Distribution Functions (DDF) f_i equal to the number of discrete speeds $(i = 1, 2, \ldots, Q)$. In the *streaming* step, each f_i is copied to the cell in the c_i direction.
- A **collision operator** Ω, which locally transforms f_i.
- An **equilibrium function**, which is used in most types of collision operators.

The cell update is

$$f_i(\boldsymbol{x} + \boldsymbol{c}_i, t + 1) = f_i^*(\boldsymbol{x}, t), \quad i = 1, \ldots, 19; \tag{1}$$

$$f_i^*(x, t) = \Omega\big(f_1(x, t), f_2(x, t), \ldots, f_{19}(x, t)\big). \tag{2}$$

For the purpose of performance benchmarking, we considered the most common variation of the LBM, with 19 speeds, a collision term in the Bhatnagar–Gross–Krook (BGK) form, and a polynomial of degree 2 as the equilibrium function. The considered collision operator is one of the most computationally cheap, so the memory-bound property of the method becomes quite pronounced.

3 The Performance Analysis

The roofline model [14] describes performance limitations based on the operational intensity of the algorithm. Operational intensity indicates how many operations can be performed per byte of data throughput. If it is high, the performance is limited by the horizontal roof of the peak performance (compute-bound). Memory-bound problems are limited by the inclined slope. Thus, the central task for the implementation of memory-bound problems is the increase in operational intensity since at some point it affects the performance more than the reduction of computational overhead and hardware optimization.

Locally Recursive non-Locally Asynchronous (LRnLA) algorithms [4] represent one efficient approach to amplify the operational intensity. In its essence, the ideal limit of conventional LBM implementations (no more than one load/save operation per f_i per cell update) may be surpassed by far.

3.1 LRnLA Algorithms

LRnLA algorithms are built as a hierarchical recursive decomposition of the dependency graph (DG). The DG can be aligned with coordinates in $(d+1)$-dimensions, where d is the dimensionality of the simulation space. This dimensionality is denoted by dD1T to emphasize the cases where the time axis is included. In the implementation discussion, we take $d = 3$; smaller values of d are used only for illustration of the concepts. The recursive definition of the LRnLA algorithm states that it is a shape in dD1T space with a rule for its decomposition into smaller shapes. The shapes should have only unilateral dependencies after the decomposition. A shape represents the computation of all DG points inside it, in the order that is determined by the dependencies between its parts. The ability to parallelize portions of operations is determined by tracing these dependencies.

The LBM stencil is closely fitted by a cube, so the ConeFold (CF) class is chosen for LBM implementations (see Fig. 1) [9].

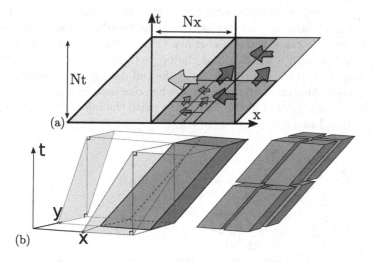

Fig. 1. (a) ConeFold projection in 1D1T. (b) ConeFold projection in 2D1T

The largest shape or a set of similar shapes contains the whole DG between two chosen synchronization time layers. The ConeFold prism may be decomposed into 2^{d+1} similar shapes (see Fig. 1a). Then, the size of its base is parameterized by the decomposition rank R and is equal to 2^{dR} cells.

Alternatively, it may be decomposed into prisms with the same height but a smaller base size, equal to DTS^d cells. The base side size DTS is a parameter of the algorithm. The high prisms are called ConeTorre (CT) and are decomposed into ConeFolds by several horizontal time slices. They are used for parallel implementations at the thread level parallelism (TLP).

By tracing the dependencies between CFs, large portions of asynchronous computations can be found. This is used for implementing the parallel code. As an example, if the CF base size is DTS, then the 2D1T CF at $(x, y, t) = (0, 0, 0)$ depends on CFs at $(DTS, 0, 0)$, $(0, DTS, 0)$, and $(-DTS, -DTS, -DTS)$. The CFs at $(0, 0, 0)$ and $(-DTS, DTS, 0)$ are independent and may be processed asynchronously. Generally, if the dD1T ConeFold is presented as a parallelepiped, the plane of asynchronous CFs is perpendicular to its longest 3D1T diagonal.

To contrast the usual implementations, where only collision and streaming steps are merged, the non-LRnLA algorithm will hereafter be called stepwise. It amounts to some kind of loop over the whole simulation domain at each time step.

ConeFold with $R = 0$ is one full cell update, which may cover the whole DG by translation. We refer to it as LRnLA cell. To decide which operations are in the LRnLA cell, we plot the DG on the smallest scale.

Firstly, we decompose one step of the LBM method into Q separate streaming operations and one collision operation Ω, which reads Q post-stream f_i^* and its output are Q post-collision f_i.

In Fig. 2, we give a visual illustration of the D1Q3 case. When we plot the DG, there is some ambiguity when assigning the coordinates in dD1T space to the operations. We can use this ambiguity to our benefit. The coordinates are often assigned in such a manner that the projection of the graph onto the dD space corresponds to the discrete space of the cell data grid. However, complex streaming algorithms are better visualized when one departs from this tradition.

The template is plotted in this manner to make the information propagation slant angle equal to $45°$.

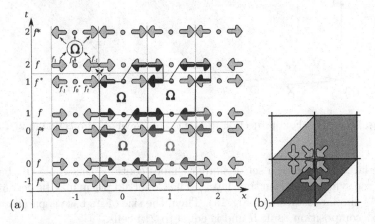

Fig. 2. (a) The 1D1T template for the DG construction (grayscale). Four LRnLA cells are shown as colored shapes. (b) LRnLA cell in 2D projection (Color figure online)

Figure 2a portrays the template that is the base for the DG plot. By thick arrows and circles, we represent the values that are described in the LBM. Note

Fig. 3. Various streaming algorithms: (a) push, (b) AA [2] (two steps), (c) EsoTwist [3]. In (a), the blue shape is the streaming of f_{-1}, the red one is the streaming of f_1, and the green one is the collision operation for all f_i. The streaming of f_0 is not represented anywhere in this text as this operation usually does not present a problem. (Color figure online)

that, at this step, there is no definite correspondence of the plotted arrows neither to the operations nor to the stored data. Same shapes represent same i, and there are two values within one time interval: post-stream f_i and post-collision f_i^*.

The operations with these values are plotted with thin arrows. Let us denote the unit operations by colored shapes as in Fig. 3a. For each basic operation, the input f_i is filled in the top half, the output f_i is filled in the bottom half.

Any unit algorithm is presented as a merging of these unit operations. If we merge the shapes that fill with color the bottom and the top of the same f_i arrow, then this value is not stored. The basic streaming algorithm "push" is shown in Fig. 3a. According to this illustration, in the "push" algorithm, the post-collision f_i^* are not stored. Analogously, in the "pull" algorithm the post-streaming f_i are not stored. The more interesting cases are the two distinct steps in the "AA" pattern (see Fig. 3b) [2] and EsoTwist (see Fig. 3c) [3]. The arrows that are not present in any read/write operation are not colored and thus omitted.

If the LRnLA cell is chosen as in Fig. 2a, it contains only one operation of every kind, and they all use the data of two neighboring cells. That is why it was naturally implemented in the first ConeFold LBM code [9].

Any other streaming algorithm may be used in ConeFold as well. The LRnLA cell for the AA algorithm would contain the update of 2^D cells in the first step and two shifted cells in the second step, owing to the fact that its two steps are different. The "pull" and "push" algorithms require three cell data in one LRnLA cell. However, this distinction might possibly be subdued by rearranging the data structure, and, moreover, at higher ranks, the difference in the used data is less visible.

The ConeFold-swap algorithm seems similar to the original swap streaming algorithm [6]. However, the appropriate order of computations is provided by higher levels of subdivision. Another distinction can be seen in a 2D illustration (see Fig. 2b; compare with illustrations in [6]).

In terms of 1D DG, EsoTwist looks more efficient since it contains less input data at any level of subdivision. The distinction is seen in higher dimensions: the pattern of access of some diagonal streaming, such as the $(1, -1)$ direction, is different.

3.2   Roofline Limitations

In [5], it is shown that the construction of the LRnLA algorithms simplifies the estimations of the location of the algorithm under the roofline. This can be summarized as follows.

By varying the prism parameters, the operational intensity can be adjusted to be arbitrarily large. It may be compared to the fact that if we take the whole DG of the problem, the operational intensity is the ratio of the total number of operations required for the simulation to the total data of the simulation. This ratio can also be arbitrarily large. On the other hand, under some assumptions, the algorithm execution time is the accumulated time of execution of its parts. Thus, the performance is limited by the minimal operational intensity which is encountered at some level of subdivision. The next level of subdivision of the stepwise algorithm is a loop over the domain in one time step. This represents the arithmetic intensity that is plotted as ideal for stepwise algorithms, and it is limited by the bandwidth of the main memory storage (RAM for CPUs, device memory for GPUs).

For LRnLA algorithms, there are several steps where a task is subdivided into subtasks. If at some decomposition level, the task data are localized in the higher level of the memory hierarchy, then its subtasks are limited by the bandwidth of this level:

$$\Pi \leq \min\left(\Pi_{\text{CPU}}, \Theta_{\text{RAM}}\frac{O(R_{\text{L3}}, d)}{S(R_{\text{L3}}, d)}, \Theta_{\text{L3}}\frac{O(R_{\text{L2}}, d)}{S(R_{\text{L2}}, d)}, \Theta_{\text{L2}}\frac{O(R_{\text{L1}}, d)}{S(R_{\text{L1}}, d)}, \Theta_{\text{L1}}\frac{o}{s}\right). \quad (3)$$

Here, O is the number of arithmetic operations in the task, S is the amount of data in the load and store operations in the task, Π_{CPU} is the peak performance of the computer, Θ_{RAM} is the memory throughput, o and s are the numbers of operations and data of one cell update. The formula is derived in [5] through the procedure of plotting arrows under the roofline from right to left. Here, we plot it in 3D, adding the data localization axis (see Fig. 4). The roofline shows the memory hierarchy of a node of the K60 cluster [1]. The marker shows the maximum performance achieved in our implementation on this system. The heat map at the bottom of the plot represents the compute-bound domain with yellow and memory bound with darker shades. The color of the arrow shows what roofline it is limited by, but it cannot be higher than any arrow to its right. Starting at the RAM level and the maximum-size CF that fits it, and progressing to lower ranks, there are two performance drops, in RAM and L3 cache. The final limit is equal to 1.6×10^9 cell updates per second, which is shown by the red arrow in Fig. 4.

The GPU peak performance is typically much larger than that of CPU, but it has a less developed memory hierarchy. The peak performance of one nVidia RTX2070 card is higher than that of the CPU-based K60 cluster node.

Since LRnLA algorithms conceal the host-device communication [15], we start the GPU roofline estimation from the host RAM (see Fig. 5). At $R = 8$, the data fit the device memory, and the decomposition rule switches to ConeTorre. Due to technical reasons (see Sect. 5), the DTS cannot presently be made greater

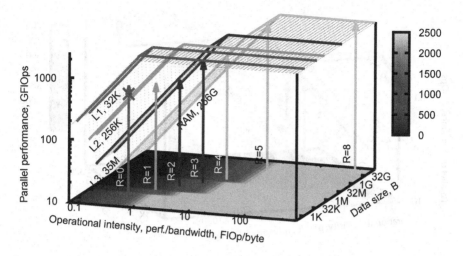

Fig. 4. The roofline for the K60. The arrows correspond to the ConeFold algorithm for D3Q19 LBM. (Color figure online)

than 8, which corresponds to $2^R = 8$ cells in its base. This leads to a large decline in estimated performance, the only one for this system. This means that higher DTS should be made possible for a substantial increase in performance.

The final estimation (red arrow) for nVidia RTX2070 is 3.0×10^9, which is about twice higher than the same result for the K60. Both overcome the main memory bandwidth limitation (RAM for CPU, device memory for GPU).

4   The Data Structure

The data layout provides the base for the algorithmic approach and for the software implementation. Thus, its choice is one of the most important steps to maximum efficiency.

On CPU, for the efficient use of AVX2/AVX512 vectorization, data should be aligned in a manner that most operations can address them in a uniform way.

On GPU, there is a similar requirement. Data are read in aligned blocks. If the required data are not aligned, then there will be more memory operations than are necessary. The thread-coalescing requirement can be expressed by the condition that the threads performing in parallel should access a compact piece of adjacent data.

For an efficient caching on CPU, the data layout should follow the patterns of access of the chosen algorithm. On GPU, the same condition enables to localize more data in the register file.

As a result, the data structure in high-performance codes is essentially hierarchical (see, for example, [11,13]). The levels of data structure hierarchy satisfy the conditions that come from different memory levels and levels of parallelism.

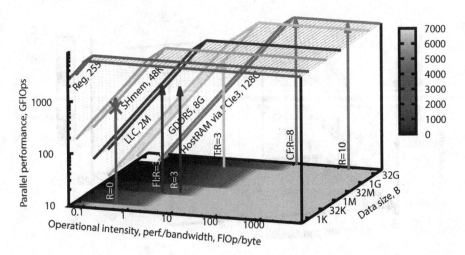

Fig. 5. The roofline for nVidia RTX2070. The arrows correspond to the TorreFold algorithm for D3Q19 LBM.

In the LRnLA approach, the data structure follows the projection of the chosen algorithm.

At higher levels, the data structure shows the links between the storage of parallel nodes. Inside one node, data may be organized in a multidimensional array, a Z-curve [9], a Peano curve [7], and so on.

4.1   Data Structure for CPU

We implemented the LBM with the ConeFold algorithm, as reported in [9].

The recursive call of the ConeFold algorithm suggests the use of the Z-curve-based array. This data structure is convenient on most CPU hardware.

The hierarchy of CPU cache has several layers with different capacities, and the Z-curve is cache-oblivious in this case. Since it fits each cache hierarchy level at some level of recursivity, there is no need to manually adjust the data structure tiles to the memory subsystem of any specific hardware.

The ConeFold subdivision terminates in eight cell updates, which are vectorized with AVX2 (float) or AVX512 (double). This is an example of non-local vectorization since the eight cells are located in different sub-domains. The element of data structure contains the vector for each f_i of the eight cells.

The implementation proved the efficiency of the data structure and the chosen algorithms. The results of some performance tests are shown in Fig. 6. The maximum achieved performance is more than 1.4×10^9 cell updates per second. It is shown as a marker under the roofline (see Fig. 4).

This result proves our theoretical estimation and approaches the efficiency of some high-performance GPU implementations. We see that the performance gap between CPU and GPU implementations of the LBM can be reduced.

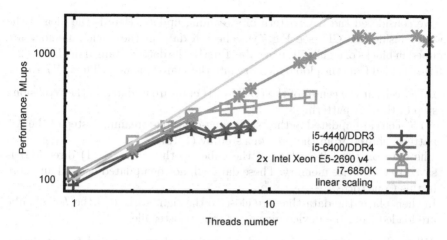

Fig. 6. Performance results for a desktop computer (i5-4440, i5-6400), a high-end desktop workstation (i7-6850K), and a node of the K60 cluster (2x Intel Xeon E5-2690 v4)

4.2 Data Structure for GPU

With GPU, it is most important to take into account the thread-coalescing requirement and data alignment. The cells should be merged into groups of sufficiently large size. Thus, cubes of size $2 \times 2 \times 2$ cells seem a good compromise between thread coalescence and data localization. Data localization is beneficial for sparse fluid domains. In these cases, the f_i data is not required in the non-fluid cells, but the memory is still allocated for the whole memory block. With smaller cell blocks, space usage is more efficient.

The f_i should be sorted inside the cell block so that they are accessed in blocks of no less than 128 B for each collision or streaming operation. This idea is implemented for stepwise algorithms in [11,13].

5   The Implementation

The CPU ConeFold implementation was discussed in [9].

The ConeTorre variation is preferred for the GPU implementation since its memory subsystem is not as developed as for CPU, and the main localization site is the register file.

The DTS parameter is chosen so that the data contained in $(DTS + 2)^3$ cells fit the shared memory. One CUDA-block is assigned to one ConeTorre and performs time layers in it one by one. The recursive subdivision down to $R = 0$ is not performed. The cells are assigned to CUDA-threads and processed in parallel in each separate time layer.

We considered the AA pattern of streaming update. Here is the idea of the updates inside one CF (see Fig. 7a). The cell data in the device memory are grouped in blocks of eight cells in a cube. Firstly, the data contained in $(DTS+2)^3$ cells are loaded from the global memory into the shared memory. Then, NT times:

- DTS^3 cells at the center of the $(DTS+2)^3$ cube are updated in the "collision" step of the AA pattern.
- DTS^3 cells are updated in the "streaming-collision-streaming" step of the AA pattern, shifted from the previous step by $(1,1,1)$.
- The data on the left surface of the cube (in the $(-1,-1,-1)$ direction) is saved to the device memory. These data will not be updated or used anymore in this ConeTorre.
- In their place, the data that are close to the right surface of the DTS^3 cube are loaded from the device memory to the register file.

When the ConeTorre is finished, all register data are saved to the device memory.

Other LRnLA algorithms performed well on GPU [5]. However, the ConeFold implementation on GPU presented some difficulties.

One of them is the fact that the dependencies between parallel CFs on CPU are ensured with a system of semaphores, which are supported in POSIX thread models. Semaphores are not supported in CUDA, and the pairwise synchronization of CUDA blocks presents a problem. We developed a manual solution, introducing an integer variable for each semaphore:

```
sem_val = AtomicSub(semaphore, 1);
volatile int* sem_ptr = &semaphore;
if (sem_val<=0) {
  while (*sem_ptr<0) {}
}
```

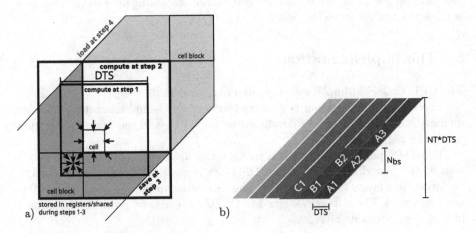

Fig. 7. (a) ConeTorre slice for LBM with AA streaming. (b) 1D projection of the parallel synchronization of ConeTorre on GPU

Here, `semaphore` is an integer stored in the global device memory. It is unlocked in another CUDA-block by an `atomicADD` operation.

The TorreFold synchronization algorithm in this case is parametrized by `CFsteps`, i.e. the number of steps between CUDA-block synchronizations. Figure 7b shows a 1D1T illustration of the concept. The CUDA-block that is processing ConeTorre A unlocks the semaphore for B1 after `CFsteps` LBM steps. Then, B1 and A2 are processed asynchronously.

According to our test, each iteration of the `while` loop requires 390 clocks on a Maxwell GPU, 400 clocks on a Pascal GPU, and 375 clocks on a Volta GPU. These figures are relatively high and may be interpreted as L2 cache latency.

The portion of the overhead for the block synchronization is shown in Fig. 8.

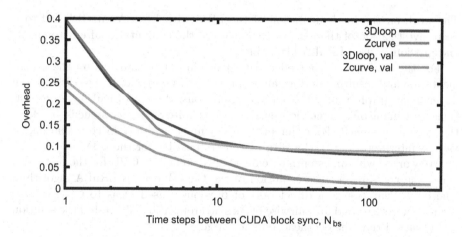

Fig. 8. The portion of the overhead for the block synchronization (dependence on the number of time steps between CUDA-block synchronizations). The 'val' mark represents the results where the `sem_val` variable stores the number of unlocked parts.

There are two more important distinctions of GPU that make the semaphore communication more complicated. Firstly, there are generally more SMs in a GPU than cores in a CPU. The number of asynchronous CFs at the start and at the end of the execution is quite low, so the parallel ability is sufficiently utilized.

Secondly, the order in which the ConeTorre are called exerts a greater influence when it comes to GPU. In the CPU case, if the ConeTorre that is locked by the surrounding semaphores is started, then the core that it is assigned to may store the context in the lower levels of the cache and continue to execute another thread. On GPU, however, since the register file is large, the context store operation would be costly and is not performed. It may even result in freezing of the simulation if the number of ConeTorre is greater than the number of SMs (which is generally the case) and some wrong CUDA-block has already begun the wait. The order of calling the corresponding CUDA blocks should be made consistent with the order of semaphore change. We tested the dependence on this looping order (see Fig. 8), and it was demonstrated that the Z-curve shows a better result than a usual 3D loop.

The dependencies between CT make possible another kind of optimization. The part of CT above the current one may be unlocked before the current one is started. The number of unlocked parts of the ConeTorre may be stored in the sem_val variable. The effect of this optimization is also plotted in Fig. 8.

The increase in DTS, which was shown to be desirable in Sect. 3, would require more advanced tools for intra-block communication. However, we do not see an adequate solution in modern GPU computing yet.

6   Conclusions

The work on the implementation of the LBM with LRnLA on GPU led to a deep investigation of the logic of this method, the computational environment, and previously used LRnLA algorithms.

For the analysis of what effect the streaming algorithm has on the implementation and resulting performance of the LRnLA code, we had to dissolve the dependency graph into unit streaming operations. Previous LRnLA decompositions were made on symmetric stencils, such as finite difference stencils [5]. Here, the streaming stencils for different f_i are symmetrical only in combination, so various interpretations of the LRnLA cell are possible for the LBM.

We worked with an essentially cube-shaped stencil on GPU for the first time, hence the necessity for the evolution of the ConeFold class LRnLA algorithm. The implementation is a direct port of previously used ideas to CUDA, so it required a workaround for pairwise block synchronization. We tested the solution and observed that the overhead remains large.

The disadvantage is brought by the underdeveloped intra-block synchronization in CUDA. If future GPUs do not support low-overhead selective block synchronization, it might be sensible to switch to another class of LRnLA algorithms that already have proved to be efficient on GPU [5].

Additionally, we plotted the roofline limitations of the proposed approach. As we can see, the estimated achievable performance ratio $\Pi_{GPU}/\Pi_{CPU} \approx 2$. If we take for comparison another high-performance implementation [10], in which $\Pi_{GPU}/\Pi_{CPU} \approx 10$, we can see that the ratio is not as high.

Moreover, for GPU computations with the LRnLA approach, the host memory can be used without a significant performance decline. Thus, this approach paves the way for high-performance large-scale simulations in CFD.

Acknowledgments. The work was supported by the Russian Science Foundation (grant No. 18-71-10004).

References

1. Computational resources of Keldysh Institute of Applied Mathematics RAS. www. kiam.ru
2. Bailey, P., Myre, J., Walsh, S.D., Lilja, D.J., Saar, M.O.: Accelerating lattice boltzmann fluid flow simulations using graphics processors. In: International Conference on Parallel Processing, ICPP 2009, pp. 550–557. IEEE (2009). https://doi.org/10.1109/ICPP.2009.38
3. Geier, M., Schönherr, M.: Esoteric twist: an efficient in-place streaming algorithmus for the lattice boltzmann method on massively parallel hardware. Computation 5(2), 19 (2017). https://doi.org/10.3390/computation5020019
4. Levchenko, V., Perepelkina, A., Zakirov, A.: Diamondtorre algorithm for high-performance wave modeling. Computation 4(3), 29 (2016). https://doi.org/10.3390/computation4030029
5. Levchenko, V.D., Perepelkina, A.Y.: Locally recursive non-locally asynchronous algorithms for stencil computation. Lobachevskii J. Math. 39(4), 552–561 (2018). https://doi.org/10.1134/S1995080218040108
6. Mattila, K., Hyväluoma, J., Rossi, T., Aspnäs, M., Westerholm, J.: An efficient swap algorithm for the lattice boltzmann method. Comput. Phys. Commun. 176(3), 200–210 (2007). https://doi.org/10.1016/j.cpc.2006.09.005
7. Neumann, P., Bungartz, H.J., Mehl, M., Neckel, T., Weinzierl, T.: A coupled approach for fluid dynamic problems using the PDE framework peano. Commun. Comput. Phys. 12(1), 65–84 (2012). https://doi.org/10.4208/cicp.210910.200611a
8. Nguyen, A., Satish, N., Chhugani, J., Kim, C., Dubey, P.: 3.5-D blocking optimization for stencil computations on modern CPUs and GPUs. In: High Performance Computing, Networking, Storage and Analysis (SC), pp. 1–13. IEEE (2010). https://doi.org/10.1109/SC.2010.2
9. Perepelkina, A., Levchenko, V.: LRnLA algorithm ConeFold with non-local vectorization for LBM implementation. Commun. Comput. Inf. Sci. 965, 101–113 (2019). https://doi.org/10.1007/978-3-030-05807-4_9
10. Riesinger, C., Bakhtiari, A., Schreiber, M., Neumann, P., Bungartz, H.J.: A holistic scalable implementation approach of the lattice Boltzmann method for CPU/GPU heterogeneous clusters. Computation 5(4), 48 (2017). https://doi.org/10.3390/computation5040048
11. Robertsén, F., Westerholm, J., Mattila, K.: Designing a graphics processing unit accelerated petaflop capable lattice boltzmann solver: read aligned data layouts and asynchronous communication. Int. J. High Perform. Comput. Appl. 31(3), 246–255 (2017). https://doi.org/10.1177/1094342016658109
12. Succi, S.: The Lattice Boltzmann Equation: for Fluid Dynamics and Beyond. Oxford University Press, Oxford (2001)
13. Tomczak, T., Szafran, R.G.: A new GPU implementation for lattice-Boltzmann simulations on sparse geometries. Comput. Phys. Commun. 235, 258–278 (2019)
14. Williams, S., Waterman, A., Patterson, D.: Roofline: an insightful visual performance model for multicore architectures. Commun. ACM 52(4), 65–76 (2009). https://doi.org/10.1145/1498765.1498785
15. Zakirov, A., Levchenko, V., Perepelkina, A., Zempo, Y.: High performance FDTD algorithm for GPGPU supercomputers. J. Phys: Conf. Ser. 759, 012100 (2016). https://doi.org/10.1088/1742-6596/759/1/012100. IOP Publishing

GPU-Accelerated Learning
of Neuro-Fuzzy System Based on Fuzzy
Truth Value

Sergey Vladimirovich Kulabukhov[(✉)] [ID] and Vasily Grigorievich Sinuk

Belgorod State Technological University Named After V.G. Shukhov,
Belgorod, Russian Federation
qlba@ya.ru, vgsinuk@mail.ru

Abstract. The article is devoted to the problem of the computational
complexity of fuzzy inference and neuro-fuzzy system learning in the case
of fuzzy inputs. We resort to parallel computations to reduce computation time. In the article, we suggest an algorithm using GPU to efficiently
perform fuzzy inference based on fuzzy truth values and the extension of
this algorithm to neuro-fuzzy system learning by evolution strategy. We
demonstrate the importance of the algorithm and include a benchmark
to compare the computation time on CPU against GPU.

Keywords: Fuzzy inference systems · Neuro-fuzzy systems ·
Fuzzy truth value · Evolution strategies · GPGPU ·
Parallel computations

1 Introduction

Fuzzy inference systems are gaining popularity. They are used in many fields of
practical interest. As a matter of fact, they are more consistent with the nature
of human thinking than systems of traditional formal logic, since they allow for
building models that reflect various aspects of uncertainty in a more adequate
manner [6]. Such models are defined via fuzzy rule bases. Fuzzy inference systems
have applications in such fields as control of technical systems, speech and image
recognition, and diverse expert systems.

Fuzzy inference systems do not address the formation of a rule base. Being a
formal representation of knowledge, fuzzy rule bases can be constructed manually. Nevertheless, some applications also provide training sets, i.e. data consisting of pairs "input values–desired output values" that can be used to adjust the
parameters of the inference system. Such situations are the subject of machine
learning and are common in applications of artificial neural networks.

The interpretation of neural network parameters is generally difficult. This
fact prevents the explicit use of knowledge of domain experts in a network and
the extraction of knowledge from a trained network. The combination of basic

This work was supported by the RFBR (grant No. 19-07-00133).

L. Sokolinsky and M. Zymbler (Eds.): PCT 2019, CCIS 1063, pp. 152–167, 2019.
https://doi.org/10.1007/978-3-030-28163-2_11

methods of fuzzy inference systems and artificial neural networks led to the creation of neuro-fuzzy systems. Various options for combining these methods have been presented in the literature. For example, parameters of membership functions, triangular norms, or even the whole rule base can undergo learning.

The parameters of fuzzy inference systems have a clear meaning, and their values after learning may reveal previously unknown knowledge about a subject area. In addition, if a fuzzy inference system can handle fuzzy input values, then they can also be used for neuro-fuzzy system learning, provided that the training set contains uncertainty. In many applications, input data contain either nonnumerical (linguistic) assessments [2,7] or input signals that are received with noise [8,9].

In this paper, we consider logical-type neuro-fuzzy systems based on fuzzy truth values [5,10]. Such systems allow for inference in case of multiple fuzzy inputs of polynomial computational complexity by using any triangular norm [11]. On the other hand, this method of fuzzy inference leads to discretization of membership functions since they undergo complex transformations which are difficult to implement in analytical form.

An evolutionary algorithm was used in [8,9] for neuro-fuzzy system learning. This algorithm assumes that the computation of objective function values is simultaneous for all elements of the offspring population created at each generation. Neuro-fuzzy system learning can be very time-consuming even in simple tasks. The parallel implementation of both inference and learning processes in such systems by using GPU is the subject of the present article. The implementation is based on NVIDIA CUDA technology.

2   Neuro-Fuzzy System Based on Fuzzy Truth Values

The problem that is to be solved by using a fuzzy inference system is formulated as follows. Consider a system with n inputs $\boldsymbol{x} = [x_1, \ldots, x_n]$ and a single output y. The relationship between inputs and the output is defined using N fuzzy rules expressed as

$$R_k\colon \text{If } x_1 \text{ is } A_{1k} \text{ and } \ldots \text{ and } x_n \text{ is } A_{nk}, \text{ then } y \text{ is } B_k, \quad k = \overline{1, N}, \quad (1)$$

where $\boldsymbol{x} \in \boldsymbol{X} = X_1 \times X_2 \times \cdots \times X_n$, $y \in Y$, and $\boldsymbol{A_k} = A_{1k} \times A_{2k} \times \cdots \times A_{nk} \subseteq \boldsymbol{X}$, $B_k \subseteq Y$ are fuzzy sets.

According to the classification proposed in [12], the specific feature of logical-type systems is that the rules expressed in (1) are formalized via fuzzy implication as fuzzy $(n+1)$-ary relations $R_k \subseteq X_1 \times \cdots \times X_n \times Y$, namely

$$R_k = A_{1k} \times \cdots \times A_{nk} \times Y \to X_1 \times \cdots \times X_n \times B_k, \quad k = \overline{1, N},$$

where "$\to$" denotes a fuzzy implication expressing a causal relationship between the antecedent "x_1 is A_{1k} and $\ldots$ and x_n is A_{nk}" and the consequent "y is B_k". The task is to determine the inference result $B'_k \subseteq Y$ for a system given in the form expressed in (1), provided that the inputs are given as $\boldsymbol{A'} = A'_1 \times \cdots \times A'_n \subseteq \boldsymbol{X}$ or "x_1 is A'_1 and $\ldots$ and x_n is A'_n".

The specific feature of the considered approach to fuzzy inference is that the inference is made within a single truth space for all premises, which is achieved by transforming the relationships between premise and fact into a so-called fuzzy truth value. By using the truth modification rule (see [1]), we can write

$$\mu_{A'}(x) = \tau_{A|A'}(\mu_A(x)),$$

where $\tau_{A|A'}(\cdot)$ is the fuzzy truth value of the fuzzy set A relative to A', which represents the compatibility $CP(A, A')$ of the term A with respect to A' [3,13]:

$$\tau_{A|A'}(t) = \mu_{CP(A,A')}(t) = \sup_{\substack{\mu_A(x)=t \\ x \in X}} \{\mu_{A'}(x)\}, \quad t \in [0,1]. \tag{2}$$

Denote $t = \mu_A(x)$. Then

$$\mu_{A'}(x) = \tau_{A|A'}(\mu_A(x)) = \tau_{A|A'}(t). \tag{3}$$

Thus, the generalized fuzzy modus ponens rule for single-input systems can be written as

$$\mu_{B'_k}(y) = \sup_{t \in [0,1]} \{\tau_{A|A'}(t) \, \mathrm{T} \, I(t, \mu_{B_k}(y))\}, \quad k = \overline{1, N},$$

where T is a t-norm and I is a fuzzy implication.

In systems with n inputs ($n > 1$), the convolution of fuzzy truth values $\tau_{A_i|A'}$ is performed for all inputs $i = \overline{1, n}$. For rules of the form (1), the fuzzy truth value of the antecedent $\boldsymbol{A_k}$ with respect to the inputs $\boldsymbol{A'}$ is defined as

$$\tau_{\boldsymbol{A_k}|\boldsymbol{A'}}(t) = \mathop{\mathrm{T}}_{i=\overline{1,n}} \tau_{A_{ki}|A'_i}(t), \quad t \in [0,1], \tag{4}$$

where T is an n-ary t-norm extended by the extension principle (see [9]). With this in mind, the inference of the output value B'_k based on the fuzzy truth value, for systems with n inputs, can be written in the form

$$\mu_{B'_k}(y) = \sup_{t \in [0,1]} \{\tau_{\boldsymbol{A_k}|\boldsymbol{A'}}(t) \, \mathrm{T} \, I(t, \mu_{B_k}(y))\}, \quad k = \overline{1, N}. \tag{5}$$

The fuzzy set B' (the output of the system as a whole) is obtained by accumulation, and in a logical approach, it is defined as an intersection operation [9]:

$$B' = \bigcap_{j=\overline{1,N}} B'_j. \tag{6}$$

Accordingly, the membership function B' is defined by means of the t-norm:

$$\mu_{B'}(y) = \mathop{\mathrm{T}}_{j=\overline{1,N}} \mu_{B'_j}(y). \tag{7}$$

The center-of-gravity defuzzification method is used in the neuro-fuzzy system to define the crisp output $\overline{y}$ of the system:

$$\overline{y} = \frac{\int_Y y \cdot \mu_{B'}(y)\, dy}{\int_Y \mu_{B'}(y)\, dy}. \tag{8}$$

In addition, a transformation is introduced to regulate the effect of each rule j on the accumulation result in accordance with its weight w_j. This transformation is pointwise. Now (7) can be written in the form

$$\mu_{B'}(y) = \mathop{\mathrm{T}}_{j=\overline{1,N}} f\left(\mu_{B'_j}(y), w_j\right), \tag{9}$$

where $f(a, w)$ is an arbitrary function that associates a membership-function value a to a new value according to the rule weight w. The function $f(a, w)$ must be nondecreasing on the first argument and, for logical-type systems, nonincreasing on the second argument.

Within this study, neuro-fuzzy system learning is performed by means of an evolution strategy (μ, λ) [9]. As it was already noted, this algorithm assumes that the computation of objective function values $R(\mathbf{p})$ is simultaneous for all forms of the offspring population O at each generation. An objective function is a function of neuro-fuzzy system parameters that reflects how far the results obtained deviate from the training set. The lower $R(\mathbf{p}_l)$ is, the more $\mathbf{p}_l$ corresponds to the training set. Let us denote the training set as

$$T = \{T_r = \langle A_1'^{(r)}, \ldots, A_n'^{(r)}, \overline{y}^{(r)} \rangle\}_{r=\overline{1,M}},$$

where $A_i'^{(r)}$ is the value of the i-th input for the r-th element of the training set, $\overline{y}^{(r)}$ is the desired output value for these input values, and M is the training set size.

The objective function for the neuro-fuzzy system is defined as

$$R(\mathbf{p}) = \frac{1}{|Y|} \sqrt{\frac{1}{M} \sum_{r=1}^{M} \left(F_{\mathbf{p}}\left(A_1'^{(r)}, \ldots, A_n'^{(r)}\right) - \overline{y}^{(r)} \right)^2}, \tag{10}$$

where $|Y|$ stands for the width of the domain of definition of the output variable, while

$$F_{\mathbf{p}}\left(A_1'^{(r)}, \ldots, A_n'^{(r)}\right)$$

is the inference result $\overline{y}$ of the neuro-fuzzy system with parameter values $\mathbf{p}$ and input values $A_1'^{(r)}, \ldots, A_n'^{(r)}$.

3   Parallel Implementation of the Learning Process

In this section, we consider the implementation of the learning process of the neuro-fuzzy system, in which all computations associated with fuzzy inference are performed on graphics processing units (GPUs).

3.1   Learning Process Overview

From a computational point of view, the evaluation of the objective function $R(\mathbf{p})$ for each element of the offspring population O is the most complex phase of neuro-fuzzy system learning. To compute $R(\mathbf{p})$ for a single $\mathbf{p}$ (a single vector of parameter values), we must compute the value of the expression under the summation symbol in (10) for each element T_r of the training set T, each time computing the inference result for the given input values. This means that we have to perform a fuzzy inference for each combination $\langle \mathbf{p}, T_r \rangle \in O \times T$, i.e. $|O| \cdot M$ times, and each of these inferences is independent of the others. Thus, all pairs $\langle \mathbf{p}, T_r \rangle \in O \times T$ can be processed in parallel. In the implementation of the algorithm using CUDA technology, every pair is processed in an individual block.

Each block is allocated its own memory, a part of which is allocated for storing intermediate results; output results are placed in another part of the memory and a third part contains a sequence of commands defining fuzzy inference operations. Before launching the kernel, command sequences that are executed by the fuzzy system to compute output values are built for each block and written to the corresponding memory location. The values of inputs and parameters are embedded explicitly in these sequences. Since commands are specified separately for each block, the developed algorithm does not impose restrictions on the set of adjusted parameters of the neuro-fuzzy system. In general, it allows parallel inference to be done for several completely different fuzzy systems. This fact expands the possibilities of neuro-fuzzy system learning.

As previously mentioned, this fuzzy inference method leads to discretization of membership functions into arrays of samples. These samples are distributed over the threads and then processed. The implementation of individual fuzzy inference operations will be described below in this section.

3.2   Data Distribution in GPU Memory

Each block is allocated the same amount of memory, which consists of three parts (Fig. 1). Let N_b be the number of blocks. The first part of the allocated memory is used for storing the results of the execution of intermediate commands; these results are arrays of samples of membership functions. The part consists of N_c arrays of N_d elements each, where N_d is the number of samples and N_c is the maximum number of commands that result in a membership function. The array elements are single-precision real numbers. The second part of the allocated memory stores a real number which is the final result of the computation, i.e. the crisp output $\bar{y}$ of the system. The third part stores the sequence of fuzzy inference commands. The size of this part is S_c, which is an estimated maximum size for command sequences (in bytes) as the size of sequences for different blocks in the general case can vary, for example, due to rule base changes during training. If the command sequence for a block is shorter than S_c, then this part is aligned with unused memory to size S_c.

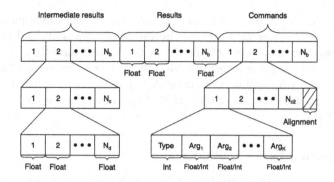

Fig. 1. Data distribution in GPU memory

Intermediate results for every block are stored first in GPU memory. Final inference results for each block are placed next, followed by command sequences. Such a distribution of parts in memory has an advantage over their grouping in blocks, namely when transferring the results from the device memory into the host memory, a continuous memory segment containing $N_b * 4$ bytes is copied in the first case, whereas, in the second, N_b four-byte segments are copied. The same happens when transferring command sequences into the device memory. Storage of all intermediate results in memory is not required; however, it provides detailed information when designing and debugging neuro-fuzzy systems.

Each command is represented as a sequence of numbers. The first number denotes the operation type; subsequent numbers are its arguments and parameters. The operation type determines the amount of numbers. If arguments include membership functions, then the command size is also determined by their type. Each argument or parameter is either a 32-bit integer or a single-precision real number.

3.3   The Main Kernel Function

The main kernel function accepts four parameters: the start address of the allocated memory a_0; the amounts of memory $N_c * N_d * 4$ and S_c, which are needed for storing, respectively, the intermediate results and the commands; and finally, the number of samples N_d. Kernel launch parameters are also specified: the grid size is set to $|O| \cdot M$ (single dimension); the block size is set to the maximum possible number not exceeding N_d; the amount of shared memory is $N_d * 4$ bytes. When the main kernel function is started, each thread calculates the addresses of each memory part for the block $b = $ blockIdx.x to which the thread belongs. The first part has the address $a_1 = a_0 + b * N_c * N_d * 4$, the second $a_2 = a_0 + N_b * N_c * N_d * 4 + b * 4$, and the third $a_3 = a_0 + N_b * N_c * N_d * 4 + N_b * 4 + b * S_c$. This function also declares the variable dp which contains both the number of saved results of intermediate commands (initially equal to zero) and a pointer to the beginning of the next command ip (initially set to a_3). Then a loop begins, in whose body the current command is read and executed, and the pointer ip is increased by the size of the command. Implementations of individual operations are allocated to separate subroutines; the values of arguments and parameters that are passed to them are determined in the main subroutine. If the result of the current operation is a membership function, then its array of samples is placed at $a_3 + \text{dp} * N_d$ and dp is incremented. The loop ends when ip points to a dummy command that marks the end of the command sequence. We describe each operation below.

3.4 Computing Fuzzy Truth Values

This operation is analytically defined by (2). Its arguments are membership functions of the term $\mu_A(x)$ and the fact $\mu_{A'}(x)$; the result is the fuzzy truth value $\tau_{A|A'}(t)$ of the term with respect to the fact. In the discrete case, we calculate sample values of the function $\tau_{A|A'}(t)$. Since this function is defined in the numerical range $[0; 1]$, we split it into N_d samples and define the value of each sample as follows:

$$\tau_{A|A'}(t_i) = \sup_{\substack{\mu_A(x) \in [t_i; t_{i+1}] \\ x \in X}} \{\mu_{A'}(x)\}, \quad t_i = \frac{i}{N_d}, \quad i = \overline{0, N_d - 1}, \tag{11}$$

whence, taking into account (3), it follows that

$$\tau_{A|A'}(t_i) = \sup_{t \in [t_i; t_{i+1}]} \{\tau_{A|A'}(t)\}. \tag{12}$$

The procedure for computing $\tau_{A|A'}(t_i)$ for each t_i is divided into three subroutines. Flowcharts of two of them at the level of individual threads are shown in Fig. 2. The main subroutine of the operation is FTV(dst, $\mu_A(x)$, $\mu_{A'}(x)$, N_d), which fills an array of N_d elements at dst with values according to (11). Membership functions $\mu_A(x)$ and $\mu_{A'}(x)$ are expressed as numerical sequences in the same way as the commands: the first number determines the type of the membership function, the other numbers are its parameters. The destination address dst points to an array in the memory space allocated for the results of the execution of intermediate commands for this block.

The computation of individual samples of $\tau_{A|A'}(t_i)$ is distributed among all threads of the block. FTV starts with getting the thread index threadIdx.x and the number of threads blockDim.x in the block. They will be henceforth referred to as variables x and h, respectively. Then a loop is executed for $i = \overline{0, N_d - 1}$, each thread performs every h-th iteration starting with the x-th. Since the value of the i-th sample equals the upper boundary of the fuzzy truth value in the range $[t_i; t_{i+1}]$ (see (12)), the ends of this range, denoted by $t_{\min}$ and $t_{\max}$, are calculated in the loop body. Then the function FTV1($\mu_A(x)$, $\mu_{A'}(x)$, $t_{\min}, t_{\max}$) is invoked, and it returns the value of (11) for the given range $[t_{\min}; t_{\max}]$, which is then assigned to the i-th element of the array at dst.

Depending on both the type of the membership function $\mu_A(x)$ and the values of its parameters, FTV1 determines all ranges such that $\mu_A(x) \in [t_{\min}; t_{\max}]$.

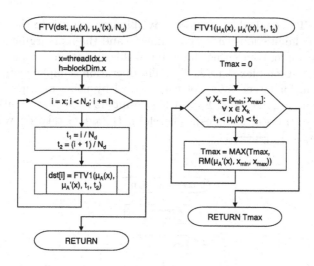

Fig. 2. Flowcharts of fuzzy truth value computation

FTV1 features algorithms to determine these ranges for all supported types of membership functions. For example, the ranges for a Gaussian membership function with center in m and standard deviation σ are

$$\left[m - \sigma\sqrt{-\ln t_{\min}}; m - \sigma\sqrt{-\ln t_{\max}} \right], \left[m + \sigma\sqrt{-\ln t_{\max}}; m + \sigma\sqrt{-\ln t_{\min}} \right].$$

For each of these ranges, FTV1 invokes the function $\mathrm{RM}(\mu_{A'}(x), x_{\min}, x_{\max})$, which returns the maximum membership degree of the fact within this range. FTV1 returns the maximum of the values returned by RM.

The flowchart of RM is not shown in the figure as it contains only the formulas that express the maximum value within a given range $[x_{\min}; x_{\max}]$ for all supported types of membership functions. In the aforementioned case of a Gaussian function, this value is given as

$$\sup_{x \in [x_{\min}; x_{\max}]} \{\mu_{A'}(x)\} = \begin{cases} 1, & \text{if } x_{\min} \leq m \leq x_{\max}, \\ \exp\left((x_{\max} - m)^2 / \sigma^2\right), & \text{if } x_{\max} < m, \\ \exp\left((x_{\min} - m)^2 / \sigma^2\right), & \text{if } x_{\min} > m. \end{cases}$$

3.5 Convolution of Fuzzy Truth Values

If a rule contains more than one subcondition, then the fuzzy truth value of the entire antecedent A_k with respect to the inputs A' is computed according to (4). This formula contains an n-ary t-norm extended by the extension principle. For $n = 2$, it is defined as

$$\mathbf{T}_{i=\overline{1,2}} \tau_{A_{ki}|A_i'}(t) = \sup_{\substack{t_1 \, \mathbf{T} \, t_2 = t \\ t_1, t_2 \in [0;1]}} \{\tau_{A_{k1}|A_1'}(t_1) \, \mathbf{T} \, \tau_{A_{k2}|A_2'}(t_2)\}, \quad t \in [0, 1]. \tag{13}$$

If $n > 2$, then $\mathbf{T}$ is applied as an associative binary operator to the result of the convolution of the previous $(n-1)$ arguments and the n-th argument.

The algorithm for computing the result of this operation is depicted in Fig. 3. Its computational complexity is $O\left(N_d^2\right)$. The operation is implemented for $n = 2$; the convolution for larger values of n is defined by using multiple convolution commands. The arguments A, B, and the result dst are fuzzy truth values in discrete form (arrays of samples). Therefore, t_1 and t_2 take on values in the discrete set $\{i/N_d\}_{i=\overline{0,N_d-1}}$.

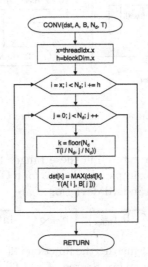

Fig. 3. Flowchart for the convolution of fuzzy truth values

The enumeration of all values of t_1 and t_2 is implemented by a double loop for the variables i and j, respectively. The iterations of the external loop for the variable i are distributed among all threads of the block in the same manner as in the case of FTV (Fig. 2). The internal loop for j is entirely executed by a single thread. The values of i and j correspond to $t_1 = i/N_d$ and $t_2 = j/N_d$, respectively. Then $t = t_1\ \mathbf{T}\ t_2$ is computed. The index of the result sample corresponding to t is $k = \lfloor t * N_d \rfloor$. Let us denote by $\hat{\tau}_t$ the argument of the supremum in (13), i.e. $\hat{\tau}_t = \tau_{A_{k1}|A_1'}(t_1)\ \mathbf{T}\ \tau_{A_{k2}|A_2'}(t_2)$, which is calculated as T(A[i], B[j]), where T is the implementation of the t-norm. If $\hat{\tau}_t$ is greater than the current value of dst[k], then dst[k] is assigned the value of $\hat{\tau}_t$.

The index k is calculated using the t-norm. In the implemented distribution of iterations for i and j, threads may process elements of the destination array with the same index. If multiple threads read the value of dst[k] and then conditionally update it, then data races may occur. However, distributing the iterations in such a way that no pair of threads obtains similar values of k depends on the t-norm and is computationally inefficient since threads will get different numbers of payloads. Data races can be eliminated by using the atomic maximum operation provided by the CUDA framework, which can be invoked by calling the function **atomicMax**. This function accepts **address** and the value **val** as arguments. If the value at **address** is less than **val**, then **val** is written at **address**. Since the function accepts only integer numbers, $\lfloor \hat{\tau}_t * \text{INT\_MAX} \rfloor$ is passed instead of $\hat{\tau}_t$. Given that $\hat{\tau}_t \in [0;1]$, no overflow can occur.

During the execution of the algorithm, the integer representation of the resulting array is located in the shared memory. This memory is used since it has shorter access time than global memory. When the execution is completed, the result is transferred to global memory at **dst** with each element converted into a floating-point number and divided by **INT_MAX**. This process is also iterative and involves all threads of the block in the same way as in previous cases.

3.6   Computation of the Result of Rule Inference

This operation is performed according to (5) after obtaining the truth value of the antecedent of the rule with respect to the inputs. The algorithm is shown in Fig. 4.

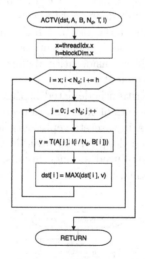

Fig. 4. Flowchart of the computation of rule inference result

The fuzzy truth value of the antecedent A is represented in discrete form, whereas the membership function of the consequent term B is given in analytical form (type and values of parameters). The discrete representation of the result is placed at **dst**. According to (5), it is necessary to have the values of B in the same points in which the result $\mu_{B'_k}$ must be computed. The discrete representation of B is prepared by the main kernel function and stored in shared memory; this process is parallelized by distributing the samples among threads. Similarly to the previous operation, a double loop is executed for two variables, i and j, to enumerate the samples of y and t, respectively. The computational complexity of this algorithm is $O\left(N_d^2\right)$.

The iterations of the loop for i are distributed among all threads; the loop for j is entirely executed by a single thread. The values of i and j correspond to $y = \mathtt{i}/N_d$ and $t = \mathtt{j}/N_d$, respectively. The values of $T\left(\mathtt{A[j]}, I\left(\mathtt{j}/N_d, \mathtt{B[i]}\right)\right)$ are computed in the body of the inner loop; the maximum of these values is recorded at dst[i] after leaving the loop. Unlike the convolution of fuzzy truth values, this algorithm does not cause data races since every sample dst[i] of the result array is processed by a single thread.

3.7   Applying Rule Weights

The transformation of the membership function of the rule inference result is pointwise, i.e. it transforms every sample value separately (see (9)). The function $f\left(a, w\right)$ is implemented in the program as

$$f\left(a, w\right) = a * w + \left(1 - w\right).$$

This function was designed by analogy with [4], where $f\left(a, w\right)$ was defined as $f\left(a, w\right) = a * w$, which is equivalent to $f\left(a, w\right) = a * w + 0 * \left(1 - w\right)$. Thus, w determines the proportion of the original function in a linear combination with $\mu_{B'_j}\left(y\right) = 0$, which is the neutral element for the union of fuzzy sets. The logical-type fuzzy system under consideration defines the accumulation through intersection, whose neutral element is $\mu_{B'_j}\left(y\right) = 1$.

The flowchart of the algorithm is depicted in Fig. 5. Its computational complexity is $O\left(N_d\right)$. The arguments of the operation are the membership function of the rule inference result A in discrete representation and the rule weight w. The algorithm contains a single loop for the variable i which denotes the sample index; the iterations are distributed among threads. The argument A and the result are both located in global memory.

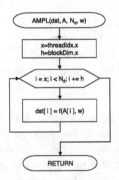

Fig. 5. Flowchart for application of rule weights

3.8   Accumulation

Accumulation is the computation of the inference B' of the fuzzy system as a whole. For logical-type systems, it is defined through the intersection of the inference results of all rules B'_j, $j = \overline{1, N}$ (see (6)). The intersection is performed pointwise; the membership degree of each sample is determined by applying the t-norm to the membership degrees of the samples in each set B'_j. The flowchart of this algorithm is not given here since it is completely similar to the one shown in Fig. 5, except that the operator in the loop body is `dst[i]=T(A[i],B[i])`. The operation is implemented for $N = 2$; accumulation for greater numbers N is programmed at command sequence level, similarly to the convolution of fuzzy truth values. The arguments `A`, `B` and the result are membership functions in discrete form, located in global memory. The enumeration of samples is implemented by a loop for the variable `i` (sample index); the iterations are distributed among threads.

3.9   Defuzzification

Defuzzification determines the crisp output of the system by accumulation result. In this research, the center-of-gravity defuzzification method was used. In the continuous case, it has the form given in (8), whereas in the discrete case, it can be defined as

$$\bar{y} = \sum_{i=0}^{N_d-1} y_i \cdot \mu_{B'}(y_i) \bigg/ \sum_{i=0}^{N_d-1} \mu_{B'}(y_i).$$

Both summations, on the numerator and on the denominator, can be distributed among threads; every thread computes its own partial sums, then they are reduced and divided. Barrier synchronization must be done before division; the division itself must be executed by a single thread. The algorithm flowchart is depicted in Fig. 6. The complexity of this algorithm is $O(N_d)$. The argument of the operation is `A`, the membership function of the system inference result in discrete form, which is stored in global memory. The result is placed at `dst`; it points to the memory part for final results. The boundaries `a` and `b` of the base set of the output variable are also passed to the function that implements the algorithm.

In the implementation of the algorithm, the whole sums `numTotal` and `denTotal` of the numerator and the denominator, respectively, are stored in shared memory. Initially, they are assigned zero values by one of the threads; then barrier synchronization is performed (not shown in the flowchart). Next, every thread computes its own partial sums of the numerator and the denominator, executing its iterations of the loop for the sample index `i`. The values y_i are calculated regardless of the original domain of definition $[a; b]$ of the output variable; it is assumed to be $[0; 1]$ instead. After completion of the loop, the threads increase the values of the whole sums by the values of their partial sums. Since all threads are modifying the same memory locations, we use the atomic addition operation `atomicAdd` which is provided by the CUDA framework. This function

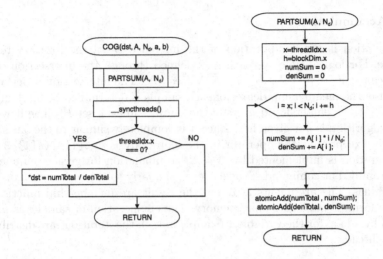

Fig. 6. Algorithm flowchart of the center-of-gravity defuzzification method

has `address` and value `val` as its arguments and increases the value at `address` by `val`. Then barrier synchronization is performed (`__syncthreads()`). Finally, the zeroth thread divides the whole sums of the numerator and the denominator, casts the quotient y_0 to the boundaries of the base set $[a; b]$, according to the formula $\overline{y} = a + (b - a) y_0$, and records the result as $\overline{y}$ at `dst`, while other threads do nothing.

4   Benchmark

Let us consider a neuro-fuzzy system that approximates the analytical dependency $f(x_1, x_2) = (x_1 - 7)^2 \cdot \sin(0.61x_2 - 5.4)$. The plot of this dependency is shown in 7. The rule base compiled according to the shape of the plot is given in Table 1.

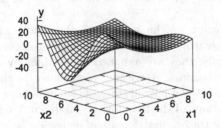

Fig. 7. Approximate dependency plot

Table 1. Rule base

x_1	x_2	y
Low	Low	High
Low	Medium	Low
Low	High	High
Medium	—	Medium
High	Low	Above medium
High	Medium	Below medium
High	High	Above medium

The neuro-fuzzy system has two inputs, each having three terms, six rules with two subconditions, and one rule with one subcondition. The computation of fuzzy truth values must be performed six times, for each term of each input variable. The convolution of fuzzy truth values occurs six times since the rule base contains six rules having two subconditions. The rule inference result is computed seven times, then each of them is applied a weight factor transformation. Finally, accumulation is performed $N - 1 = 6$ times, followed by defuzzification. The neuro-fuzzy system performs 33 operations in total; 13 of them have complexity $O\left(N_d^2\right)$. All terms are defined by Gaussian membership functions; each of them has two parameters for adjustment. Moreover, each rule has its own adjusted weight. The system contains altogether $(3 + 3 + 5) * 2 + 7 = 29$ parameters.

Let us estimate the computational complexity of system learning. A common number of generations for evolution strategies is $N_g = 300$; $\mu = 40$, $\lambda = 4 * \mu = 160$. Let the size of the training set be $M = 80$. In this situation, fuzzy inference is executed $N_g * \lambda * M = 3.84$ million times. Every inference requires $13 * N_d^2 + 20 * N_d$ iterations of loops. For $N_d = 1024$, it results in 13.6 millions of iterations; so the entire learning process requires approximately $5.2 * 10^{13}$ iterations. Figure 8

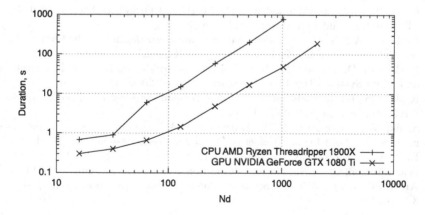

Fig. 8. A comparison of the learning process duration on CPU and GPU

portrays the results of an experiment comparing the duration of the learning process on CPU and GPU for different values of N_d. We used a parallel CPU-based implementation of the neuro-fuzzy system. For $N_d = 1024$, learning on CPU took 769.33 seconds, whereas on GPU it completed in 48.06 seconds. Thus, the GPU-based implementation accelerated the learning process by a factor of 16 approximately.

5   Conclusions

The parallel algorithms we have developed make it possible to significantly accelerate neuro-fuzzy system learning, which makes them more useful in practical applications. Owing to its high time complexity, the problem of acceleration becomes essential. The benchmark we provided demonstrates that the GPU-accelerated implementation of learning shortens its duration by a factor of 16. The suggested neuro-fuzzy system also allows for handling fuzzy input values, in contrast to most modern fuzzy modeling frameworks (see [6]).

References

1. Borisov, A., Alekseev, A., Krumberg, O., et al.: Decision Making Models Based on Linguistic Variable. Zinatne, Riga (1982). (in Russian)
2. Borisov, V., Kruglov, V., Fedulov, A.: Fuzzy Models and Networks. Hot Line - Telecom, Moscow (2007). (in Russian)
3. Dobuis, D., Prade, H.: Possibility Theory. Applications to the Representation of Knowledge in Informatics. Radio and Communication, Moscow (1990). (in Russian)
4. Programmable controllers - part 7: Fuzzy control programming. International Standard Electrotechnical Commission, Geneva, Switzerland (2000)
5. Kutsenko, D., Sinuk, V.: Inference method for systems with multiple fuzzy inputs. Bull. Russ. Acad. Sci. Control Theory Syst. **3**, 48–56 (2015). https://doi.org/10.7868/S0002338815030129. (in Russian)
6. Leonenkov, A.: Fuzzy Modeling in MATLAB and FuzzyTech Environment. BHV - Petersburg, Saint Petersburg (2003). (in Russian)
7. Rothstein, A., Shtovba, S.: Identification of nonlinear dependence by fuzzy training set. Cybern. Syst. Anal. **2**, 17–24 (2006). (in Russian)
8. Rutkowska, D., Pilinsky, M., Rutkowsky, L.: Neural Networks, Genetic Algorithms and Fuzzy Systems. Hot Line - Telecom, Moscow (2004). (in Russian)
9. Rutkowsky, L.: Methods and Techniques of Computational Intelligence. Hot Line - Telecom, Moscow (2010). (in Russian)
10. Sinuk, V.G., Polyakov, V.M., Kutsenko, D.A.: New fuzzy truth value based inference methods for non-singleton MISO rule-based systems. In: Abraham, A., Kovalev, S., Tarassov, V., Snášel, V. (eds.) Proceedings of the First International Scientific Conference "Intelligent Information Technologies for Industry" (IITI' 16). AISC, vol. 450, pp. 395–405. Springer, Cham (2016). https://doi.org/10.1007/978-3-319-33609-1_36

11. Sinuk, V.G., Kulabukhov, S.V.: Neuro-fuzzy system based on fuzzy truth value. In: Kuznetsov, S.O., Osipov, G.S., Stefanuk, V.L. (eds.) RCAI 2018. CCIS, vol. 934, pp. 91–101. Springer, Cham (2018). https://doi.org/10.1007/978-3-030-00617-4_9

12. Zadeh, L.: Outline of a new approach to the analysis of complex systems and decision processes. IEEE Trans. Syst. Man Cybern. 3(1), 28–44 (1973)

13. Zadeh, L.: PRUF - a meaning representation language for natural language. Intern. J. Man-Mach. Stud. 10, 395–460 (1978)

Time Series Discord Discovery on Intel Many-Core Systems

Mikhail Zymbler[1]([✉]) [iD], Andrey Polyakov[1], and Mikhail Kipnis[2]

[1] South Ural State University, Chelyabinsk, Russia
mzym@susu.ru, avpgenium@gmail.com
[2] South Ural State Humanitarian and Pedagogical University, Chelyabinsk, Russia
mmkipnis@gmail.com

Abstract. A discord is a refinement of the concept of an anomalous subsequence of a time series. The task of discovering discords is applied in a wide range of subject areas involving time series: medicine, economics, climate modeling, and others. In this paper, we propose a novel parallel algorithm for discord discovery using Intel MIC (Many Integrated Core) accelerators in the case when time series fit in the main memory. We achieve parallelization through thread-level parallelism and OpenMP technology. The algorithm employs a set of matrix data structures to store and index the subsequences of a time series and to provide an efficient vectorization of computations on the Intel MIC platform. Moreover, the algorithm exploits the ability to independently computing Euclidean distances between subsequences of a time series. The algorithm iterates subsequences in two nested loops; it parallelizes the outer and the inner loops separately and differently, depending on both the number of running threads and the cardinality of the sets of subsequences scanned in the loop. The experimental evaluation shows the high scalability of the proposed algorithm.

Keywords: Time series · Discord discovery · OpenMP · Intel Xeon Phi · Data layout · Vectorization

1 Introduction

The problem of finding an anomalous subsequence in a time series (i.e. a subsequence with the least similarity to any other subsequences) is one of the topical issues in time-series data mining and has applications in a wide range of subject areas: medicine, economics, climate modeling, predictive maintenance, energy consumption, and others.

In [9], Keogh *et al.* introduced the term *discord* to refine the concept of an anomalous subsequence. A discord of a time series can informally be defined as a subsequence that has the largest distance to its nearest non-self match neighbor. Discords are attractive as anomaly detectors because they only require one intuitive parameter (the length of the subsequence), unlike most anomaly detection algorithms, which typically require many parameters [10]. The HOTSAX

© Springer Nature Switzerland AG 2019
L. Sokolinsky and M. Zymbler (Eds.): PCT 2019, CCIS 1063, pp. 168–182, 2019.
https://doi.org/10.1007/978-3-030-28163-2_12

algorithm [9] employs SAX (Symbolic Aggregate ApproXimation) [14] transformation of subsequences and Euclidean distance for discord discovery. HOTSAX iterates subsequences according to an effective heuristics that allows for pruning large amounts of unpromising candidates for discord without calculating the distance.

In this paper, we address the task of accelerating the HOTSAX algorithm on Intel MIC (Many Integrated Core) systems [4,17]. MIC accelerators are based on the Intel x86 architecture and provide a large number of computing cores with 512-bit wide vector processing units (VPU) while supporting the same programming methods and tools as a regular Intel Xeon CPU. Intel provides two generations of MIC systems (under the codename of Intel Xeon Phi), namely Knights Corner (KNC), featuring 57 to 61 cores, and Knights Landing (KNL), featuring 64 to 72 cores. The benefits from the use of MIC accelerators usually manifest in applications where the processing of large amounts of data (at least hundreds of thousands of elements) can be arranged as loops that may be submitted to vectorization by a compiler [18]. Vectorization means the compiler's ability to transform the loops into sequences of vector operations [2] of VPUs.

In this study, we propose a parallel algorithm for discord discovery on Intel MIC systems, assuming that all the data involved in the computations fit into the main memory. The paper is structured as follows. In Sect. 2, we give the formal definitions along with a brief description of HOTSAX. Section 3 contains a short overview of related work. Section 4 presents the proposed parallel algorithm. In Sect. 5, we give the results of the experimental evaluation of our algorithm. Finally, Sect. 6 summarizes the results obtained and suggests directions for further research.

2   Problem Statement and the Serial Algorithm

2.1   Notations and Definitions

Below, we follow [9] to give some formal definitions and the statement of the problem.

A *time series* T is a sequence of real-valued elements: $T = (t_1, \ldots, t_m)$, $t_i \in \mathbb{R}$. The length of a time series is denoted by $|T|$.

A *subsequence* $T_{i,n}$ of a time series T is its contiguous subset of n elements that starts at position i: $T_{i,n} = (t_i, t_{i+1}, \ldots, t_{i+n-1})$, $1 \le n \ll m$, $1 \le i \le m - n + 1$. We denote by S_T^n the set of all subsequences of length n in T. Let N denote the number of subsequences in S_T^n, i.e. $N := |S_T^n| = m - n + 1$.

A *distance function* for any two subsequences is a nonnegative and symmetric function Dist: $\mathbb{R}^n \times \mathbb{R}^n \to \mathbb{R}$.

We say that two subsequences $T_{i,n}, T_{j,n} \in S_T^n$ are a *non-self match* to each other at distance $\mathrm{Dist}(T_{i,n}, T_{j,n})$ if $|i - j| \ge n$. A non-self match of a subsequence $C \in S_T^n$ is denoted by M_C.

A subsequence $D \in S_T^n$ is said to be a *discord* of T if it has the largest distance to its nearest non-self match. That is,

$$\forall C, M_C \in S_T^n \quad \min(\mathrm{Dist}(D, M_D)) > \min(\mathrm{Dist}(C, M_C)). \qquad (1)$$

As the distance function, we use the ubiquitous Euclidean distance measure, defined as follows. Given two subsequences $X, Y \in S_T^n$, the Euclidean distance between them is calculated as

$$ED(X, Y) := \sqrt{\sum_{i=1}^{n}(x_i - y_i)^2}. \tag{2}$$

2.2 The Serial Algorithm

The HOTSAX algorithm [9] consists of two stages. At the first stage, HOTSAX converts each subsequence of the input time series into its SAX representation [14]. Then the algorithm forms an array of SAX words and counts how often each word occurs. Afterward, it builds a prefix trie with each leaf containing a list of all array indices that map to that terminal node. At the second stage, the algorithm scans subsequences via the trie and discovers a discord. We below describe these stages in more detail.

The algorithm consequentially applies z-normalization, PAA transformation, and SAX transformation to a given subsequence to produce its SAX word.

We define the *z-normalization* of a subsequence $C \in S_n^T$ as a subsequence $\hat{C} = (\hat{c}_1, \dots, \hat{c}_n)$ in which

$$\hat{c}_i = \frac{c_i - \mu}{\sigma}, \ 1 \leq i \leq n;$$
$$\mu = \frac{1}{n}\sum_{i=1}^{n}c_i, \ \sigma^2 = \frac{1}{n}\sum_{i=1}^{n}c_i^2 - \mu^2. \tag{3}$$

The *PAA (Piecewise Aggregate Approximation)* [14] represents a subsequence $C = (c_1, \dots, c_n)$ as a vector $\overline{C} = (\overline{c}_1, \dots, \overline{c}_w)$ in a w-dimensional space, for a certain parameter $w \leq n$; moreover, the i-th coordinate of $\overline{C}$ is calculated as follows:

$$\overline{c}_i = \frac{w}{n} \cdot \sum_{j=\frac{n}{w}\cdot(j-1)+1}^{\frac{n}{w}\cdot j} c_j. \tag{4}$$

Next, the PAA representation is coded through the *SAX (Symbolic Aggregate approXimation)* [14] transformation. For a given subsequence $C = (c_1, \dots, c_n)$, its SAX word $\hat{C} = (\hat{c}_1, \dots, \hat{c}_w)$ is produced as follows. Assume we have an alphabet $\mathcal{A} = (\alpha_1, \dots, \alpha_{|\mathcal{A}|})$ to map a vector $\overline{C}$ into a word $\hat{C}$; here, $|\mathcal{A}|$ is the alphabet cardinality, and $\alpha_1 = $ 'a', $\alpha_2 = $ 'b', $\alpha_3 = $ 'c', and so on. Then,

$$\hat{c}_i = \alpha_i \Leftrightarrow \beta_{j-1} \leq \hat{c}_i < \beta_j, \tag{5}$$

where β_i are *breakpoints* [14] defined as a sorted list of numbers, $\mathcal{B} := (\beta_0, \beta_1, \dots, \beta_{|\mathcal{A}|-1}, \beta_{|\mathcal{A}|})$, such that $\beta_0 := -\infty$, $\beta_{|\mathcal{A}|} := +\infty$, and the area under the $N(0, 1)$

Gaussian curve between β_i and β_{i+1} equals $\frac{1}{|\mathcal{A}|}$. The breakpoints may be determined by looking them up in a statistical table [9]. It has been confirmed empirically that $w = 3, 4$ and $|\mathcal{A}| = 3, 4$ are suitable values for the discovery of time-series discords in a wide spectrum of subject areas [9].

Further, the algorithm produces an array of SAX words, counts the frequency of each word, and builds a prefix trie to store this information. The *prefix trie* [6] is a tree in which each edge is labeled with a symbol from the $\mathcal{A}$ alphabet in such a manner that all the edges connecting a node with its children are labeled with different symbols. The SAX word that corresponds to a leaf of the prefix trie is obtained by concatenation of all the characters that label the edges from the root to the leaf. Each leaf stores a sorted list of all array indices of the respective SAX word.

Alg. 1 HOTSAX(IN T, n; OUT $pos_{bsf}, dist_{bsf}$)

1: $dist_{bsf} \leftarrow 0$; $dist_{min} \leftarrow \infty$
2: **for** $C_i \in (NearDiscords \cdot Others)$ **do**
3:     **for** $C_j \in (Neighbors(C_i) \cdot Strangers(C_i))$ **do**
4:         $dist \leftarrow \text{ED}(C_i, C_j)$
5:         **if** $dist < dist_{bsf}$ **then**
6:             **break**
7:         $dist_{min} \leftarrow \min(dist, dist_{min})$
8:     $dist_{bsf} \leftarrow \max(dist_{min}, dist_{bsf})$; $pos_{bsf} \leftarrow i$
9: **return** $\{pos_{bsf}, dist_{bsf}\}$

Algorithm 1 presents the HOTSAX pseudo code. The algorithm takes a time series and a discord length and returns the index of the discord as well as the distance to its nearest neighbor subsequence. The algorithm looks through all the pairs of subsequences of the time series, calculates the Euclidean distance between them, and finds the maximum among the distances to nearest neighbors. The subsequence having the maximum distance to its nearest neighbor is a discord. The subsequences are iterated through two nested loops. Throughout the iterations, unpromising subsequences are pruned without calculating their distances. A subsequence is said to be unpromising if some of its neighbors is closer to it than the current maximum of the distances to all the nearest neighbors.

HOTSAX iterates subsequences according to a heuristic rule that enables pruning large amounts of unpromising subsequences. For this, the algorithm employs the following four sets of subsequences. The *NearDiscords* set contains subsequences with the least frequent SAX words; the *Others* set contains the rest of the subsequences. For a given subsequence C, the $Neighbors(C)$ set contains the subsequences whose SAX words match the SAX word of C. Conversely, the $Strangers(C)$ set contains the subsequences whose SAX words differ from that of C. The heuristics prescribes the following order for the iteration of subsequences. In the outer loop, subsequences from the *NearDiscords* set should be considered first, and then those from the *Others* set. In the inner loop, subsequences from the *Neighbors* set are considered first, and then those from the *Strangers* set.

3   Related Work

Following their introduction in [9], time-series discords and the HOTSAX algorithm have motivated considerable interest and follow-up work. Discords are applied for discovering abnormal heart rhythm in ECG [5], weird patterns of electricity consumption [1], unusual shapes [20], and others.

Among the attempts made to improve HOTSAX, we can mention the following. The *iSAX* algorithm [16] and the *HOTiSAX* algorithm [3] are the most direct enhancements to HOTSAX based on the indexable SAX transformation. A different approach is used in the *WAT* algorithm [7], in which Haar wavelets are employed instead of SAX transformations for time-series approximation. Another worth-noting example is the *Hash_DD* algorithm [19], which makes use of a hash table as an alternative to the prefix trie. We should also mention the *BitClusterDiscord* algorithm [13], which resorts to clustering of the bit representation of subsequences.

With reference to parallel discord discovery, we can draw attention to the following developments for computing systems with distributed memory. The *PDD (Parallel Discord Discovery)* algorithm [8] employs a Spark cluster [24] to split time series into fragments that are processed separately. PDD puts forward the Distributed Discord Estimation (DDE) method, which estimates the discord's distance to the nearest neighbor and minimizes the communication between computing nodes. Next, for each subsequence, PDD calculates the distance to its nearest neighbor and updates both pos_{bsf} and $dist_{bsf}$. A bulk of continuous subsequences is transmitted and calculated in batch mode to reduce message passing overhead. During this process, the early abandon technique is used to reduce computational complexity. Although PDD outruns HOTSAX, message passing between cluster nodes is a potential cause of significant degradation of the algorithm's performance as the number of nodes increases.

In [22], Yankov *et al.* propose a disk-aware algorithm (for brevity, referred to as *DADD, Disk-Aware Discord Discovery*) based on the concept of a *range discord*. For a given range r, the algorithm finds all discords at a distance of at least r from their nearest neighbor. The algorithm performs in two phases, namely candidate selection and discord refinement, with each phase requiring one linear scan through the time series on disk. By running HOTSAX for a uniformly random sample of the whole time series, which fits in the main memory, one can obtain the r parameter as the $dist_{bsf}$ value of the discord found [22]. Later, in [23], Yankov *et al.* presented a parallel version of DADD based on the MapReduce paradigm (for brevity, we denote the corresponding algorithm by *MR-DADD*). We should also mention the *DDD (Distributed Discord Discovery)* algorithm [21], which parallelizes DADD through a Spark cluster [24]. As opposed to DADD and MR-DADD, DDD computes the distance without taking advantage of an upper bound for early abandoning, which would increase the algorithm's performance.

To the best of our knowledge, no attempts have been made to parallelize HOTSAX for multi-core CPUs or many-core accelerators. Such an algorithm might be useful, though, both when time series fit in the main memory (e.g., to discover discords in ECG time series with tens of millions of elements) and in the case of disk-aware discord discovery as a sub-algorithm to obtain the r parameter quickly.

4 Accelerating Discord Discovery on Intel MIC Systems

The parallelization of the HOTSAX algorithm for the Intel MIC platform employs thread-level parallelism and OpenMP technology [15]. It is based on the following ideas: separate parallelization of outer and inner loops that iterate subsequences, use the square of the Euclidean distance, and use matrix representation of data.

At the first stage of HOTSAX, a *matrix data layout* enables an effective parallelization of computations in (3), (4), and (5) through OpenMP. In addition, matrix data structures are aligned in the main memory, so the calculations are organized with as many vectorizable loops as possible. We avoid unaligned memory access since it can cause inefficient vectorization due to time overhead for loop peeling [2]. Since OpenMP is more suitable to process matrices than trees, we employ a set of matrix data structures that store the same information as the prefix trie.

At the second stage of HOTSAX, distances between subsequences can be calculated independently by different threads of the parallel application. The parallel iteration of subsequences from the *NearDiscords*, *Others*, *Neighbors*, and *Strangers* sets can be performed *separately* and *differently* for the outer and the inner loops, depending on the number of running threads and the cardinality of the above-mentioned sets. To speed up the computations in (2), *the square-root calculation can be omitted* since this does not change the relative ranking of potential discords (indeed, the ED function is monotonic and concave).

In Sects. 4.1 and 4.2, we will show an approach to the implementation of these ideas.

4.1 Parallel Implementation of the Algorithm

To parallelize HOTSAX, we split the algorithm into two steps, namely *finding* and *refinement* (see Algorithm 2 and Algorithm 3, respectively; cf. Algorithm 1). In the finding step, the iteration in the outer loop involves only subsequences from the *NearDiscords* set. At the same time, only the inner loop is parallelized since the number of least frequent SAX words (i.e. the cardinality of the *NearDiscords* set) is potentially less than the number of threads the algorithm is running on.

In the refinement step, the iteration in the outer loop involves only subsequences from the *Others* set; the outer loop is parallelized, in view of the fact

Alg. 2 PhiDiscordDiscovery-Find(IN T, n; OUT pos_{bsf}, $dist_{bsf}$)

1: $dist_{bsf} \leftarrow 0$; $dist_{min} \leftarrow \infty$
2: **for** $C_i \in NearDiscords$ **do**
3:      #pragma omp parallel for
4:      **for** $C_j \in (Neighbors(C_i) \cdot Strangers(C_i))$ **do**
5:          $d \leftarrow \mathrm{ED}^2(C_i, C_j)$
6:          **if** $d < dist_{bsf}$ **then**
7:              **break**
8:          $dist_{min} \leftarrow \min(d, dist_{min})$
9:      $dist_{bsf} \leftarrow \max(dist_{min}, dist_{bsf})$
10:     **if** $dist_{bsf} < dist_{min}$ **then**
11:         $pos_{bsf} \leftarrow i$
12: **return** $\{pos_{bsf}, dist_{bsf}\}$

Alg. 3 PhiDiscordDiscovery-Refine(IN T, n; OUT pos_{bsf}, $dist_{bsf}$)

1: $dist_{bsf} \leftarrow 0$; $dist_{min} \leftarrow \infty$
2: #pragma omp parallel for
3: **for** $C_i \in Others$ **do**
4:      **for** $C_j \in (Neighbors(C_i) \cdot Strangers(C_i))$ **do**
5:          $d \leftarrow \mathrm{ED}^2(C_i, C_j)$
6:          **if** $d < dist_{bsf}$ **then**
7:              **break**
8:          $dist_{min} \leftarrow \min(d, dist_{min})$
9:      $dist_{bsf} \leftarrow \max(dist_{min}, dist_{bsf})$
10:     **if** $dist_{bsf} < dist_{min}$ **then**
11:         $pos_{bsf} \leftarrow i$
12: **return** $\{pos_{bsf}, \sqrt{dist_{bsf}}\}$

that the cardinality of the *Others* set is potentially greater than the number of threads the algorithm is running on.

In both steps, the loop is parallelized by the standard OpenMP compiler directive #pragma omp parallel for. Pruning unpromising subsequences results in uneven computational loading of threads. Therefore, to increase the efficiency of the parallel algorithm, we add to the above-mentioned #pragma the schedule (dynamic) parameter, which dynamically distributes loop iterations among threads. The statements in the loop body that calculates the squared Euclidean distances are vectorized by the compiler.

4.2   Internal Data Layout

The parallel algorithm employs the data structures depicted in Fig. 1. The time series is stored as a matrix of aligned subsequences to enable computations over aligned data with as many auto-vectorizable loops as possible. Let us denote by $width_{VPU}$ the number of floats stored in the VPU. If n (i.e. the length of the discord to be discovered) is not a multiple of $width_{VPU}$, then the subsequence

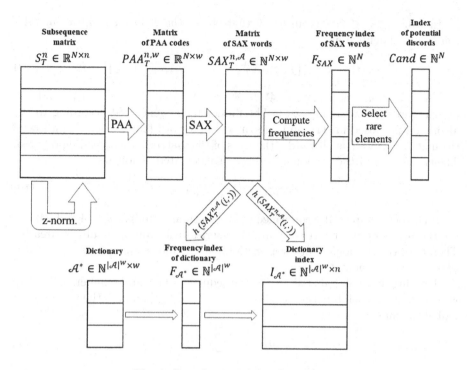

Fig. 1. Data layout of the algorithm

is padded with zeroes, with a number of zeroes $pad := width_{VPU} - (n \bmod width_{VPU})$. The aligned subsequence $\tilde{T}_{i,n}$ is defined as follows:

$$\tilde{T}_{i,n} := \begin{cases} t_i, t_{i+1}, \ldots, t_{i+n-1}, \overbrace{0, 0, \ldots, 0}^{pad}, & \text{if } n \bmod width_{VPU} > 0; \\ t_i, t_{i+1}, \ldots, t_{i+n-1}, & \text{otherwise.} \end{cases} \qquad (6)$$

The *subsequence matrix* $S_T^n \in \mathbb{R}^{N \times (n+pad)}$ is defined as

$$S_T^n(i, j) := \tilde{t}_{i+j-1}. \qquad (7)$$

The *matrix of PAA codes* $PAA_T^{n,\,w} \in \mathbb{R}^{N \times w}$ contains the data calculated in accordance with (4).

The *matrix of SAX words* $SAX_T^{n,\,\mathcal{A}} \in \mathbb{N}^{N \times w}$ contains the data calculated according to (5).

The *index of potential discords* is an ascending-ordered array $Cand \in \mathbb{N}^N$ that contains the indices of those subsequences of S_T^n having the least frequent SAX words in $SAX_T^{n,\,\mathcal{A}}$. The index of potential discords determines the order of iteration of subsequences that can be discords and is formally defined in the following way:

$$Cand(i) = k \Leftrightarrow F_{SAX}(k) = \min_{1 \le j \le N} F_{SAX}(j) \wedge$$
$$\forall i < j \quad Cand(i) < Cand(j). \qquad (8)$$

In (8), $F_{SAX} \in \mathbb{N}^N$ is an array that stores the *frequency indices of SAX words* and is defined as

$$F_{SAX}(i) = k \Leftrightarrow |\{j \mid SAX_T^{n,\mathcal{A}}(j,\cdot) = SAX_T^{n,\mathcal{A}}(i,\cdot)\}| = k. \tag{9}$$

The *dictionary* is a matrix $\mathcal{A}^* \in \mathbb{N}^{dict_{size} \times w}$ that contains all possible w-length words in the alphabet $\mathcal{A}$. The dictionary is generated according to the algorithm proposed in [11], and all symbols in a word (the elements in a row of the matrix), as well as all words (the rows of the matrix), are ascending-ordered. The cardinality of the dictionary $dict_{size}$ is calculated as follows:

$$dict_{size} := \bar{A}_{|\mathcal{A}|}^w = |\mathcal{A}|^w. \tag{10}$$

It has been empirically confirmed that such small values as $w = 4$ and $|\mathcal{A}| = 4$ are the best suited for virtually any time-series task from any subject area [9]. Therefore, the dictionary fits well in the main memory (indeed, $dict_{size} \times w = 4^4 \times 4 = 256$ elements).

Treating the $\mathcal{A}$ alphabet as an ordered sequence of natural numbers $1, \ldots, |\mathcal{A}|$, we can define a hash function $h \colon \mathbb{N}^w \to \{1, \ldots, dict_{size}\}$ to map the words of the alphabet, namely

$$h(a_1, \ldots, a_w) := \sum_{j=1}^{w+1} a_j \cdot w^{w-j-1}. \tag{11}$$

Next, the *dictionary index* is a matrix $I_{\mathcal{A}^*} \in \mathbb{N}^{dict_{size} \times N}$ comprising the indices of alphabet words contained in the matrix of SAX words. The dictionary index is defined as

$$I_{\mathcal{A}^*}(i,j) = k \Leftrightarrow \mathcal{A}^*(i,\cdot) = SAX_T^{n,\mathcal{A}}(k,\cdot). \tag{12}$$

Finally, the *frequency index of the dictionary* is an array $F_{\mathcal{A}^*} \in \mathbb{N}^{|\mathcal{A}|^w}$ whose elements are the numbers of occurrences of the words of the dictionary in the matrix of SAX words. The frequency index is defined in the following manner:

$$F_{\mathcal{A}^*}(i) = k \Leftrightarrow k = |\{j \mid \mathcal{A}^*(i,\cdot) = SAX_T^{n,\mathcal{A}}(j,\cdot)\}|. \tag{13}$$

5   The Experiments

5.1   The Experimental Setup

We evaluated the proposed algorithm in experiments conducted on Intel many-core systems at the Siberian Supercomputing Center[1] and the South Ural State University [12] (see Table 1 for a summary of the hardware involved).

In the first series of experiments, we assessed the performance and scalability of the algorithm while varying the discord length. We measured the run time (after deduction of the I/O time required for reading input data and writing the

---

[1] Hardware specifications of the Siberian Supercomputing Center.

Table 1. Hardware environment for the experiments

Characteristic	Intel Xeon CPU		Intel Xeon Phi Accelerator	
Model	E5-2630v4	E5-2697v4	SE10X (KNC)	7290 (KNL)
Physical cores	2×10	2×16	60	72
Hyperthreading factor	2×	2×	4×	4×
Logical cores	40	64	240	288
Frequency, GHz	2.2	2.6	1.1	1.5
VPU size, bit	256	256	512	512
Peak performance, TFLOPS	0.390	0.600	1.076	3.456

results) and calculated the algorithm's speedup and parallel efficiency, which are defined as follows. The speedup and the parallel efficiency of a parallel algorithm employing k threads are calculated, respectively, as $s(k) = \frac{t_1}{t_k}$ and $e(k) = \frac{t_1}{k \cdot t_k}$, where t_1 and t_k are the run times of the algorithm when one and k threads, respectively, are employed. We used two synthetic time series considered in [8] for evaluation of the PDD algorithm, namely SCD-1M and SCD-10M (with 10^6 and 10^7 elements, respectively).

In the second series of experiments, we compared the performance of our algorithm against analogs we have already considered: PDD, DDD, and MR-DADD (see Sect. 3). Throughout the experiments, we used the same datasets that were employed for the evaluation of the competitors. We ran our algorithm on an Intel Xeon Phi KNL system (see Table 1) with a reduced number of cores to make the peak performance of our accelerator approximately equal to that of the system on which the corresponding competitor was evaluated. For comparison purposes, we used the run times reported in the respective papers [8,21,23] (for the DDD and MR-DADD algorithms, we excluded the run time needed to calculate the r parameter).

5.2 Results and Discussion

Figure 2 depicts experimental results regarding the algorithm's scalability on the Intel Xeon Phi KNL system. The algorithm showed a 40 to 60× speedup and a 50 to 85% parallel efficiency (with respect to the length of the discord that is to be found) when the number of threads matches the number of physical cores the algorithm runs on. As expected, the algorithm performs at its best with greater values of the m and n parameters (time-series length and discord length, respectively) because this provides a higher computational load. When more than one thread per physical core is employed, the algorithm shows sub-linear speedup and an accordingly diminished parallel efficiency, but without a tendency to stagnate or degrade.

Experimental results concerning the algorithm's performance on various Intel many-core systems are depicted in Fig. 3. The algorithm performs better on systems with greater numbers of cores. At the same time, when the algorithm

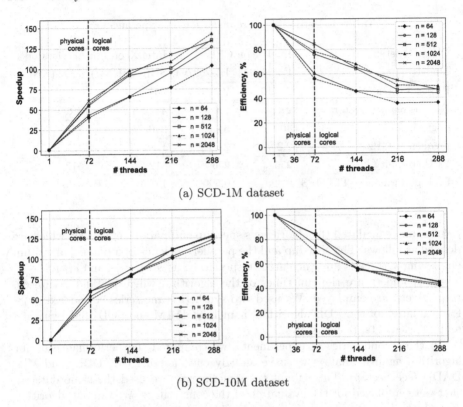

(a) SCD-1M dataset

(b) SCD-10M dataset

Fig. 2. Scalability of the algorithm

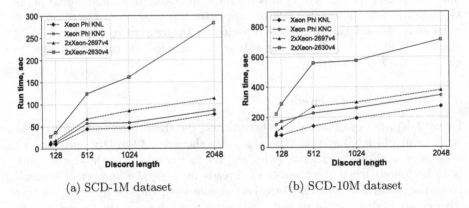

(a) SCD-1M dataset   (b) SCD-10M dataset

Fig. 3. Performance of the algorithm

runs on the Intel MIC accelerator, it performs better than when it runs on a
node equipped with two Intel Xeon CPUs.

Summing up, the proposed algorithm efficiently utilizes the vectorization
capabilities of the many-core system and shows high scalability, especially in

Table 2. Comparison of the proposed algorithm with analogs

Experimental setup				Performance, s	
Analog		Time series length	# cores (threads) of Intel MIC	Our algorithm	Competitor
Competitor	Hardware				
MR-DADD [23]	8 CPU 3.0 GHz	10^6	8 (32)	101.6	240
DDD [21]	4 CPU 2.13 GHz	10^7	4 (16)	1 745.3	5 382
PDD [8]	10 CPU 1.2 GHz	10^7	10 (40)	833.3	399 600

case of high computational load due to greater time-series length and discord length (tens of millions and tens of thousands of elements, respectively).

Table 2 summarizes the performance of the proposed algorithm compared with analogs. We can see that our algorithm outruns its competitors. PDD is far behind due to a significant overhead caused by message passing among cluster nodes. The reason for DDD and MR-DADD being inferior to our algorithm is disk I/O overhead, which can amount to a half of the whole run time of the algorithm. In addition to the use of the main memory rather than the disk, our algorithm also takes advantage of the vectorization capabilities of the Intel MIC accelerator. We may conclude that the parallel algorithm we have proposed should be preferred over similar disk-aware parallel algorithms for discord discovery in the case when time series fit in the main memory.

6 Conclusions

In this paper, we addressed the task of accelerating the discovery of time series discords on Intel MIC (Many Integrated Core) systems. A discord is a refinement of the concept of an anomalous subsequence (i.e. a subsequence that is the least similar to all the other subsequences) of a time series. Discord discovery is applied in a wide range of subject areas involving time series: medicine, economics, climate modeling, and others.

We proposed a novel parallel algorithm for discord discovery on Intel MIC systems in the case of time series that fit in the main memory. Our algorithm parallelizes the serial HOTSAX algorithm by Keogh et al. [9]. The parallelization makes use of thread-level parallelism and OpenMP technology and is based on the following ideas: separation of parallelization of iteration loops, use of the squared Euclidean distance, and application of matrix layout for algorithm's data.

The proposed algorithm showed high scalability throughout the experimental evaluation, especially in the case of high computational load due to greater time series length and discord length (tens of millions and tens of thousands of elements, respectively). Moreover, the experiments showed that our algorithm outperforms analogous disk-aware parallel algorithms for discord discovery when time series fit in the main memory.

In further studies, we plan to elaborate versions of the algorithm for other hardware platforms, namely GPU accelerators and cluster systems with nodes based on Intel MIC or GPU accelerators.

Acknowledgments. This work was financially supported by the Russian Foundation for Basic Research (grant No. 17-07-00463) and by the Ministry of Science and Higher Education of the Russian Federation (government orders 2.7905.2017/8.9 and 14.578.21.0265).

The authors thank the Siberian Supercomputer Center (Novosibirsk, Russia) and the South Ural State University (Chelyabinsk, Russia) for the computational resources provided.

References

1. Ameen, J., Basha, R.: Mining time series for identifying unusual sub-sequences with applications. In: First International Conference on Innovative Computing, Information and Control, ICICIC 2006, Beijing, China, 30 August–1 September 2006, pp. 574–577. IEEE Computer Society (2006). https://doi.org/10.1109/ICICIC.2006.115

2. Bacon, D.F., Graham, S.L., Sharp, O.J.: Compiler transformations for high-performance computing. ACM Comput. Surv. **26**(4), 345–420 (1994). https://doi.org/10.1145/197405.197406

3. Buu, H.T.Q., Anh, D.T.: Time series discord discovery based on iSAX symbolic representation. In: 3rd International Conference on Knowledge and Systems Engineering, KSE 2011, Hanoi, Vietnam, 14–17 October 2011, pp. 11–18. IEEE Computer Society (2011). https://doi.org/10.1109/KSE.2011.11

4. Chrysos, G.: Intel® Xeon Phi coprocessor (codename Knights Corner). In: 2012 IEEE Hot Chips 24th Symposium (HCS), Cupertino, CA, USA, 27–29 August 2012, pp. 1–31 (2012). https://doi.org/10.1109/HOTCHIPS.2012.7476487

5. Chuah, M.C., Fu, F.: ECG anomaly detection via time series analysis. In: Thulasiraman, P., He, X., Xu, T.L., Denko, M.K., Thulasiram, R.K., Yang, L.T. (eds.) ISPA 2007. LNCS, vol. 4743, pp. 123–135. Springer, Heidelberg (2007). https://doi.org/10.1007/978-3-540-74767-3_14

6. Fredkin, E.: Trie memory. Commun. ACM **3**(9), 490–499 (1960). https://doi.org/10.1145/367390.367400

7. Fu, A.W., Leung, O.T.-W., Keogh, E., Lin, J.: Finding time series discords based on haar transform. In: Li, X., Zaïane, O.R., Li, Z., et al. (eds.) ADMA 2006. LNCS (LNAI), vol. 4093, pp. 31–41. Springer, Heidelberg (2006). https://doi.org/10.1007/11811305_3

8. Huang, T., et al.: Parallel discord discovery. In: Bailey, J., Khan, L., Washio, T., Dobbie, G., Huang, J.Z., Wang, R. (eds.) PAKDD 2016. LNCS (LNAI), vol. 9652, pp. 233–244. Springer, Cham (2016). https://doi.org/10.1007/978-3-319-31750-2_19

9. Keogh, E.J., Lin, J., Fu, A.W.: HOT SAX: efficiently finding the most unusual time series subsequence. In: Proceedings of the 5th IEEE International Conference on Data Mining, ICDM 2005, Houston, Texas, USA, 27–30 November 2005, pp. 226–233. IEEE Computer Society (2005). https://doi.org/10.1109/ICDM.2005.79

10. Keogh, E.J., Lonardi, S., Ratanamahatana, C.A.: Towards parameter-free data mining. In: Kim, W., Kohavi, R., Gehrke, J., DuMouchel, W. (eds.) Proceedings of the 10th ACM SIGKDD International Conference on Knowledge Discovery and Data Mining, Seattle, Washington, USA, 22–25 August 2004, pp. 206–215. ACM (2004). https://doi.org/10.1145/1014052.1014077

11. Knuth, D.: The Art of Computer Programming, Volume 4, Fascicle 3: Generating All Combinations and Partitions. Addison-Wesley Professional, Boston (2005)
12. Kostenetskiy, P., Semenikhina, P.: SUSU supercomputer resources for industry and fundamental science. In: 2018 Global Smart Industry Conference (GloSIC), Chelyabinsk, Russia, 13–15 November 2018, p. 8570068 (2018). https://doi.org/ 10.1109/GloSIC.2018.8570068
13. Li, G., Bräysy, O., Jiang, L., Wu, Z., Wang, Y.: Finding time series discord based on bit representation clustering. Knowl.-Based Syst. **54**, 243–254 (2013). https:// doi.org/10.1016/j.knosys.2013.09.015
14. Lin, J., Keogh, E.J., Lonardi, S., Chiu, B.Y.: A symbolic representation of time series, with implications for streaming algorithms. In: Zaki, M.J., Aggarwal, C.C. (eds.) Proceedings of the 8th ACM SIGMOD Workshop on Research Issues in Data Mining and Knowledge Discovery, DMKD 2003, San Diego, California, USA, 13 June 2003, pp. 2–11. ACM (2003). https://doi.org/10.1145/882082.882086
15. Mattson, T.: Introduction to OpenMP. In: Proceedings of the ACM/IEEE SC 2006 Conference on High Performance Networking and Computing, Tampa, FL, USA, 11–17 November 2006, p. 209. ACM Press (2006). https://doi.org/10.1145/ 1188455.1188673
16. Shieh, J., Keogh, E.J.: iSAX: indexing and mining terabyte sized time series. In: Li, Y., Liu, B., Sarawagi, S. (eds.) Proceedings of the 14th ACM SIGKDD International Conference on Knowledge Discovery and Data Mining, Las Vegas, Nevada, USA, 24–27 August 2008, pp. 623–631. ACM (2008). https://doi.org/10. 1145/1401890.1401966
17. Sodani, A.: Knights Landing (KNL): 2nd generation Intel® Xeon Phi processor. In: 2015 IEEE Hot Chips 27th Symposium (HCS), Cupertino, CA, USA, 22–25 August 2015, pp. 1–24. IEEE (2015). https://doi.org/10.1109/HOTCHIPS.2015. 7477467
18. Sokolinskaya, I., Sokolinsky, L.: Revised pursuit algorithm for solving non-stationary linear programming problems on modern Computing clusters with manycore accelerators. In: Voevodin, V., Sobolev, S. (eds.) RuSCDays 2016. CCIS, vol. 687, pp. 212–223. Springer, Cham (2016). https://doi.org/10.1007/978-3-319-55669-7_17
19. Thuy, H.T.T., Anh, D.T., Chau, V.T.N.: An effective and efficient hash-based algorithm for time series discord discovery. In: 2016 3rd National Foundation for Science and Technology Development Conference on Information and Computer Science (NICS), Danang, Vietnam, 14–16 September, pp. 85–90 (2016). https:// doi.org/10.1109/NICS.2016.7725673
20. Wei, L., Keogh, E.J., Xi, X.: SAXually explicit images: finding unusual shapes. In: Proceedings of the 6th IEEE International Conference on Data Mining, ICDM 2006, Hong Kong, China, 18–22 December 2006, pp. 711–720. IEEE Computer Society (2006). https://doi.org/10.1109/ICDM.2006.138
21. Wu, Y., Zhu, Y., Huang, T., Li, X., Liu, X., Liu, M.: Distributed discord discovery: Spark based anomaly detection in time series. In: 17th IEEE International Conference on High Performance Computing and Communications, HPCC 2015, 7th IEEE International Symposium on Cyberspace Safety and Security, CSS 2015, and 12th IEEE International Conference on Embedded Software and Systems, ICESS 2015, New York, NY, USA, 24–26 August 2015, pp. 154–159. IEEE (2015). https:// doi.org/10.1109/HPCC-CSS-ICESS.2015.228

22. Yankov, D., Keogh, E.J., Rebbapragada, U.: Disk aware discord discovery: finding unusual time series in terabyte sized datasets. In: Proceedings of the 7th IEEE International Conference on Data Mining, ICDM 2007, Omaha, Nebraska, USA, 28–31 October 2007, pp. 381–390. IEEE Computer Society (2007). https://doi.org/10.1109/ICDM.2007.61
23. Yankov, D., Keogh, E.J., Rebbapragada, U.: Disk aware discord discovery: finding unusual time series in terabyte sized datasets. Knowl. Inf. Syst. **17**(2), 241–262 (2008). https://doi.org/10.1007/s10115-008-0131-9
24. Zaharia, M., Chowdhury, M., Franklin, M.J., Shenker, S., Stoica, I.: Spark: cluster computing with working sets. In: Nahum, E.M., Xu, D. (eds.) 2nd USENIX Workshop on Hot Topics in Cloud Computing, HotCloud 2010, Boston, MA, USA, 22 June 2010. USENIX Association (2010)

Supercomputer Simulation

The Use of Parallel Technologies in the Solution of a Problem of Geophysics Concerned with the Detection of Weakly Scattering Objects

Alexander Danilin[1(✉)], Gennadiy Erokhin[1], Maksim Kozlov[1], and Alexander Bugaev[2]

[1] Immanuel Kant Baltic Federal University, Kaliningrad, Russia
{ADanilin,GErokhin,MKozlov}@kantiana.ru
[2] Moscow Institute of Physics and Technology, Dolgoprudniy, Russia
Bugaev@cplire.ru

Abstract. In this research, we study the use of multiprocessor systems for the numerical solution of a problem of geophysics, namely the determination of weak scattering objects in complex geological media through the solution of acoustic tasks in direct and inverse time. We use an explicit finite-difference scheme for the numerical solution of the problem. The scheme is based on shifted grids and fits well with multiprocessor systems. In addition, we address the solution of the problem for multiple sources, a situation that frequently arises in geophysics. This paper proposes an optimal approach to the implementation of computations on a computational cluster using the MPI and OpenMP libraries.

Keywords: Parallelization · Applied mathematics · Geophysics · Reverse time migration · Numerical simulation

1 Introduction

A need to use modern supercomputers with shared memory always arises in a natural manner when solving major tasks of applied mathematical modeling of geophysical processes. Supercomputers are used not only to overcome the memory limitations involved with the use of a single computing node but also to accelerate the whole execution of the algorithm by parallelizing computations on all available cores. One of the most common approaches consists in applying MPI technology for performing computations on systems with distributed memory. At the same time, modern computing nodes in fact always consist of several cores with shared memory. It seems therefore natural to use "hybrid" algorithms based on both MPI and OpenMP, in which communications between nodes are carried out by taking advantage of MPI, while OpenMP streams work on shared memory within nodes. It is quite well known that the "pure" MPI implementation of algorithms in which the distribution of data over processes occurs

© Springer Nature Switzerland AG 2019
L. Sokolinsky and M. Zymbler (Eds.): PCT 2019, CCIS 1063, pp. 185–196, 2019.
https://doi.org/10.1007/978-3-030-28163-2_13

naturally (e.g., explicit schemes for parabolic equations, the method of separation of variables for elliptic equations, domain decomposition methods, etc.) is highly effective (almost linear time scaling on the number of MPI processes). To achieve a comparable performance in OpenMP implementations within a single computing node, a significant modification of the code is usually necessary, with the introduction of local arrays for each OpenMP flow. In this paper, we suggest a hybrid approach to solving the geophysical problem of determining weak scattering objects with a large number of sources.

In this article, we describe the use of parallel technologies for implementing a new acoustic-media visualization method based on an accurate statistical analysis of the amplitude and phase of interrelated vectors [1–4].

2   Vector-Pair Reverse-Time Migration

2.1   Forward Problem

To determine weak scattering objects, we consider two direct acoustic problems with direct and inverse time. We deduce acoustic waves from the pair $(p, \overrightarrow{u})$, where p is pressure and $\overrightarrow{u}$ is the particle velocity vector field [5–7]. The elements of the pair satisfy first-order linear differential equations. The forward wave $(p^{\mathrm{f}}, \overrightarrow{u}^{\mathrm{f}})(x, t; x_s)$, $t \in [0, T]$, satisfies the Cauchy problem

$$
\begin{cases}
\dfrac{1}{c^2} p_t = \operatorname{div} u + \delta(x - x_s) f(t), \\
\overrightarrow{u}^{\mathrm{f}}_t = \nabla p^{\mathrm{f}}, \\
p^{\mathrm{f}}\big|_{t=0} = 0, \ \overrightarrow{u}^{\mathrm{f}}\big|_{t=0} = 0.
\end{cases}
\tag{1}
$$

Here, $\delta(x - x_s)$ is the Dirac delta function, which models a point source located at $x_s \in \Gamma = \{x \in R^n \mid x^n = 0, \ n = 2, 3\}$, and $f(t)$ is a Ricker impulse (see Fig. 1).

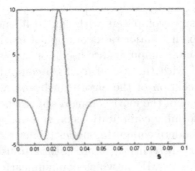

Fig. 1. Ricker impulse (f is the peak frequency, $f = 40\,\mathrm{Hz}$)

Let $p_0 = p^f|_{\Gamma \times [0,T]}$ be the "measured" pressure. Then the adjoint problem for (1) can be written as

$$\begin{cases} \frac{1}{c^2} p_t^b - c^2 \mathrm{div}(\nabla \overrightarrow{u}^b) = 0, \\ \overrightarrow{u}_t^b = \nabla p^b + p_0 \delta(x^n) \overrightarrow{\nu}_\Gamma, \\ p^b|_{t=T} = 0, \quad \overrightarrow{u}^b|_{t=T} = 0, \end{cases} \tag{2}$$

where $\overrightarrow{\nu}_\Gamma = (0,\ldots,0,1)$ is the unit normal vector to Γ.

Ideally absorbing layers are used as boundary of the computational domain. Let ν be the outward unit normal vector to the PML (Perfectly Matched Layer) layer (its boundaries can be curvilinear) [9]. More precisely, the vector ν is orthogonal to the level surfaces of the damping function $\alpha(x)$. Let us decompose the differential operator ∇ into a sum, as follows:

$$\nabla = \nabla^\perp + \nabla^\parallel, \tag{3}$$
$$\nabla^\perp = (\nu, \nabla)\nu, \quad \nabla^\parallel = \nabla - \nabla^\perp. \tag{4}$$

Thus, we have the following decomposition of the divergence operator:

$$\mathrm{div} = \mathrm{div}^\perp + \mathrm{div}^\parallel, \tag{5}$$
$$\mathrm{div}^\perp f = \nabla_i^\perp f^i, \quad \mathrm{div}^\parallel f = \nabla_i^\parallel f^i, \tag{6}$$

Then, in the PML layer, the acoustic equations are replaced by the following system of equations for $p^\perp, p^\parallel, u^\perp, u^\parallel$:

$$\frac{1}{c^2}(p_t^\perp + \alpha p^\perp) = \mathrm{div}^\perp u, \tag{7}$$

$$\frac{1}{c^2} p_t^\parallel = \mathrm{div}^\parallel u, \tag{8}$$

$$u_t^\perp + \alpha u = \nabla^\perp p, \tag{9}$$

$$u_t^\parallel = \nabla^\parallel p, \tag{10}$$

where $p = p^\perp + p^\parallel$, $u = u^\perp + u^\parallel$, and $\alpha(x)$ is the damping function. The damping function in the direction $\nu(x)$ is a smooth nonnegative function (see Fig. 2), equal to zero on the inner boundary of the PML layer and monotonously increasing in the direction of the outer boundary. We use a well-known empirical formula for plane-parallel layers, namely

$$\alpha(x) = \frac{(n+1)c_{max}}{2\delta} \log\left(\frac{1}{R}\right)\left(\frac{x - x_0}{\delta}\right)^n. \tag{11}$$

2.2   Vector-Pair Reverse-Time Migration Method

We propose new methods of visualization of the environment that take advantage of a statistical approach and are based on both the inversion of the wave field

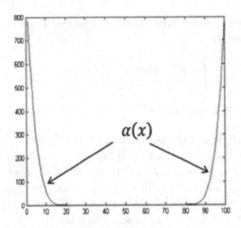

Fig. 2. The damping function

and the ideas underlying the Reverse-Time Migration method. Since we are not considering a wave equation but a system of first-order acoustics equations, it is possible to construct various formulas taking into account the interaction of the vector fields of the velocity of displacement of particles in direct and inverse time. The obtained formulas give additional information enabling one to directly determine the type and magnitude of the acoustic-impedance variation for objects of a class of reflecting domains and diffraction objects. By using this method, we can determine the reflection coefficient taking into account the sign of the impedance jump for the local reflective site and other additional information.

We call $(p^b, \vec{u}^b)$ the back waves, which include two vector fields: $\vec{u}^f(x,t;x_s)$ and $\vec{u}^b(x,t;x_s)$. In what follows, we only use these vector fields (see Fig. 3).

By considering the statistics of interrelated vector pairs for a certain set of time samples, we can analyze each point of a given space and determine some characteristics of the acoustic field. Let us introduce interconnected image conditions for vector pairs [5],

$$I(x) = R_Q(\vec{u}^f, \vec{u}^b), \tag{12}$$

where $(\vec{u}^f, \vec{u}^b)$ are defined in some admissible set Q,

$$Q \subseteq \{t_k, x_s \mid t_k = \tau(x,x_s) + k\delta t, \ k = 0, \ldots, N_k, \ s = 1, \ldots, N_s\}, \tag{13}$$

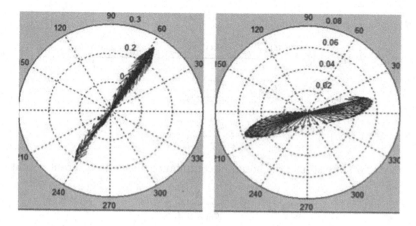

Fig. 3. Vector field for a fixed point x and a fixed source x_s. Left: $\overrightarrow{u}^{\,\mathrm{f}}$; right: $\overrightarrow{u}^{\,\mathrm{b}}$

and R_Q is an operator that is applied to the pair of vectors $(\overrightarrow{u}^{\,\mathrm{f}}, \overrightarrow{u}^{\,\mathrm{b}})$ in Q. The operator R_Q may take any of the following forms:

$$R_Q(\overrightarrow{u}^{\,\mathrm{f}}, \overrightarrow{u}^{\,\mathrm{b}})(x) = \sum_{s=1}^{N_s}\sum_{k=0}^{N_k}(|\overrightarrow{u}^{\,\mathrm{f}}||\overrightarrow{u}^{\,\mathrm{b}}|)(x; t_k, x_s), \tag{14}$$

$$R_Q(\overrightarrow{u}^{\,\mathrm{f}}, \overrightarrow{u}^{\,\mathrm{b}})(x) = \sum_{s=1}^{N_s}\sum_{k=0}^{N_k}(|\overrightarrow{u}^{\,\mathrm{f}}|/|\overrightarrow{u}^{\,\mathrm{b}}|)(x; t_k, x_s), \tag{15}$$

$$R_Q(\overrightarrow{u}^{\,\mathrm{f}}, \overrightarrow{u}^{\,\mathrm{b}})(x) = M(|\overrightarrow{u}^{\,\mathrm{f}}||\overrightarrow{u}^{\,\mathrm{b}}|)/M(|\overrightarrow{u}^{\,\mathrm{b}}|^2), \tag{16}$$

where M is the mathematical expectation.

3   The Numerical Solution

3.1   Finite-Difference Obvious Economic Scheme

In the discrete representation, the system is described, obviously, by a finite explicit cost-effective scheme with shifted grids. In the two-dimensional case, the scheme can be written as

$$\left(\frac{1}{c^2}\right)^{i,j}\frac{p_{i,j}^{k+1/2} - p_{i,j}^{k-1/2}}{\Delta t} = (D_x u_1)_{i,j}^k + (D_z u_2)_{i,j}^k, \tag{17}$$

$$\frac{u_1{}_{i+1/2,j}^{k+1} - u_1{}_{i+1/2,j}^k}{\Delta t} = D_x u_{i,j}^{k+1/2}, \tag{18}$$

$$\frac{u_2{}_{i,j+1/2}^{k+1} - u_2{}_{i,j+1/2}^k}{\Delta t} = D_z u_{i,j}^{k+1/2}, \tag{19}$$

where

$$(D_x u_1)_{i,j}^k + (D_z u_2)_{i,j}^k = \frac{u_1{}_{i+1/2,j}^k - u_1{}_{i-1/2,j}^k}{h} + \frac{u_2{}_{i,j+1/2}^k - u_2{}_{i,j-1/2}^k}{h}, \quad (20)$$

$$(D_x u)_{i,j}^{k+1/2} = \frac{u_{i+1,j}^{k+1/2} - u_{i,j}^{k+1/2}}{h}, \quad (21)$$

$$(D_z u)_{i,j}^{k+1/2} = \frac{u_{i,j+1}^{k+1/2} - u_{i,j}^{k+1/2}}{h}. \quad (22)$$

The space and time steps are chosen in compliance with the Courant condition:

$$\Delta t < \frac{h}{c_{max}\sqrt{2}}. \quad (23)$$

3.2   OpenMP Parallelization

We consider an algorithm for the parallel implementation of an explicit finite-difference scheme with shifted grids. A switch to a new temporal layer is only possible after obtaining the result on the previous layer. Thus, parallelization can be accomplished only when switching between temporal layers (see Fig. 4).

The computational process is parallelized at the stage of formation and solution of an individual system of linear equations from the set, since these systems are independent of each other. After completing the solution of the whole set of systems of linear equations, it is necessary to synchronize the flows inasmuch as the calculation of the next layer depends on the results on the previous one. There are several manners in which we can allocate memory for data storage of arrays:

– by declaring private variables for the thread that will be stored on the computational flow stack;
– by allocating dynamic memory in each thread;
– by allocating dynamic memory in the main thread for all threads and using in child threads part of the memory allocated in the main stream.

From the point of view of program execution, the first method is the fastest. However, if static memory allocation is used when compiling the program, then the number of grid nodes is still unknown, which leads to redundancy of allocated memory. The allocation and release of dynamic memory during the execution of a parallel block requires a considerable amount of computer time. Moreover, memory operations in different threads can block each other since threads use a common "heap" of the process. Given all the above factors, it is advisable to allocate dynamic memory in advance for all parallel threads and use the corresponding part of the memory in a parallel block, depending on the thread ID [8].

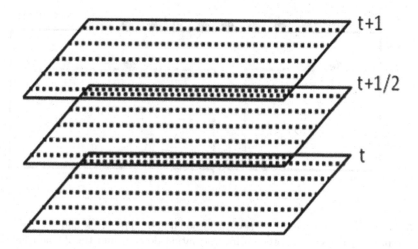

Fig. 4. Time layers

3.3 MPI Parallelization

The general formulation of the problem considers a large number of sources. A certain domain is limited for each source (see Fig. 5). Assuming that the domain for one source fits in the RAM of one node, it would be optimal to use the MPI library to distribute the sources among the cluster nodes [8].

Thus, we chose a data distribution approach that consists in selecting the main node which distributes the input data to other nodes by means of the message exchange functions MPI::Send/MPI::Recv.

3.4 Numerical Experiments

The simulation and test of the programs were carried out on the HPC "RocsCluster", which consists of 128 nodes, each containing eight-core Intel Server BoardS 5400SF servers. Each server is equipped with two Quad-Core Intel Xeon E5472 four-core processors (3.00 GHz, 2×6 MB of shared L2 cache). Each node has 32 GB RAM. The nodes communicate over a Fast Ethernet network. The operating system is OS Linux x86-64 + MPILibrary.

The parameters for the test models were the following (see Fig. 6):

– calculated area: 17000 m $\times 3500$ m;
– number of sources: 200, with a step of 50 m;
– calculated area for one source: 7000 m $\times 3500$ m; number of receivers: 701, with a step of 10 m;
– grid step: 5 m;
– time step: 0.4 ms;
– Ricker impulse frequency: 40 Hz.

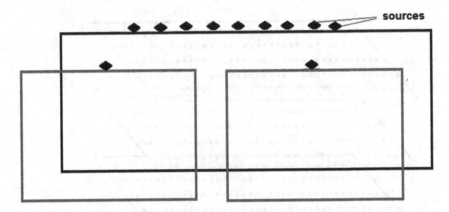

Fig. 5. Data for one source in the two-dimensional case. The black and the red rectangles represent, respectively, the entire calculation domain and the domain for one source (Color figure online)

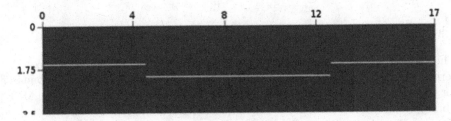

Fig. 6. Model with three borders; speed in the environment: 3000 m/s; speed in the borders: 4000 m/s

During the algorithm execution, the background speed was considered constant and equal to 3000 m/s. In Fig. 7 we can see the angles α and β obtained for any point $(t, s) \in Q$. These angles determine a point in the square $-180 < \alpha, \beta < 180$. The color corresponds to the magnitude of the vector $\vec{u}^b(t, s)$. We call this representation the angle distribution of the vector $\vec{u}^b$ magnitudes at point x. We can see that the angle distribution strongly depends on whether the point x is either a reflection or a diffraction point. This difference makes it possible to choose the operator R_Q so as to correspond to either reflection or diffraction points. Thus, filtering is performed on the basis of the angle distribution of the vector $\vec{u}^b$ magnitudes.

The results of such filtering (reflection and diffraction filters) for a model of three thin layers (see Fig. 6) are shown in Figs. 8 and 9.

The bar chart in Fig. 10 depicts how the speed of solving a problem depends on the number of cluster nodes used.

In the case of MPI, only one core per node is used; in the case of MPI + OpenMP, eight cores per node are used. It can be seen from the bar chart that

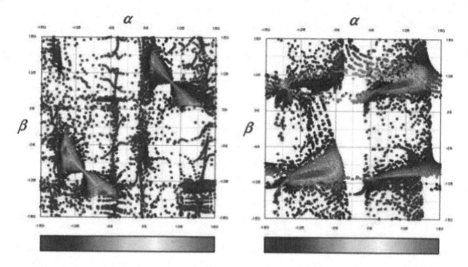

Fig. 7. (α, β): angle distribution of $|\vec{u}^{b}|$ at a reflection point (a) and at a diffraction point (b) (Color figure online)

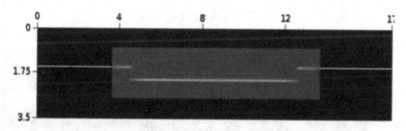

Fig. 8. Source data processing results. Isolation of reflectors

Fig. 9. Source data processing results. Extraction of diffracting parts

the time required to solve the problem depends on the number of cluster nodes used and also on the number of cores on each node.

A model of two diffractors in an homogeneous medium (second model) is shown in Fig. 11. Each diffractor is generated by a velocity jump in only one node. The left diffractor (white) reduces the velocity by 0.1 km/s (background velocity: 3 km/s), while the right one (red) increases the velocity by 0.1 km/s (Figs. 12, 13 and 14).

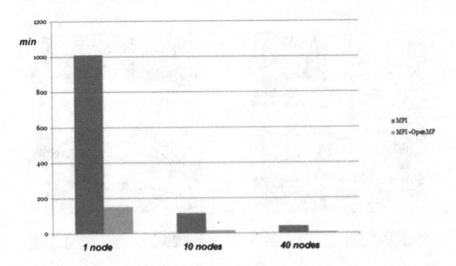

Fig. 10. The bar chart of solution time versus number of nodes used

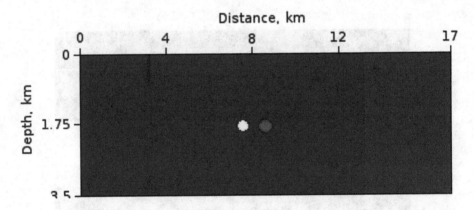

Fig. 11. Model of "soft" and "hard" diffractors (Color figure online)

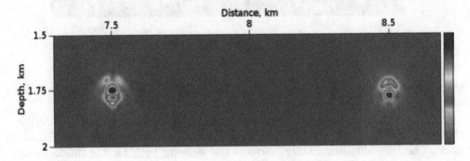

Fig. 12. Result VPRTM: Soft diffractor filter (Color figure online)

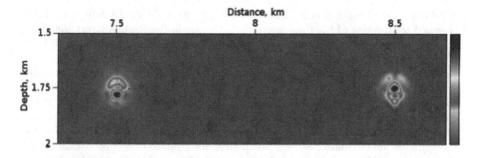

Fig. 13. Result VPRTM: Hard diffractor filter (Color figure online)

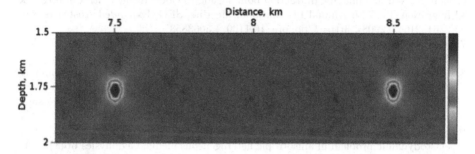

Fig. 14. VPRTM results: Soft + Hard diffractor filter (Color figure online)

4   Conclusions

We were able to ascertain that the vector-pair reverse-time migration method with phase filtering exhibits a rather high sensitivity to speed variations. This is a promising method for analyzing the amplitude and angle as well as the speed of migration and can be the basis for a new analysis of scattering and angle. However, this is a computationally complex method which can be executed only on high-performance clusters. We demonstrated that this task is optimally parallelized using the "hybrid" MPI + OpenMP approach.

Acknowledgments. This research was supported by the Russian Science Foundation (grant 16 - 11 - 10027).

References

1. Alekseev, A.S., Erokhin, G.N.: Integation in geophysical inverse problems (Integrated Geophysics): USSR Academy of Sciences Proceedings, UDC 550.3:517.97, vol. 308, pp. 1327–1331 (1989)
2. Baysal, E., Kosloff, D.D., Sherwood, J.W.C.: Reverse time migration. Geophysics **48**, 1514–1524 (1983). https://doi.org/10.1190/1.1441434

3. Erokhin, G.N., Kremlev, A.N., Starikov, L.E., Maltcev, V.V., Zdolnik, S.E.: CSP-method prospecting of fracture-cavernous reservoirs in the Bazhen formation of the Salym Oilfield. In: 74th Annual International Conference and Exhibition, EAGE, Extended Abstracts, Y028 (2012)
4. Kremlev, A.N., Erokhin, G.N., Starikov, L.E., Rodin, S.V.: Fracture and cavernous reservoirs prospecting by the CSP prestack migration method. In: 73rd Annual International Conference and Exhibition, EAGE, Extended Abstracts, B024 (2011). https://doi.org/10.3997/2214-4609.20148996
5. Erokhin, G., Pestov, L., Danilin, A., Kozlov, M., Ponomarenko, D.: Interconnected vector pairs image conditions: new possibilities for visualization of acoustical media. SEG Technical Program Expanded Abstracts 2017, pp. 4624–4629 (2017). https://doi.org/10.1190/segam2017-17587902.1
6. Zhu, X., Wu, R.: Imaging diffraction points using the local image matrix in prestack migration. In: 78th Annual International Meeting, SEG, Expanded Abstracts, pp. 2161–2165. (2008). https://doi.org/10.1190/1.3063853
7. Yoon, K., Marfurt, K.J.: Reverse-time migration using the Poynting vector. Explor. Geophys. **37**, 102–107 (2006)
8. Rabenseifner R., Hager G., Jost G.: Hybrid MPI/OpenMP parallel programming on clusters of multi-core SMP nodes. In: Proceedings of the 2009 17th Euromicro International Conference on Parallel, Distributed and Network-based Processing, pp. 427–436 (2009)
9. Colino, F., Tsogka, C.: Application of PML absorbing layer model to the linear elastodynamic problem in anisotropic heterogeneous media. Geophysics **66**(1), 294–307 (2001)

Use of Parallel Technologies for Numerical Simulations of Unsteady Soil Dynamics in Trenchless Borehole Drilling

Yury Perepechko$^{(\boxtimes)}$ ⓘ, Sergey Kireev ⓘ, Konstantin Sorokin ⓘ,
and Sherzad Imomnazarov ⓘ

N. A. Chinakal Institute of Mining, SB RAS, Novosibirsk, Russia
`perep@igm.nsc.ru`

Abstract. The work focuses on modeling unsteady dynamics of saturated soil in problems related to trenchless technology for laying underground communications. We devise parallel algorithms for solving the
corresponding nonlinear nonstationary problems of continuum mechanics. A special feature of trenchless technology is the extraction of portions
of the soil core which forms inside an open-end hollow pipe while it is
pushed into the ground by pneumatic percussive blows. Wave processes
are to be taken into account as the pipe is "hammered" through the
ground. The impact on the physical and mechanical properties of the
soil must be considered as well. We apply a finite-difference approximation procedure to thermodynamically consistent equations of saturated-
soil dynamics. For this, we resort to the control-volume method, whose
parallel implementation is based on domain-decomposition methods. In
addition, we perform numerical simulations using high-performance computing for various modes of pulsation of the pipe and periodic injections
of an ejecting medium so as to compact the core and extract it.

Keywords: Mathematical modeling ·
Parallel computing technologies · Trenchless technologies ·
Granular media · Two-velocity hydrodynamics ·
Control-volume method · Decomposition method

1   Introduction

The ongoing urbanization and compaction of urban surface infrastructure lead
to the need for development of new facilities for water, gas, and electricity supply, communication lines, made mainly by using underground methods. Currently, the following are the most frequent methods used in the construction
of underground infrastructure: directional drilling of boreholes of various diameters, piercing boreholes into the ground with a closed front-end pipe using a
pneumatic hammer, and pushing into the ground an open front-end pipe [1].

© Springer Nature Switzerland AG 2019
L. Sokolinsky and M. Zymbler (Eds.): PCT 2019, CCIS 1063, pp. 197–210, 2019.
https://doi.org/10.1007/978-3-030-28163-2_14

The last method causes only a minimal damage to geomaterials and is the least energy-consuming. A feature of this technology is that the soil is directed into the pipe interior and a ground plug (core) is formed. The motion of the pipe with the ground plug leads to a sharp increase in deformation of surrounding soil, reduction in penetration rate, and as a result, an increase in energy consumption. Timely removal of the ground core becomes an important factor determining the effectiveness of this technology. Existing methods for core removal, such as soil collecting devices, screw conveyors, and high-pressure jet washout, also require additional technical tools and energy. A promising method consists in the removal of portions of the ground core while it forms, by injecting an ejecting medium to the downhole pipe end via an additional pipeline [1]. This method does not require the use of auxiliary equipment.

The efficiency of this method of laying underground engineering communications can be improved by using modern methods of mathematical and computer modeling. The results of numerical and physical modeling of processes related to dynamic insertion of a hollow tube with an open end, including a discussion on the problems of core formation and its properties, can be found, for example, in [2–5]. In the present paper, we consider the use of trenchless technologies for the development of an efficient method of drilling in underground constructions, with removal of portions of the ground core. This approach is considered as a problem related to the movement of a heterophasic granular deformable medium in the framework of two-speed hydrodynamics. We examine an unsteady non-isothermal flow of a two-phase mixture of viscous compressible fluids in the presence of an impurity, additionally taking into account thermal and diffusion effects. The determining equations of motion for such a two-phase medium are obtained by the method of conservation laws and are thermodynamically consistent [6–8]. The finite-difference implementation of the model equations is based on the control-volume method [9,11,12], which guarantees the integral preservation of the main parameters of the model. The parallel implementation resorts to domain-decomposition methods [17]. An advantage of this approach is that the method of conservation laws ensures that the mathematical model fulfills the basic physical principles, while the method of finite volumes ensures the fulfillment of conservation laws on an arbitrary volume, and hence the physical correctness of the results. The dynamic models of saturated porous media, granular media, and two-phase mixtures of compressible fluids were successfully considered previously within the framework of this approach [7–10]. In this study, we conduct numerical simulations of a two-phase flow of a saturated granular medium under a pulsating action exerted on the pipe end with periodic injections of an ejecting medium through an additional channel. In the simulations, we examine various model parameters. In addition, we consider the stability of the numerical algorithm and the efficiency of the parallel algorithm.

2   The Mathematical Model

The technology consists in pushing a pipe into the ground under the action of a pneumatic hammer; as a result, the loose soil enters the internal cavity of the pipe, compacts gradually in there, and forms a core [5]. Portions of the core are extracted by injecting an ejecting medium (air, water, or a water-air mixture) to the pipe end through a separate pipeline (see Fig. 1). After the soil core forms, the ejection medium is fed through an upper opening into the internal cavity of the pipe, a portion of the soil core is cut away and moved out through the pipe to the pit. In the mathematical model, unlike the physical one, the periodic blow of the pneumatic hammer is simulated by setting the conditions corresponding to a periodic injection of soil in the left end of the pipe (see Fig. 1). Pipe deformations in this statement of the problem are not considered.

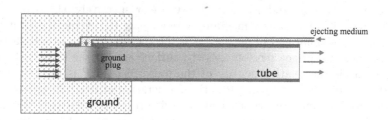

Fig. 1. Statement of the problem of a pipe pushed into the soil under the action of periodic injections of an ejecting medium near the downhole face

Saturated soil was considered as a two-phase granular medium consisting of soil particles and liquid or air phases. An individual volume of such a two-phase medium is characterized by densities and velocities of the phases, entropy, and specific content of soil granules.

The governing equations of such a medium can be derived in the framework of the method of conservation laws [6]. This method is based on the harmonization of the laws of thermodynamics, the laws of conservation of mass, energy, and momentum, and the Galilean invariance of the equations. The full system of constitutive equations of hydrodynamics for a two-phase mixture with an impurity [10,17] has the form

$$\frac{\partial \rho}{\partial t} + \operatorname{div} \mathbf{j} = 0, \quad \frac{\partial J}{\partial t} + \operatorname{div}(J\mathbf{u}_1) = 0, \tag{1}$$

$$\frac{\partial \rho_\alpha}{\partial t} + \operatorname{div}(c_1\rho_1\mathbf{u}_1 + c_2\rho_2\mathbf{u}_2 + \mathbf{L}_\alpha) = 0, \tag{2}$$

$$\frac{\partial j_i}{\partial t} + \partial_k\left(\rho_1 u_{1i}u_{1k} + \rho_2 u_{2i}u_{2k} + P\delta_{ik}\right. \\ \left. - (\eta_1 + \eta_{12})\, u_{1ik} - (\eta_2 + \eta_{12})\, u_{2ik}\right) = \rho \mathbf{g}, \tag{3}$$

$$\frac{\partial u_{2i}}{\partial t} + (\mathbf{u}_2, \nabla)\, u_{2i} = -\frac{1}{\rho}\partial_i P + \varsigma\frac{J}{\rho}\partial_i \sigma + \frac{\rho_1}{2\rho}\partial_i \mathbf{w}^2 - \frac{1}{\rho_2}b\,(j_i - \rho u_{1i}) \\ + \frac{1}{\rho_2}\partial_k\left(\eta_2 u_{2ik} + \eta_{12} u_{1ik}\right) + g_i, \tag{4}$$

$$\frac{\partial S}{\partial t} + \operatorname{div}\left(S_1\mathbf{u}_1 + S_2\mathbf{u}_2 + \frac{\mathbf{L}_q}{T} - \frac{\mu_\alpha \mathbf{L}_\alpha}{T}\right) = \frac{R}{T}. \tag{5}$$

The dissipative function R is determined by the formula

$$
R = \frac{1}{\rho_2} b \left(j_i - \rho u_{1i} \right)^2 + \kappa \left(\frac{\nabla T}{T} \right)^2 + 2\nu \nabla \left(\frac{\mu_\alpha}{T} \right) \nabla T
$$
$$
+ DT^2 \left(\nabla \left(\frac{\mu_\alpha}{T} \right) \right)^2 + \frac{1}{2} \eta_1 \partial_k u_{1i} + \frac{1}{2} \eta_2 \partial_k u_{2i} + \eta_{12} \partial_k u_{1i} \partial_k u_{2i}. \tag{6}
$$

The dissipative flows associated with thermal and diffusion phenomena have the form

$$
\mathbf{L}_q = -\kappa \frac{1}{T} \nabla T - \nu T \nabla \left(\frac{\mu_\alpha}{T} \right), \quad \mathbf{L}_\alpha = -\nu \frac{1}{T} \nabla T - DT \nabla \left(\frac{\mu_\alpha}{T} \right). \tag{7}
$$

Here, $\rho = \rho_1 + \rho_2$ and $\mathbf{j} = \rho_1 \mathbf{u} + \rho_2 \mathbf{v}$ are, respectively, the density and momentum of the two-velocity medium; μ and μ_α are the chemical potentials of the two-phase medium and impurities; J is the specific content of soil particles; σ is the tensor of surface tension; ς is the surface area of the dispersed phase; $\mathbf{j}_0 = \mathbf{j} - \rho \mathbf{u}_2$ is the relative density of the pulse phases; P is the pressure; S is the density of entropy; finally, $\mathbf{g}$ is the acceleration of free fall (gravitational acceleration). The impurity concentration is determined by the ratio $c = \rho_\alpha / \rho$: $c_1 \rho_1 = c\rho_1 + 2\lambda_1 \rho_2$, $c_2 \rho_2 = c\rho_2 - 2\lambda_1 \rho_2$ [10]. The density of the entropy of the phases can be expressed as

$$
S_2 = \frac{\rho_2}{\rho} S - 2 \left(\lambda_2 - \lambda_1 \frac{\rho \mu_\alpha}{T} \right) \frac{\rho_2}{\rho},
$$
$$
S_1 = \frac{\rho_1}{\rho} S + 2 \left(\lambda_2 - \lambda_1 \frac{\rho \mu_\alpha}{T} \right) \frac{\rho_2}{\rho}.
$$

The kinetic coefficients of interfacial friction (b), shear viscosity of the phases (η_i), diffusion (D), thermal conductivity of the two-phase medium (κ), and the coefficient ν are functions of thermodynamic parameters. Moreover,

$$
u_{1ik} = \partial_k u_{1i} + \partial_i u_{1k} - \frac{2}{3} \delta_{ik} \, \mathrm{div} \, \mathbf{u}_1,
$$
$$
u_{2ik} = \partial_k u_{2i} + \partial_i u_{2k} - \frac{2}{3} \delta_{ik} \, \mathrm{div} \, \mathbf{u}_2.
$$

Bulk viscosity effects are not taken into account.

The internal energy of the two-phase medium, $E_0 = E_0 \left(\rho, \rho_\alpha, J, \mathbf{j}_0, S \right)$, determines its thermodynamic properties. The equation of state of the two-phase medium is given in a linear approximation:

$$
\delta \rho = \rho \alpha \, \delta p - \rho \beta \, \delta T, \quad \delta s = \frac{c_p}{T} \delta T - \frac{1}{\rho} \beta \, \delta p. \tag{8}
$$

The coefficients of volumetric compression (α), thermal expansion (β), and specific heat capacity (c_p) are regarded as additive in the phases. The presence of impurities is considered in the approximation of the ideal solution. The surface tension and chemical potential of the impurity are determined by the ratios

$$
\sigma = \sigma_0 \frac{T_c - T}{T_c - T_{\mathrm{ref}}} - a_2 \ln \left(1 + a_3 c \right), \quad \mu_\alpha = d_1 P + d_2 T + \bar{R} T \ln (c), \tag{9}
$$

where $\bar{R}$ is the universal gas constant.

3   The Numerical Algorithm

The discretization of the original master equations relies on the control-volume method on a uniform rectangular mesh [11,12]. We use an implicit time scheme to determine the values of dependent variables in each time layer [9]. A nonlinear second-order HLPA scheme [13] is introduced to calculate the flows on the faces of the control volumes. In addition, a central difference scheme is applied to approximate the diffusion terms. The numerical solution of the systems of equations for computing the pressure field consistent with the flow field is achieved by applying the IPSA iteration procedure [14]. Nonlinearities present in all equations are eliminated thanks to a general iterative approach to construct a computational algorithm that is based on a simple iteration method, involves local iterations for solving each equation containing nonlinear terms and is closed by a global iterative procedure for computing the pressure with recalculation of thermodynamic parameters at each iteration. The scheme of the algorithm can be found in [17].

The main computational complexity of the numerical algorithm arises from the solution of a large number of systems of linear algebraic equations (SLAE) that are obtained as a result of the discretization of the differential equations. Therefore, the choice and effective implementation of solvers that take into account subsequent parallelization are the key factors affecting performance. Three methods were used to solve the SLAE:

- alternating direction method (ADM): three-diagonal run over alternate coordinate directions until the convergence condition is reached;
- stabilized biconjugate gradient method (BiCGStab): the method is implemented in the PETSc library [15];
- direct solution method: is based on the LDU decomposition of the matrix and implemented in the PARDISO solver as part of the Intel MKL library [16].

Since a rectangular mesh is used for space discretization, the ADM seems to be the most suitable for an efficient implementation and subsequent parallelization, owing to its simple structure and high degree of parallelism. Moreover, the study showed that, for the class of problems under consideration, the execution of the method in one of the coordinate directions has almost no effect on the convergence process. Therefore, the mentioned direction was excluded, which reduced the computational complexity of the corresponding solver to a half. The ADM can be successfully used to solve the equations of motion and transport. However, this method may diverge when is applied for solving the pressure-correction equation. It was therefore necessary to use other methods.

The choice of the PETSc solver library and the Intel MKL PARDISO solver was motivated by their efficient parallel implementation on distributed memory systems. A comparison of the methods implemented in the PETSc library showed that the solver based on the BiCGStab method produces the shortest computation time for the class of problems under consideration. Both solvers can be successfully used to solve all types of equations of the problem: equations of motion, transport, and pressure-correction.

4 Parallel Implementation

We devised a parallel program based on the above-mentioned numerical algorithm. The program was written in Fortran and parallelized using MPI standard and OpenMP API. Previously, we reported (see [17]) a version of the program using only MPI. The parallel implementation takes advantage of the domain-decomposition method. The 2D simulation domain is divided into subdomains in both directions. All arrays of mesh values are appropriately cut up and distributed among parallel processes. The number of subdomains in each direction is an implementation parameter. The shadow edges of the subdomains have a width of two cells, which corresponds to the size of the difference schemes used. The parallel program makes it possible to conduct simulations of the considered processes on distributed memory supercomputers.

To solve the SLAE arising in the simulation, the program uses three parallel solvers:

- our own implementation of ADM, adapted to the considered class of problems;
- the implementation of the BiCGStab method from the PETSc library [15],
- the PARDISO solver from the Intel MKL library [16].

The solver based on the ADM does not require that the sparse matrix be explicitly built and works with the original distributed 2D arrays. Therefore, no data redistribution among parallel processes is necessary. To reduce a possible imbalance caused by the sequential nature of the sweep algorithm, we used a pipelined algorithm [18,19]. The essence of the algorithm lies in the fact that, when the sweep runs for a large number of rows, each process computes a row and sends the coefficients in the sweep direction immediately, thereby providing work for the next process; only after that, it starts to compute the next row.

Unlike the previous implementation [17], the ADM solver was parallelized using OpenMP and a number of optimization procedures were applied to it. The OpenMP parallelization is based on the fact that several rows in a process can be computed simultaneously using different threads. So, the rows that are local to the process are divided into blocks, whose size is a parameter of the implementation. The rows within each block are processed in parallel by OpenMP threads. The resulting sweep coefficients of the whole block are sent to the next process using a single data transfer, which is another improvement (optimization). The local subdomain must be divided into at least two blocks to gain some effect from the use of the pipelined algorithm. One more improvement is a sequential bypass of mesh elements in memory when processing a block of rows, which should give an effect on large meshes.

The ADM solver implies that a number of iterations are to be performed until the convergence criterion is met. When we applied it to the solution of the motion equations, we noticed that most computations for the calculation of the sweep coefficients do not depend on the iteration of the ADM. Therefore, the sweep coefficients may be partially precalculated at the start of the ADM solver and saved in a set of arrays. These arrays can be used throughout the ADM iterations to finish the calculation of the sweep coefficients. This optimization allowed to drastically reduce the computational complexity of the ADM solver. In addition, the precalculation was parallelized using OpenMP. This variant of the ADM solver will henceforth be referred to as "optimized". Applying the same optimization to the solution of the other equations is much more complicated and does not obviously lead to an improvement. This was therefore left for future consideration.

The PETSc is an MPI-based parallel solver library [15]. At low level, it uses the optimized Intel MKL library. This to some extent ensures that all the necessary low-level optimizations, such as efficient use of caches and vectorization, are fulfilled. Of all the solvers implemented in the PETSc library, the best solution time was shown by the BiCGStab solver. We therefore chose it to be used in the program. In the BiCGStab solver, the sparse matrix of the SLAE is built explicitly, i.e. some sort of redistribution of data among processes is supposed. To minimize data exchanges, the PETSc library supports various methods of distribution of the elements of the global SLAE matrix among processes, including those consistent with the used 2D decomposition of the solution vector. This consistent distribution was used in the parallel program. The PETSc library does not support parallelization with threads. However, unlike the previous version of the program [17], OpenMP is now used for the parallel computation of the matrix of coefficients within each MPI process.

The Intel MKL PARDISO solver is an optimized parallel direct solver recently added to the parallel program. It supports MPI as well as OpenMP parallelization and requires the sparse matrix to be built explicitly. The computation of the matrix of coefficients within each MPI process is parallelized with OpenMP. For maximum performance, the MKL documentation strongly recommends using OpenMP parallelization rather than MPI, that is, to run one MPI process per physical node and set the number of OpenMP threads per node equal to the number of physical cores on the node.

5   Performance Evaluation

The resources listed below were used to evaluate the performance of the parallel program:

- A workstation with one Intel Core i7-7700 3.6 GHz CPU (four cores, two threads per core), 32 GB RAM. In what follows, it will be referred to as the Workstation.
- MVS-10P cluster [20], consisting of several partitions. The following two partitions were used:
    - Broadwell partition with nodes containing two 16-core Intel Xeon E5-2697Av4 2.6 GHz (32 cores per node, 2 threads per core), 128 GB RAM. These nodes will henceforth be referred to as the Broadwell nodes.
    - KNL partition with nodes containing one 72-core Intel Xeon Phi 7290 1.5 GHz (72 cores per node, 4 threads per core), 96 GB RAM, 16 GB MCDRAM. In what follows, these nodes will be referred to as the KNL nodes.

Moreover, we used the following software:

- On the Workstation: GNU Fortran v.5.5.0, Intel MKL v.2018.0 update 1, PETSc v.3.10.2 (built with Intel MKL), MPICH v.3.2.
- On the MVS-10P cluster: Intel Fortran Compiler v.17.0.4, Intel MKL v.2017.0 update 3, PETSc v.3.10.2 (built with Intel MKL), Intel MPI library v.2017 update 3.

Extreme pressure and velocity values and the complexity of the mathematical model in this problem make it necessary to use very fine meshs to increase stability, up to 10 000 nodes in each direction. To investigate the properties of parallel programs and determine optimal implementation parameters, large numbers of test runs are required. Accordingly, coarse meshes of size 400 × 400 and 800 × 800 were used to study the performance of the parallel program.

As a test problem, we considered a pressure flow of compressible fluids in a vertical channel under the action of a gravity field. The computational domain dimensions are $LX = 0.15$ m, $LY = 2.0$ m. The computation time for ten time steps is given in all cases.

The problem solution involves three computationally intensive stages: solution of the equations of motion, solution of the pressure-correction equation, and solution of the transfer equations. The program contains three different solvers which may be applied for solving the equations: the ADM solver (normal and optimized versions), the PETSc BiCGStab solver, and the Intel MKL PARDISO solver. First of all, one needs to determine which solver should be applied for the solution at each stage. Figure 2 portrays the evaluation results for the used solvers on the Workstation, with a mesh of size 400 × 400. The results show the execution time of the solvers applied to the solution of different equations with various parallelization parameters. Eight hardware threads were used on the Workstation since this number turned out to be the most effective. The test showed that the ADM optimized solver is the fastest when solving the equations of motion; the PETSc BiCGStab proved to be the fastest in the solution of the pressure-correction equation, while the ADM and the PETSc BiCGStab solvers showed the best time when solving the transfer equation. The Intel MKL PARDISO solver in all cases turned out to be slower than the others.

We note from the results that the optimal parallelization parameters are distinct for different solvers. The Intel MKL PARDISO solver works much faster when using parallelization with threads, the PETSc solver when using MPI processes, and the ADM solvers can show comparable results in both cases. The optimal parameters of the topology of MPI processes also differ. As a result, the following set of solvers was planned for further use: the ADM optimized solver for the solution of the equations of motion, the PETSc BiCGStab solver for the solution of the pressure-correction and transfer equations. The block sizes in the ADM solvers were set to 64 because this size gives a solution time close to the minimum in all cases of parallelization parameters. A subsequent study proved that this set of solvers was also optimal for the Broadwell and the KNL clusters.

The next step in the study was to determine the optimal parameters for the execution of the parallel program on a single cluster node. The parameters whose

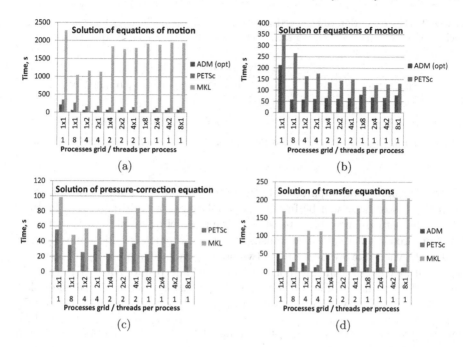

Fig. 2. Times required for the solution of the equations using different solvers with various parallelization parameters (Workstation, mesh size 400×400): equations of motion (a, b), pressure-correction equation (c), and transfer equations (d)

optimal values needed to be determined are the ratio of processes to threads, the topology of the MPI process grid, and the number of hardware threads to be used. Test results on both the Broadwell and the KNL nodes showed that parallelization only by MPI processes is better than parallelization only by OpenMP threads or any combination of them for the selected set of solvers. It was further determined that the optimal number of hardware threads for the Broadwell node is 32, which coincides with the number of cores in the node. Furthermore, the optimal number of hardware threads for the KNL node turned out to be 144, i.e. two threads per core. The study of the optimal topology of the grid of MPI processes showed that the best time is achieved when using a grid that is close to a square, which is the standard Cartesian grid used in the MPI library by default. This is determined by the ADM optimized solver, which is the one that makes the greatest contribution to the program execution time. The PETSc solver works faster on a more elongated grid of processes but its contribution to the total execution time is less significant.

The best execution time, speedup, and parallelization efficiency for a single Broadwell or KNL cluster node are shown in Table 1. The parallelization efficiency is calculated regarding the number of cores (not hardware threads) per cluster node. The results show that the Broadwell node runs on a given program approximately three times faster than the KNL node.

Table 1. Best execution time, speedup, and parallelization efficiency for a single Broadwell or KNL cluster node (mesh sizes: 400×400 and 800×800, 10 time steps)

	Broadwell node		KNL node	
Mesh size	400×400	800×800	400×400	800×800
Execution time	10.9 s	54.2 s	35.4 s	136.5 s
Speedup	17	16.8	30	34.6
Parallelization efficiency	53.1%	52.4%	41.7%	48.1%

The implementation parameters that provide a close-to-best performance on a single cluster node will be further extrapolated to a multi-node run. Figure 3 shows the performance characteristics of the parallel program for different numbers of Broadwell nodes. The unusual increase in efficiency for a mesh of size 800×800 when moving from one to two nodes may be explained by an increased aggregate cache size and memory bandwidth. The results show that the program can be scaled to four cluster nodes when using a mesh of size 400×400, and to eight nodes when using a mesh of size 800×800. The main factor that limits the scalability of the program, as it can be seen from Figs. 3a and b, is the PETSc solver, which is used for solving the pressure-correction equation. It can be assumed that the scalability of the PETSc solver can be improved by changing the parameters of the grid of processes. But alas, that would worsen the performance characteristics of the ADM optimized solver to a much greater extent. The tests showed that changing the grid topology or the ratio of processes to threads when working on several nodes as compared to those used on one node does not reduce the total execution time. Table 2 summarizes the best performance characteristics obtained on the Broadwell cluster for meshes of sizes 400×400 and 800×800.

Table 2. Best execution time, speedup, and parallelization efficiency for the Broadwell cluster (mesh sizes: 400×400 and 800×800, 10 time steps)

Mesh size	400×400	800×800
Number of Broadwell nodes, cores	4 nodes, 128 cores	8 nodes, 256 cores
Execution time	6.03 s	14.5 s
Speedup	30.7	62.6
Parallelization efficiency	24%	24.5%

6   Simulation Results

Figure 4 depicts the results of test computations for the problem with periodic injection of a saturated granular medium (soil) through the left boundary of

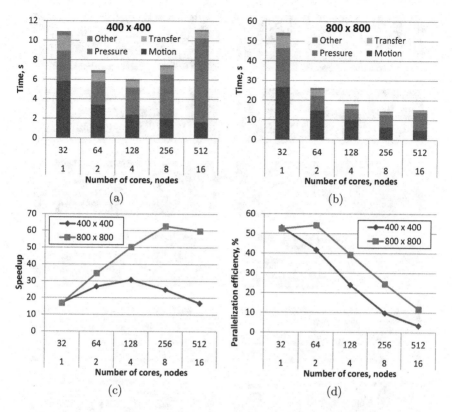

Fig. 3. Execution time (a, b), speedup (c), and parallelization efficiency (d) for meshes of sizes 400×400 and 800×800, depending on the number of Broadwell nodes

the pipe and periodic injection of an ejecting medium. The parameters of the computational domain were specified as follows (see Fig. 1): pipe size: $0.2 \times 1\,\text{m}$, hole diameter: $0.03\,\text{m}$, distance from the left edge of the pipe to the hole: $0.15\,\text{m}$. The pipe was initially filled with a saturated granular medium having a volumetric water content of 10%. A gas-water mixture with a water content of 90% is injected through the hole on the upper wall. The thermodynamic parameters of the phases correspond to the parameters of water and sand. The following boundary conditions were set (see Fig. 1): no leakage and no slippage conditions were assumed on the lower and upper walls (excluding the hole); the right boundary was considered free. Saturated soil is injected in a pulse regime through the left boundary. The injection of the gas-water mixture through the upper opening occurs in antiphase with soil injection into the pipe. The distribution of the specific content of soil particles and the velocity field of the solid phase are depicted in Fig. 4 for different regimes of periodic injection of saturated soil through the left boundary at different moments. We considered various problem statements, when the frequency of the injection was $0.01\,\text{Hz}$ (left diagrams) and $0.035\,\text{Hz}$ (right diagrams). In both problem statements, the

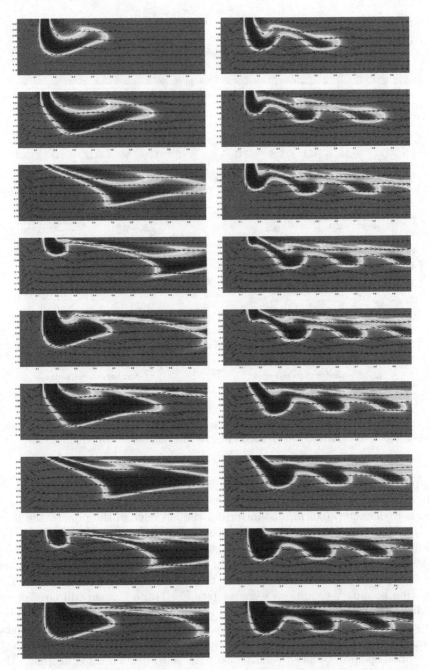

Fig. 4. Specific content of soil particles, J (shown in color), and the velocity field $\mathbf{u}_1$ (shown by vectors) of the solid phase at various time points: $50, 75, 100, 125, 150, 175, 200, 225$, and $250\,\mathrm{s}$. The left diagrams correspond to a frequency of injection of $0.01\,\mathrm{Hz}$, while the right diagrams correspond to $0.035\,\mathrm{Hz}$. (Color figure online)

two-phase flow velocity at the left boundary and at the upper opening were assumed to be equal to 0.01 m/s and 0.1 m/s, respectively.

The results for the frequency 0.01 Hz (left diagrams) demonstrate the potential of the periodic distribution (along the channel) of soil density. Under the given problem parameters, the characteristic size of the denser layer of soil is about 45 cm. An increase in the frequency of soil injection (action of the pneumatic hammer) leads to the formation of a continuous permeable longitudinal layer in the upper part of the channel (right diagrams), which reduces the probability of formation of a compacted transverse layer of soil and, subsequently, a soil plug. For a more realistic description of the process of formation of a compacted soil layer and its removal from the pipe by the action of an ejecting medium, it is necessary to expand the model by means of a mechanism of soil compaction in a plug.

7   Conclusions

We constructed a model to simulate the movement of a saturated granulated soil core inside a hollow rigid cylinder. Further, we conducted simulations of the process of introduction of a pipe into the soil and extraction of portions of the ground core. It was shown that a significant acceleration of the parallel implementation of the program can be achieved on supercomputers. As a part of the present work, we managed to optimize a previously developed parallel program for simulation of a nonstationary nonisothermal flow of a mixture of viscous compressible fluids. The improvement of the program was based on optimization, parallelization with OpenMP, and incorporation of the Intel MKL PARDISO solver. We identified an optimal set of solvers and their parameters. The corresponding performance results for a single Broadwell or KNL node, as well as for a Broadwell cluster are given in the paper. We studied the scalability of the program to four Broadwell nodes for a mesh of size 400×400 and to eight Broadwell nodes for a mesh of size 800×800. The best speedups obtained were, respectively, 30.7 and 62.6. A comparison of the results of numerical simulations with experimental data will be exposed in a forthcoming article, after obtaining the necessary results from field experiments.

Acknowledgments. The work was supported by the Russian Scientific Foundation (grant No. 17-77-20049).

References

1. Smolyanitsky, B.N., et al.: Modern Technologies for the Construction of Extended Wells in Soil Massifs and Technical Means for Controlling Their Trajectories. Publishing House of the Siberian Branch of the Russian Academy of Sciences, Novosibirsk (2016)
2. Kondratenko, A.S., Petreev, A.M.: Features of the earth core removal from a pipe under combined vibro-impact and static action. J. Min. Sci. **44**(6), 559–568 (2008). https://doi.org/10.1007/s10913-008-0063-5

3. Grabe, J., Pucker, T.: Improvement of bearing capacity of vibratory driven open-ended tubular piles. In: Meyer, V. (ed.) Proceedings of 3rd International Symposium on Frontiers in Offshore Geotechnics 2015 in Oslo (Norway), vol. 1, pp. 551–556. Taylor & Francis Group, London (2015)

4. Labenski, J., Moormann, C., Ashrafi, J., Bienen, B.: Simulation of the plug inside open steel pipe piles with regards to different installation methods. In: Kuliešius, V., Bondars, K., Ilves, P. (eds.) Proceedings of 13th Baltic Sea Geotechnical Conference, pp. 223–230, Vilnius Gediminas Technical University (2016). https://doi.org/10.3846/13bsgc.2016.034

5. Danilov, B.B., Kondratenko, A.S., Smolyanitsky, B.N., Smolentsev, A.S.: Improving the technology of drilling wells in the soil by the method of forcing. Phys. Tech. Probl. Dev. Miner. Resour. **3**, 57–64 (2017)

6. Khalatnikov, I.M.: An Introduction to the Theory of Superfluidity. W.A. Benjamin, New York (1965)

7. Dorovsky, V.N.: Mathematical models of two-velocity media. I. Math. Comput. Model. **21**(7), 17–28 (1995). https://doi.org/10.1016/0895-7177(95)00028-Z

8. Dorovsky, V.N., Perepechko, Y.V.: Mathematical models of two-velocity media. II. Math. Comput. Model. **24**(10), 69–80 (1996). https://doi.org/10.1016/S0895-7177(96)00165-3

9. Perepechko, Y.V., Sorokin, K.E.: Two-velocity dynamics of compressible heterophase media. J. Eng. Thermophys. **22**(3), 241–246 (2013). https://doi.org/10.1134/S1810232813030089

10. Dorovsky, V.N., Perepechko, Y.V., Sorokin, K.E.: Two-velocity ow containing-surfactant. J. Eng. Thermophys. **26**(2), 160–182 (2017). https://doi.org/10.1134/S1810232817020047

11. Patankar, S.V., Spalding, D.B.: A calculation procedure for heat, mass and momentum transfer in three-dimensional parabolic flows. Int. J. Heat Mass Transf. **15**, 1787–1806 (1972). https://doi.org/10.1016/0017-9310(72)90054-3

12. Date, A.W.: Introduction to Computational Fluid Dynamic. Cambridge University Press, New York (2005)

13. Wang, J.P., Zhang, J.F., Qu, Z.G., He, Y.L., Tao, W.Q.: Comparison of robustness and efficiency for SIMPLE and CLEAR algorithms with 13 high-resolution convection schemes in compressible flows. Numer. Heat Transf. Part B **66**, 133–161 (2014). https://doi.org/10.1080/10407790.2014.894451

14. Yeoh, G.H., Tu, J.: Computational Techniques for Multi-Phase Flows. Butterworth-Heinemann, Oxford (2010). https://doi.org/10.1016/B978-0-08-046733-7.00003-5

15. PETSc: Portable, Extensible Toolkit for Scientific Computation. https://www.mcs.anl.gov/petsc

16. MKL, Intel® Math Kernel Library. http://software.intel.com/en-us/intel-mkl

17. Perepechko, Y., Kireev, S., Sorokin, K., Imomnazarov, S.: Modeling of nonstationary two-phase flows in channels using parallel technologies. In: Sokolinsky, L., Zymbler, M. (eds.) PCT 2018. CCIS, vol. 910, pp. 266–279. Springer, Cham (2018). https://doi.org/10.1007/978-3-319-99673-8_19

18. Povitsky, A.: Parallelization of the Pipelined Thomas Algorithm. ICASE Report No. 98–48, NASA Langley Research Center, Hampton (1998)

19. Sapronov, I.S., Bykov, A.N.: A parallel-pipelined algorithm. Atom. **4**, 24–25 (2009)

20. MVS-10P cluster, JSCC RAS. http://www.jscc.ru

High-Performance Calculations for River Floodplain Model and Its Implementations

Boris Arkhipov[1][ID], Sergey Rychkov[2][ID], and Anatoliy Shatrov[2,3(✉)][ID]

[1] Federal Research Center of Informatics of RAS, Dorodnicyn Computer Center, Moscow, Russia
arhip@ccas.ru
[2] Vyatka State University, Kirov, Russia
rychkov@list.ru, avshatrov1@yandex.ru
[3] Kirov State Medical University, Kirov, Russia

Abstract. This work presents a parallel cluster version of a 2D hydrodynamical river flood model and its implementations to prevent environmental disasters. This problem is one of practical significance. Half encircled by a bend of the river Vyatka, a few kilometers upstream from the city of Kirov, we find the Kirovo-Chepetsk floodplain. Chemical waste dumps are located in immediate proximity to this place. In spring, flood streams wash away the pollutants, posing a threat to the water supply system of Kirov. In the present study, we describe such events considering a two-dimensional system of shallow water equations as a mathematical model of evolving processes. An equation of transport and diffusion of contaminants is added to this system. For the numerical implementation of the model, we use a system of finite-difference equations. The model is used for forecasting floods occurring in the Kirovo-Chepetsk floodplain in spring. The model was verified against satellite shots of the spreading area of floods and monitoring data on the concentration of pollutants collected by the ecological services of Kirovo-Chepetsk Chemical Works and the Ministry of Ecology and Environmental Protection of the Kirov Region. The constructed model can be also used to predict the influence of protective hydro-technical facilities before their construction, thereby diminishing the pollution hazards.

Keywords: High-performance calculations · Flood modeling ·
Transport of pollutants in flood streams

1 Introduction

In recent years, we are observing a rapid development of high-performance computer technologies, as well as the dissemination of their implementations beyond the scientific and designing spheres into industrial production and socioeconomic processes, including ecological issues. This work deals with an important environmental problem which can be described as follows.

© Springer Nature Switzerland AG 2019
L. Sokolinsky and M. Zymbler (Eds.): PCT 2019, CCIS 1063, pp. 211–224, 2019.
https://doi.org/10.1007/978-3-030-28163-2_15

A few kilometers from the city of Kirov, upstream the river Vyatka, we find the Kirovo-Chepetsk floodplain, half-encircled by a bend of the river. A map of the floodplain is given in Fig. 1. Note that, in general, the river flows from east to west. Wastes from a chemical plant are dumped in immediate proximity to the floodplain. Due to the action of rain and groundwater, some floodplain lakelets have turned into kind of accumulators of chemical contaminants. In spring, flood streams reaching these lakelets wash away the pollutants and transport them into the Vyatka riverbed, which poses a threat to the water supply system of the city of Kirov and its suburbs.

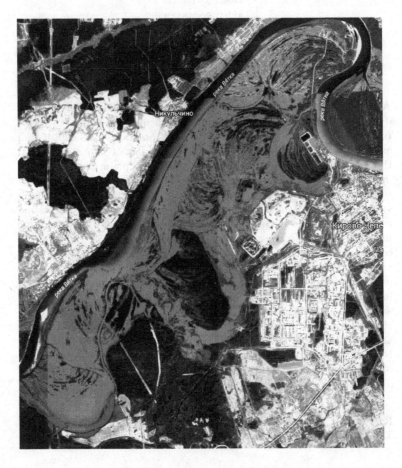

Fig. 1. Kirovo-Chepetsk floodplain

The administration of the Kirov Region set the task of creating a system of measures for the forecasting of spring floods and prevent their unfavorable aftereffects. Our work appeared in response to that request and represents an attempt to model flood processes, made in the framework of contract NIOKTR 01201275894 (22.10.2012), signed with the regional government department for environmental protection.

2   Mathematical Model of the Problem

The mathematical model of the problem is based on the well-known Navier–Stokes–Saint-Venant equations for shallow water [1]. These are a system of 2D fluid hydrodynamics equations written for the water in a riverbed taking into account some boundary conditions on air–water and water–bottom interfaces, as well as conditions on open inlet-outlet boundaries and on the riverbanks. The geometry of the problem and the main notations are shown in Fig. 2, in which a section of the riverbed is depicted with the water flow along the X-axis.

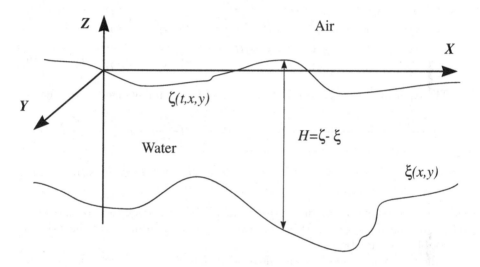

Fig. 2. Main notations

Here $H(t, x, y) = \zeta(t, x, y) - \xi(t, x, y)$, is the water depth, with $\xi(t, x, y)$ the relief height and $\zeta(t, x, y)$ the water surface height (as usual all heights are assumed to be above the Baltic Sea level [2]). Obviously, the water depth is assumed equal to zero on solid earth areas. Then the above-mentioned system can be written in the following form:

$$\frac{\partial H}{\partial t} + \frac{\partial (Hu)}{\partial x} + \frac{\partial (Hv)}{\partial y} = 0, \tag{1}$$

$$\frac{\partial (Hu)}{\partial t} + \frac{\partial (Hu^2)}{\partial x} + \frac{\partial (Huv)}{\partial y} = -\frac{\partial (gH^2/2)}{\partial x} - gH\frac{\partial \xi}{\partial x} - \alpha |\mathbf{u}| u + fHv, \tag{2}$$

$$\frac{\partial (Hv)}{\partial t} + \frac{\partial (Huv)}{\partial x} + \frac{\partial (Hv^2)}{\partial y} = -\frac{\partial (gH^2/2)}{\partial y} - gH\frac{\partial \xi}{\partial y} - \alpha |\mathbf{u}| v - fHu, \tag{3}$$

where u and v are the components of the local water velocity $\mathbf{u}$, f is a factor related to Coriolis forces, α is the water–bottom friction coefficient, and g is, as usually, the acceleration of gravity.

By introducing full flows U and V,

$$U = \int_{\xi}^{\zeta} u \cdot dz = \bar{u} \cdot H, \qquad V = \int_{\xi}^{\zeta} v \cdot dz = \bar{v} \cdot H,$$

we can write the above system in terms of full flows (it is in this form that we use it later on):

$$\frac{\partial H}{\partial t} + \frac{\partial U}{\partial x} + \frac{\partial V}{\partial y} = 0, \tag{4}$$

$$\frac{\partial U}{\partial t} + \frac{\partial (Uu)}{\partial x} + \frac{\partial (Vu)}{\partial y} = -\frac{\partial (gH^2/2)}{\partial x} - gH\frac{\partial \xi}{\partial x} - \alpha|\mathbf{u}|\frac{U}{H} + fV, \tag{5}$$

$$\frac{\partial V}{\partial t} + \frac{\partial (Uv)}{\partial x} + \frac{\partial (Vv)}{\partial y} = -\frac{\partial (gH^2/2)}{\partial y} - gH\frac{\partial \xi}{\partial y} - \alpha|\mathbf{u}|\frac{V}{H} - fU. \tag{6}$$

The equation for transport and diffusion of contaminants can be written as

$$\frac{\partial (H \cdot C)}{\partial t} + \frac{\partial (C \cdot U)}{\partial x} + \frac{\partial (C \cdot V)}{\partial y} = K\frac{\partial}{\partial x}\left(H\frac{\partial C}{\partial x}\right) + K\frac{\partial}{\partial y}\left(H\frac{\partial C}{\partial y}\right)$$
$$+ \sum_{n=1}^{N} J_n \cdot \delta(x - x_n, y - y_n), \tag{7}$$

where C is the concentration of contaminants, K is the coefficient of horizontal diffusion, and J_n, x_n, y_n are the intensity and the coordinates of the nth dot source of contamination.

When modeling riverbed and floodplain flow processes, including flooding and/or draining land patches, the main factors are the gradient of pressure and the water–bottom friction [3]. Other factors are relatively small and can be (and actually are) neglected.

As the simplest version, nonflow boundary conditions can be considered on the solid side boundaries (riverbanks):

$$U_n = 0, \tag{8}$$

where U_n is the normal full flow.

Alternatively, one can consider the following condition on the open inlet boundary:

$$U_n = q(t), \tag{9}$$

where $q(t)$ is a given river water inflow.

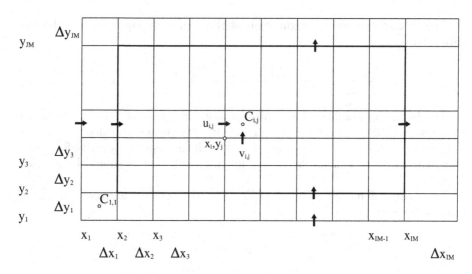

Fig. 3. The finite-difference grid

3   The Finite-Difference Scheme

To compute the solution of the problem stated above, we used an explicit scheme with a forward two-point stencil in time and a three-point stencil in space [4].

It is assumed that the flooding region of riverbed–floodplain streams has a rectangular shape with edges oriented north–south and east–west and is filled with rectangular cells, as shown in Fig. 3.

Let us consider one of the cells. At its center, the cell has a node indexed by i and j. The cell dimensions are Δx_i and Δy_j. In fact, all $\Delta x_i = \Delta x$ and all $\Delta y_i = \Delta y$. We assume that the continuous fields corresponding to the water depth $H(x,y)$, the water level $\zeta(x,y)$, the relief $\xi(x,y)$, contaminant concentrations $C(x,y)$, and the full flows $U(x,y)$ and $V(x,y)$, are digitized and represented by appropriate two-dimensional arrays, respectively $H(i,j), \zeta(i,j), \xi(i,j), C(i,j)$, $U(i,j)$, and $V(i,j)$, where $1 \leq i \leq IM$ and $1 \leq j \leq JM$. Note that the $C(i,j)$ element is located at the center of the cell, $U(i,j)$ is on the west edge of the cell, and $V(i,j)$ on the south edge. The variable $C(i,j)$ at the center of the cell corresponds thus to both the value $U(i,j)$ on the west edge of the cell and the value $V(i,j)$ on the south edge.

For the correct solution of the problem, one must define water in- and out-flows for the region under consideration. In fact, that means that the values of $U(2,i), U(IM,j)$, $V(i,2)$, and $V(i,JM)$ must be given (boundary conditions). The initial values of $\zeta(t_0, x_i, y_j)$ must be given at all nodes as well.

We consider the equations of continuity and of contaminant transport at the nodes $(2 \ldots IM - 1, 2 \ldots JM - 1)$.

The finite-difference approximation of the continuity equation has the following form:

$$\frac{H_{i,j}^{n+1} - H_{i,j}^n}{\Delta t} + \frac{U_{i+1,j}^n - U_{i,j}^n}{\Delta x} + \frac{V_{i,j+1}^n - V_{i,j}^n}{\Delta y} = 0. \tag{10}$$

The boundary conditions for this equation are required only at the nodes where water inflows to the region. This equation defines the depth in a cell deduced from the flows on the edges of the cell. If the depth is negative for whatever reason (due to calculation errors, for instance), then it should be equaled to zero.

The finite-difference equivalent of Eq. (7) with neglected horizontal diffusion terms can be written as

$$\frac{H_{i,j}^{n+1} C_{i,j}^{n+1} - H_{i,j}^n C_{i,j}^n}{\Delta t} + \frac{\tilde{C}_{i+1,j}^n U_{i+1,j}^n - \tilde{C}_{i,j}^n U_{i,j}^n}{\Delta x}$$
$$+ \frac{\tilde{C}_{i,j+1}^n V_{i,j+1}^n - \tilde{C}_{i,j}^n V_{i,j}^n}{\Delta y} = \frac{J_{i,j}^n}{\Delta x \Delta y}. \tag{11}$$

Hence we obtain

$$C_{i,j}^{n+1} = \frac{H_{i,j}^n}{H_{i,j}^{n+1}} C_{i,j}^n - \frac{\Delta t \Delta y (\tilde{C}_{i+1,j}^n U_{i+1,j}^n - \tilde{C}_{i,j}^n U_{i,j}^n)}{H_{i,j}^{n+1} \Delta x \Delta y}$$
$$- \frac{\Delta t \Delta x (\tilde{C}_{i,j+1}^n V_{i,j+1}^n - \tilde{C}_{i,j}^n V_{i,j}^n)}{H_{i,j}^{n+1} \Delta x \Delta y} + \frac{\Delta t J_{i,j}^n}{H_{i,j}^{n+1} \Delta x \Delta y}. \tag{12}$$

Here we use the following notations:

$$\tilde{C}_{i+1,j}^n = \begin{cases} C_{i,j}^n & \text{if } U_{i+1,j}^n > 0, \\ C_{i+1,j}^n & \text{if } U_{i+1,j}^n < 0; \end{cases} \qquad \tilde{C}_{i,j}^n = \begin{cases} C_{i-1,j}^n & \text{if } U_{i,j}^n > 0, \\ C_{i,j}^n & \text{if } U_{i,j}^n < 0. \end{cases} \tag{13}$$

Consequently, a conservative scheme with directed differences is used.

The dynamics equations for the U-variables are written at the U-nodes $(3 \ldots IM - 1, 2 \ldots JM - 1)$, and for the V-variables, at the V-nodes $(2 \ldots IM - 1, 3 \ldots JM - 1)$. These are called the inner nodes.

At the inner nodes, the flows are defined by the following formulae or vanish:

$$U^{n+1} = \frac{1}{\left(\dfrac{1}{\Delta t} + \dfrac{\alpha |u^n|}{H^n}\right)} \left[-gH^n \frac{\partial (H^n + \xi^n)}{\partial x} + \frac{U^n}{\Delta t}\right], \tag{14}$$

$$V^{n+1} = \frac{1}{\left(\dfrac{1}{\Delta t} + \dfrac{\alpha |u^n|}{H^n}\right)} \left[-gH^n \frac{\partial (H^n + \xi^n)}{\partial y} + \frac{V^n}{\Delta t}\right]. \tag{15}$$

To demonstrate how the calculations are performed, let us consider a $U(i, j)$-node. Each cell on the floodplain can be either wet, i.e. $H(i, j) > 0$, or dry, i.e. $H(i,j) = 0$. The $U(i,j)$-node chosen can be located between two wet cells, between two dry cells, or between a wet cell and a dry one. To determine which case applies, it is enough to test the product $P = H(i-1,j) \cdot H(i,j)$. If $P > 0$, then both cells are wet and formula (14) should be used in a slightly modified form:

$$
U_{i,j}^{n+1} \approx \frac{1}{\left(\dfrac{1}{\Delta t} + \dfrac{\alpha |\mathbf{u}^n|_{u;i,j}}{H_{u;i,j}^n}\right)} \left[-gH_{u;i,j}^n \frac{\zeta_{i,j}^n - \zeta_{i-1,j}^n}{\Delta x} + \frac{U_{i,j}^n}{\Delta t} \right], \qquad (16)
$$

where

$$
|\mathbf{u}^n|_{u;i,j} \approx \{(U_{i,j}^n)^2 + [0.25(V_{i,j}^n + V_{i,j+1}^n + V_{i-1,j}^n + V_{i-1,j+1}^n]^2\}^{1/2}/H_{u;i,j}^n, \quad (17)
$$
$$
H_{u;i,j}^n = 0.5(H_{i,j}^n + H_{i,j+1}^n). \qquad (18)
$$

If $P = 0$, then we should try the value of $S = 0.5[H(i-1,j) + H(i,j)]$. If $S = 0$, then both cells are dry and $U(i,j) = 0$.

If $P = 0$ and $S > 0$, then this means that the $U(i,j)$-node is between a wet cell and a dry one. In this case, we should compute $DH = H(i,j) - H(i-1,j)$ and $DZ = \zeta(i,j) - \zeta(i-1,j)$. If $DH \cdot DZ > 0$, then flooding of a dry cell occurs and Eq. (16) must be used. Otherwise, if $DH \cdot DZ \leq 0$, then $U(i,j) = 0$.

The remaining variables, $U(1\ldots2,j), U(IM,j), V(i,1\ldots2), V(i,JM), C(1,j)$, $C(IM,j), C(i,1), C(i,JM)$, are used for the boundary conditions.

The final stage of the problem solution produces the data arrays for water depths $H(i,j)$, flow velocities $U(i,j)$ and $V(i,j)$, and the contaminant concentration field $C(i,j)$. The vast amount of data obtained, which may be several gigabytes in size, is then processed by some utility programs specially designed to visualize them for subsequent analysis.

A similar algorithm was described in more detail in [5].

4   The Need for Parallel Processing

The relief data used were obtained by digitizing aerial photographs of the floodplain on an uniform grid with a five meter displacement between nodes. As a result, we obtained arrays containing 2122×2320 elements for $\xi(i,j)$. The other data arrays were of the same dimensions, and the finite-difference system was solved on the same grid.

To ensure the stability of the explicit scheme of computations, it was required to choose a small time step of the order of $\Delta t \approx 0.15\,\mathrm{s}$, as it had been analytically estimated at the beginning and confirmed during actual computations [6].

The substantial amount of data and the required small time step led to unfavorable results. It took six to seven days to calculate a typical three-week spring flood on a single-processor computer.

The enormous cost in processor time, on the one hand, and the natural parallelism of the data, on the other, suggested the idea of designing a parallel version of the computer program. It was implemented in the simplest way: by slicing the region and all arrays in vertical bars with overlapping edges. Then the finite-difference system was solved in each part independently in parallel, and overlapping data were exchanged after each time iteration. That is a standard well-known method [7] and there is no need for further details.

The program code was written in Fortran [8] using both MPI and OpenMP libraries. Calculations were performed on the HPC ENIGMA X000 cluster supercomputer at Vyatka State University. This supercomputer has a peak performance of 10.7 Tflops. The parallel performance gives a considerable gain in time, depending on the number of cluster nodes used.

5   Modeling Results

In addition to the floodplain relief data, some boundary conditions on the riverbed intake and the intensities of contamination sources are required to solve the problem.

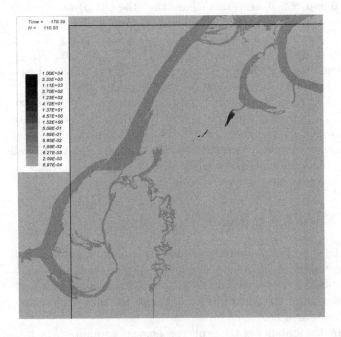

Fig. 4. Contamination before the flood

The model provides for the simulation of various versions of flood behavior by setting different time dependencies of the flow $q(t)$ on the east boundary of the floodplain. There is a large amount of such data available thanks to regular hydrological observations recorded over many years. Setting the functions $q(t)$ for forthcoming flood scenarios was consequently based on such historical statistical data, taking into account forecast data on the height of the flood, offered by the local hydrometeorological service.

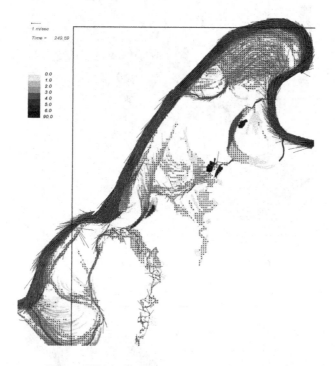

Fig. 5. Water flows on the tenth day of flood

The intensities of the sources of contamination were estimated using real samples taken from the polluted lakes just before the onset of the flood. Figure 4 shows the location of the polluted lakes on the floodplain and contaminant concentrations before the flood.

In the years 2013–2017, the model was intensely used for forecasting floods taking place on the Kirovo-Chepetsk floodplain. Simultaneously, the environmental monitoring services carried out regular gathering of samples to measure contaminant concentrations on the floodplain and in the Vyatka riverbed at the river Voloshka mouth [9].

Figures 5, 6, 7, 8, 9 and 10 show some examples of modeling results: distribution maps of water flows and contaminants during a typical three-week flood with a peak water level some higher than the average statistical level.

It should be noted that the modeling results are similar to the corresponding data of hydrological monitoring conducted by official services.

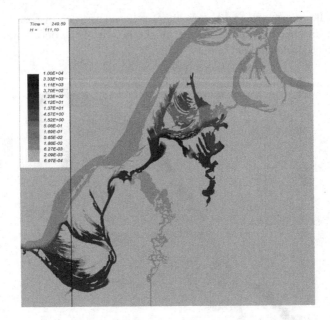

Fig. 6. Contamination on the tenth day of flood

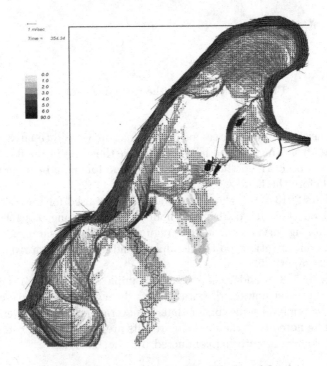

Fig. 7. Water flows on the fifteenth day of flood

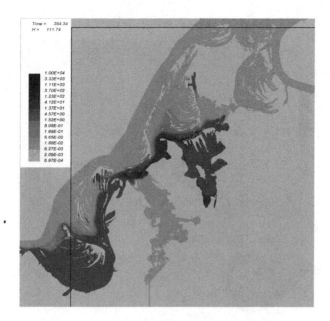

Fig. 8. Contaminations on the fifteenth day of flood

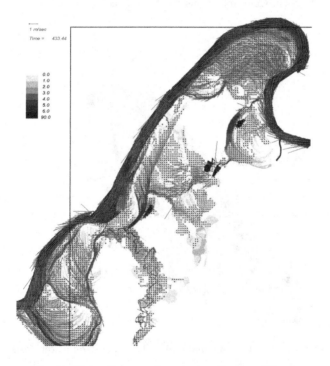

Fig. 9. Water flows on the nineteenth day of flood

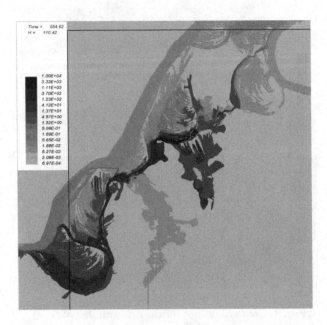

Fig. 10. Contaminations on the nineteenth day of flood

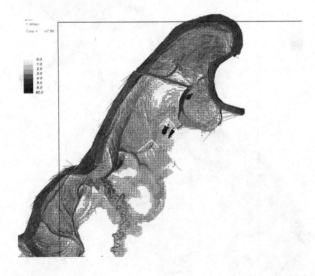

Fig. 11. Effects of a canal and embankments on water flows

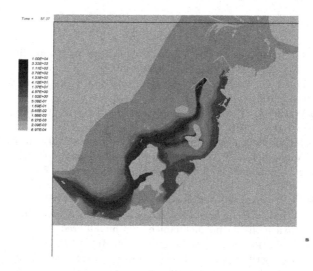

Fig. 12. Effects of a canal and embankments on contaminations

6   Extra Features of the Model

It has been suggested to build some hydro-technical facilities, such as embankments or canals, to redirect the streams on the floodplain to draw them away from the polluted lakes, thereby reducing the amount of contaminants transported into the Vyatka riverbed.

The reviewed model can be used for estimating the expediency and effectiveness of such projects against their cost. It is not difficult to simulate hydrotechnical facilities by simply modifying the relief height data. Obviously, that should be made through a utility program. The results of such simulations are shown in Figs. 11 and 12, in which a draining canal and two chevron-shaped embankments are positioned before the polluted lakes.

7   Conclusions

In this work, we constructed a mathematical model of flood events and described its cluster implementation. In addition, we demonstrated its feasibility and described some of its features.

The model was used during five years to forecast floods in the Kirovo-Chepetsk floodplain and their aftereffects. According to the conclusions of local official services, the forecasts were correct 75% of the time, which may be regarded as satisfactory.

To improve the model forecasting ability, it is necessary to introduce some modifications, for instance, to reduce flow–ground friction coefficients on various land patches and, above all, to implement a more sophisticated description of the sources of contamination.

References

1. Van Rijn, L.C.: Principles of Fluid Flow and Surface Waves in Rivers, Estuaries, Seas and Oceans. Aqua Publications, Blokzijl (1990)
2. Feistel, R., Naush, G., Wasmund, N. (eds.): State and Evolution of the Baltic Sea. Wiley, Hoboken (2008)
3. Peeters, F., Wuest, A., Piepke, G., Imboden, D.M.: Horizontal mixed in lakes. J. Geophys. Res. **101**(c8), 361–375 (1996)
4. Roach, P.: Vychislitelnaya gidrodinamika. Mir, Moscow (1980). (in Russian)
5. Arkhipov, B.V., Solbakov, V.V., Soloviev, M.B., Shapochkin, D.A.: Ekologicheskoe modelirovanie i lagranzhev podkhod. Matem. Modelirovanie **25**, 47–61 (2013). (in Russian)
6. Arkhipov, B.V., Rychkov, S.L., Solbakov, V.V., Soloviev, M.B., Shapochkin, D.A., Shatrov, A.V.: Modelirovanie pavodkovykh navodneniy v Kirovo-Chepetskoy pojme. In: XVIII Zimnyaya shkola po mekhanike sploshnykh sred, Perm', 18–22 fevralya 2013 g, Tezisy dokladov. Perm'-Ekaterinburg (2013). (in Russian)
7. Shpakovskij, G.I., Serikova, N.V.: Programirovanie dlya mnogoprotsessornykh sistem v standarte MPI. BGU, Minsk (2002). (in Russian)
8. Arkhipov, B.V., Rychkov, S.L., Solbakov, V.V., Soloviev, M.B., Shapochkin, D.A., Shatrov, A.V.: Raschyot urovney zatopleniya i perenosa zagryazneniy v rechnoy v rechnoy poyme v protsesse razvitiya pavodka. Svidetelstvo o registratsii programm dla EVM No. 2015610085 ot 12.01.2015 (2015). (in Russian)
9. Rychkov, S.L., Shatrov, A.V.: Gidrodinamicheskaya model pavodkovykh navodneniy r. Vyatka. In: II Vserossiyskaya nauchnaya konferentsiya "Okruzhayushchaya sreda i ustoychivoye razvitie regionov" Thes. dokl. 23.09–27.09 2013 KFU, Kazan' (2013). (in Russian)

The Use of Supercomputer Technologies for Predictive Modeling of Pollutant Transport in Boundary Layers of the Atmosphere and Water Bodies

Aleksandr I. Sukhinov[1], Aleksandr E. Chistyakov[1], Alla V. Nikitina[2(✉)],
Alena A. Filina[3], Tatyana V. Lyashchenko[2], and Vladimir N. Litvinov[4]

[1] Don State Technical University, Rostov-on-Don, Russia
`sukhinov@gmail.com, cheese_05@mail.ru`
[2] Southern Federal University, Rostov-on-Don, Russia
`nikitina.vm@gmail.com, t.lyashchenko@tmei.r`
[3] Supercomputers and Neurocomputers Research Center Co., Ltd.,
Taganrog, Russia
`j.a.s.s.y@mail.ru`
[4] Azov-Black Sea Engineering Institute of FSBHEEPT,
Don State Agrarian University, Rostov-on-Don, Russia
`litvinovvn@rambler.ru`

Abstract. The paper describes the development, research, and numerical implementation of interrelated mathematical models of hydrophysics and biological kinetics. These models were implemented on a supercomputer as a system for monitoring and controlling the quality of shallow waters and predicting processes of dispersion of contaminants in boundary layers of the atmosphere and water bodies. The program complex can be used as an efficient tool for monitoring the ecological situation in water bodies subjected to increasing anthropogenic pressure, climatic and industrial challenges, and emergency situations of anthropogenic or natural character. The software complex (the monitoring and control systems) comprises discrete analogs of models of water ecology based on high-order accuracy schemes. We apply the modified alternating triangular method to the solution of grid equations used for discretization of model problems in aquatic ecology. This method has the best convergence rate under the condition of asymptotic stability of difference schemes for parabolic equations with efficiency improved on the basis of updated spectral estimates. The design of effective parallel algorithms for the numerical implementation of problems of hydrophysics and biological kinetics offers an opportunity to consider processes of sediment dispersion in "air–water" systems in real and accelerated time.

Keywords: Dispersion of pollutants ·
Boundary layers of atmosphere and water bodies · Sea of Azov ·
Monitoring and control system · Predictive modeling ·
Parallel algorithms · Supercomputers

© Springer Nature Switzerland AG 2019
L. Sokolinsky and M. Zymbler (Eds.): PCT 2019, CCIS 1063, pp. 225–241, 2019.
https://doi.org/10.1007/978-3-030-28163-2_16

1   Introduction

Most pollutants emitted by industrial and transportation sources concentrate in the boundary layer of the atmosphere, are deposited on the water surface or fall onto it with precipitations. It is known that, far from river flows, more than 60% of nutrients involved in the production and destruction processes of phytoplankton get into the water from the air (see Fig. 1). It has become imperative to develop mathematical models able to predict changes in the environmental situation in shallow waters and coastal areas, namely the dispersion of pollutants occurring in the boundary atmosphere, the transport of pollutants and sediments, the formation of organic deposits, and others. These models must be implemented on high-performance computer systems so as to provide a forecast basis for sustainable development of aquatic ecosystems.

According to the Guidance Document GD.52.24.309-2016 [1], a system of stations (observation points) must be developed for collecting systematic and effective information on the ecological state of surface water bodies. The location of permanent stations involved in measurements of the concentration of pollutants in the Azov–Black Sea Basin are shown in Fig. 2, as per data of the Crimean Directorate on Hydrometeorology and Environmental Monitoring.

Fig. 1. Pollutant emissions

Fig. 2. Permanent stations

On the basis of data obtained during research expeditions (see Fig. 3) and data provided by the Unified State System of Information on the Situation in the World Ocean (ESIMO), we carried out a complex geoinformational analysis of spatial-temporal processes and phenomena. This analysis enabled us to develop models of hydrobiological processes taking place in the Sea of Azov [2].

2   Problem Statement

The spatially heterogeneous mathematical model that describes the dispersion of pollutants has the form

$$\frac{\partial S_i}{\partial t} + u\frac{\partial S_i}{\partial x} + v\frac{\partial S_i}{\partial y} + (w - w_{gi})\frac{\partial S_i}{\partial z} = \mu_i \Delta S_i + \frac{\partial}{\partial z}\left(\nu_i \frac{\partial S_i}{\partial z}\right) + \psi_i, \quad (1)$$

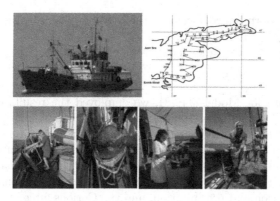

Fig. 3. Research expeditions

where S_i is the concentration of the i-th impurity, $i = \overline{1, 19}$; w_{gi} is the rate of gravitational sedimentation of the i-th component in suspended state; ψ_i is either a chemical-biological source (runoff) or a summand describing aggregation (clumping-declumping) if the corresponding component is a suspension. The correspondence between the numbers $i = \overline{1, 19}$ and the components is as follows: 1—hydrogen sulfide (H_2S), 2—elemental sulfur (S), 3—sulfates (SO_4), 4—thiosulfates (and sulfites), 5—total organic nitrogen (N), 6—ammonium (NH_4) (ammonium nitrogen), 7—nitrites (NO_2), 8—nitrates (NO_3), 9—phytoplankton, 10—zooplankton, 11—dissolved manganese (DOMn), 12—suspended manganese (POMn), 13—dissolved oxygen (O_2), 14—silicates (metasilicate (SiO_3), orthosilicate (SiO_4)), 15—phosphates (PO_4), 16—silicic acid (metasilicon (H_2SiO_3), orthosilicon (H_2SiO_3)), 17—iron (Fe^{2+}), 18—aluminum (Al), 19—molybdenum (Mo); $U = (u, v, w)$ are the velocity vector components of water flow movement; μ_i, ν_i are the diffusion coefficients in the horizontal and vertical directions of i-th substance.

The model takes into account the following factors: movement of water flows; microturbulent diffusion; interaction and gravitational sedimentation of pollutants and plankton; biogenic, temperature, and oxygen regimes; and salinity.

The computational domain $\bar{G}$ (the Sea of Azov) is a closed area limited by the undisturbed water surface Σ_0, the bottom $\Sigma_H = \Sigma_H(x, y)$, and the cylindrical surface σ for $0 < t \leq T$; moreover, $\Sigma = \Sigma_0 \cup \Sigma_H \cup \sigma$ is the sectionally smooth boundary of the G domain.

System (1) is considered under the following boundary conditions:

$$S_i = 0 \text{ on } \sigma \text{ if } U_n < 0,$$

$$\frac{\partial S_i}{\partial n} = 0 \text{ on } \sigma \text{ if } U_n \geq 0,$$

$$\frac{\partial S_i}{\partial z} = \varphi(S_i) \text{ on } \Sigma_0,$$

$$\frac{\partial S_i}{\partial z} = -\varepsilon_i S_i \text{ on } \Sigma_H, \tag{2}$$

where ε_i is the absorption coefficient of the i-th component by the bottom material, and φ is a given function.

The initial conditions for system (1) are the following:

$$S_i|_{t=0} = S_{i0}(x, y, z), \ i = \overline{1, 19}. \tag{3}$$

As input data for Model (1)–(3), we employed the results of calculations based on the motion model of a multicomponent air environment [3], as well as a hydrodynamic model of the Sea of Azov [4–6] that takes into account various factors, such as the influence of the wind, river flows (Don, Kuban, Mius, and about forty small watercourses), water exchange with other water bodies, the bottom relief, the complex coastline, the friction with the bottom, temperature, salinity, evaporation and precipitation, and the Coriolis force.

Observation models were added for the description of chemical-biological sources [7–9]. In this research, we considered various functional dependencies introduced in the observation models to account for the effects of salinity (C), temperature (T), and illumination (I) on the plankton productivity function (α) in the water. The functional dependencies of the observation models are shown in Fig. 4.

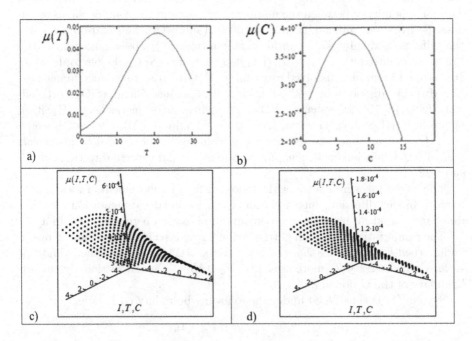

Fig. 4. Functional dependencies of observation models

In Fig. 4:

(a) $\alpha(T) = \alpha_0 \exp\left[-\{(T - T_{\text{opt}}\sigma_T)\}^2 - \mu_1 T + \mu_2\right]$, parameter values: $T_{\text{opt}} = 25$, $\sigma_T = 12$, $\alpha_0 = 0.12$, $\alpha_1 = 0.06$, $\alpha_2 = 0.43$;

(b) $\alpha(C) = \alpha_0 \exp[-\{(C-C_{opt})/\sigma_C\}^2 - \alpha_1 C + \alpha_2]$, parameter values: $C_{opt} = 12$,
  $\sigma_C = 10$, $\alpha_0 = 0.001$, $\alpha_1 = 0.1$, $\alpha_2 = 0.01$;
(c) $\alpha(I, T, C) = \alpha_0 \exp(aT)(I/I_{opt})\eta_0 \exp[-\{(C - C_{opt})/\sigma_C\}^2 - \eta_1 C + \eta_2]$,
  parameter values: $C_{opt} = 12$, $\alpha_0 = 0.8$, $a = 0.063$, $I = I_{opt} = 86$, $\sigma_C = 15$,
  $\eta_0 = 0.001$; $\eta_1 = 0.1$, $\eta_2 = 0.1$;
(d) parameter values: the same as in (c), but $I = 10$, $I_{opt} = 86$.

3   Numerical Realization of the Problem of Transport of Pollutants

Each equation of System (1)–(3) can be expressed as a diffusion-convection-reaction equation in the two-dimensional case,

$$c'_t + u c'_x + v c'_y = (\mu c'_x)'_x + (\mu c'_y)'_y + f, \tag{4}$$

with boundary conditions

$$c'_n(x, y, t) = \alpha_n c + \beta_n, \tag{5}$$

where u, v are the water velocity components; μ is the turbulent exchange coefficient; f is a function that describes the intensity and distribution of sources; and α_n, β_n are given coefficients.

3.1   Discretization of the Model

A uniform grid was defined for the numerical implementation of the discrete mathematical model [10–12]:

$$\omega_h = \{t^n = n\tau, \ x_i = i h_x, \ y_j = j h_y; \ n = \overline{0, N_t}, \ i = \overline{0, N_x}, \ j = \overline{0, N_y},$$
$$N_t \tau = T, \ N_x h_x = l_x, \ N_y h_y = l_y\}, \tag{6}$$

where τ is the time step; h_x, h_y are the spatial steps; N_t is the upper time bound; N_x, N_y are the spatial bounds; and l_x, l_y are the maximum dimensions of the computational domain.

In the case of partially filled cells, the discrete analogs of the convective, $u c'_x$, and the diffusive, $(\mu c'_x)'_x$, operators of the second order of accuracy can be written as

$$(q_0)_{i,j} u c'_x \simeq (q_1)_{i,j} u_{i+1/2,j} \frac{c_{i+1,j} - c_{i,j}}{2h_x} + (q_2)_{i,j} u_{i-1/2,j} \frac{c_{i,j} - c_{i-1,j}}{2h_x}, \tag{7}$$

$$(q_0)_{i,j} (\mu c_x) \simeq (q_1)_{i,j} \mu_{i+1/2,j} \frac{c_{i+1,j} - c_{i,j}}{h_x^2} - (q_2)_{i,j} \mu_{i-1/2,j} \frac{c_{i,j} - c_{i-1,j}}{h_x^2}$$
$$- |(q_1)_{i,j} - (q_2)_{i,j}| \mu_{i,j} \frac{\alpha_x c_{i,j} + \beta_x}{h_x}, \tag{8}$$

where $q_l, l \in \{0, 1, 2\}$ are coefficients that describe the "fullness" of control domains.

3.2   Method of Solution of the Grid Equations

Let us write the discrete analog of Model (1)–(3) in operator form [13,14]:

$$Ax = f, \tag{9}$$

where A is a non-degenerate operator defined in a real Hilbert space. We considered the implicit two-layer iterative scheme

$$B\frac{y^{k+1} - y^k}{\tau_k} + Ay^m = f, \ k = 0, 1, \ldots, \tag{10}$$

with a random initial approximation $y_0 \in H$ and B a non-degenerate operator. Any two-layer iterative method based on (10) is characterized by the operators A and B and the energy space H_D. Note that the convergence of this method has been proved and the iterative parameters were given. The main problem in the theory of iterative methods is the selection of the optimal parameter τ_k [15]. The formula for the iterative parameter τ_{k+1} of the method of minimum corrections (MMC) has the form

$$\tau_{k+1} = (A\omega_k, r_k)/(A\omega_k, A\omega_k), \ k = 0, 1, \ldots; \ r_k = Ay_k - f; \ \omega_k = B^{-1}r_k, \tag{11}$$

where r_k is the residual vector and ω_k is the correction vector [16].

4   Parallel Implementation

We describe below parallel algorithms with various types of domain decomposition for solving Problem (1)–(3).

Algorithm 1. Each processor is assigned a computational domain after the initial computational domain is partitioned in two coordinate directions (see Fig. 5). Adjacent domains overlap over two layers of nodes in the direction perpendicular to the plane of the partition [17].

   The residual vector and its uniform norm are calculated after each processor receives information for its own part of the domain [18]. Then, each processor determines the maximum module element in the residual vector and sends its value to all remaining calculators. Now, to calculate the uniform norm of the residual vector, it is enough to obtain the maximum element on each processor [19]. The parallel algorithm for calculating the correction vector has the form

$$(D + \omega_m R_1)D^{-1}(D + \omega_m R_2)w^m = r^m, \tag{12}$$

where R_1 is a lower-triangular matrix and R_2 is an upper-triangular matrix. For calculating the correction vector, we have to solve the following two equations:

$$(D + \omega_m R_1)y^m = r^m, \quad (D + \omega_m R_2)w^m = Dy^m. \tag{13}$$

   Initially, the vector y^m is calculated and the computations start in the lower left corner. Then, the correction vector w^m is calculated starting in the upper right corner. The transference of elements after calculation of two layers by the first processor is shown in Fig. 6. Only the first processor does not require additional information and can independently work with its part of the domain. Other processors wait for the results from the previous processor, while it transfers the calculated values of the grid functions at grid nodes located in the preceding positions of this line. The process continues until all layers are calculated.

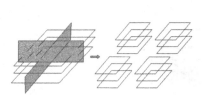

Fig. 5. Domain decomposition

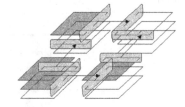

Fig. 6. Scheme of calculation of the vector y^m

Let us make a theoretical estimation of the time required by the modified alternating triangular method in the case of a system of linear algebraic equations with a seven-diagonal matrix by decomposition in two spatial directions on a cluster of distributed computations. The whole computational domain is distributed among processors (p is the total number of processors, $p = n_x \cdot n_y$, $n_x \geq n_y$), i.e. each processor is assigned a domain of size N/p, $N = N_x N_y N_z$, where N_x, N_y, and N_z are the numbers of nodes in the respective spatial directions; t_0 is the time required for the execution of one arithmetic operation; t_x is the latency; t_p is the time for transferring floating point numbers. Thus, we obtain the following theoretical estimates for the acceleration $S_{(1)}$ and the efficiency $E_{(1)}$ of parallel Algorithm 1:

$$S^t_{(1)} = \frac{p}{1 + \left(\sqrt{p} - 1\right)\left(\frac{36}{50N_z} + \frac{4p}{50t_0}\left(t_p\left(\frac{1}{N_x} + \frac{1}{N_y}\right) + \frac{t_x\sqrt{p}}{N_x N_y}\right)\right)}, \quad E^t_{(1)} = \frac{S^t_{(1)}}{p}.$$

Algorithm 2. The *k-means* method is used for the geometric partition of the computational domain for uniform loading of MCS calculators (processors) (see Fig. 7). This method is based on the minimization of the functional $Q = Q^{(3)}$ of the total variance of the element scatter (nodes of the computational grid) relative to the gravity center of subdomains,

$$Q^{(3)} = \sum_i \frac{1}{|X_i|} \sum_{x \in X_i} d^2(x, c_i) \to \min,$$

where

$$c_i = \frac{1}{|X_i|} \sum_{x \in X_i} x$$

is the center of the subdomain X_i and $d(x, c_i)$ is the distance between the calculated node and the center of the grid subdomain in the Euclidean metric. The k-means method converges only when all subdomains are approximately equal. All points on the boundary of each subdomain are required for data exchange during the computational process. We used Jarvis's algorithm for this purpose (to construct the convex hull). In addition, we made up a list of neighboring subdomains for each subdomain and created an algorithm for data transfer between subdomains.

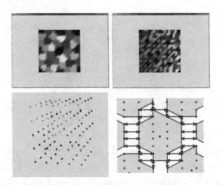

Fig. 7. Results of the k-means method for model domain decomposition into 38, 150 (for a two-dimensional domain), and 6 (for a three-dimensional domain) subdomains. Arrows indicate exchanges between subdomains.

The theoretical estimates for the acceleration and the efficiency of Algorithm 2 are the following:

$$S_{(2)}^t = \frac{p \cdot \chi}{1 + \left(\sqrt{p} - 1\right)\left(\frac{36}{50 N_x} + \frac{4p}{50 t_0}\left(t_p\left(\frac{1}{N_x} + \frac{1}{N_y}\right) + \frac{t_x \sqrt{p}}{N_x N_y}\right)\right)}, \quad E_{(2)}^t = \frac{S_{(2)}^t}{p},$$

where χ is the ratio of the number of computational nodes to the total number of nodes (computational and fictitious).

5   Results of the Experimental Studies

Parallel algorithms for the adaptive alternating-triangular method were implemented on a multiprocessor computer system (MCS) at Southern Federal University. The peak performance of this MCS is 18.8 teraFLOPS. It has eight computational racks. The computational field of the MCS is based on the HP BladeSystem c-class infrastructure with integrated communication modules, power supply, and cooling systems. The MCS uses five hundred and twelve single-type 16-core Blade servers HP ProLiant BL685c as computational nodes, each equipped with four 4-core AMD Opteron 8356 processors (2.3 GHz and 32 GB RAM).

Thus, the total number of computing cores is 2048, and the total amount of RAM is 4 TB.

We compared the parallel implementation of Algorithms 1 and 2 for solving Problem (1)–(3). The results are summarized in Table 1.

Table 1. Comparison of acceleration and efficiency of the algorithms

p	$t_{(1)}, s$	$S_{(1)}^t$	$S_{(1)}$	$E_{(1)}^t$	$E_{(1)}$	$t_{(2)}, s$	$S_{(2)}^t$	$S_{(2)}$	$E_{(2)}^t$	$E_{(2)}$
1	7.491	1.0	1.0	1	1	6.073	1.0	1.0	1	1
2	4.152	1.654	1.804	0.827	0.902	3.121	1.181	1.946	0.59	0.973
4	2.550	3.256	2.938	0.814	0.7345	1.811	2.326	3.354	0.582	0.839
8	1.450	6.318	5.165	0.7897	0.6456	0.997	4.513	6.093	0.654	0.762
16	0.882	11.928	8.489	0.7455	0.5306	0.619	8.520	9.805	0.533	0.613
32	0.458	21.482	16.352	0.6713	0.511	0.317	15.344	19.147	0.48	0.598
64	0.266	35.955	28.184	0.5618	0.4404	0.184	25.682	33.018	0.401	0.516
128	0.172	54.618	43.668	0.4267	0.3411	0.117	39.013	51.933	0.305	0.406

In Table 1, $t_{(k)}$, $S_{(k)}$, and $E_{(k)}$ are, respectively, processing time, acceleration, and efficiency of the k-th algorithm; $S_{(k)}^t$ and $E_{(k)}^t$ are, respectively, the theoretical estimates of the acceleration and the efficiency of the k-th algorithm, $k \in \{1, 2\}$. The corresponding acceleration graphs of Algorithms 1 and 2 in the solution of (1)–(3), obtained theoretically and practically, are given in Fig. 8.

The practical estimate of Algorithm 1 acceleration (graph 3 in Fig. 8), in contrast to the practical estimate of Algorithm 2 acceleration (graph 2), does not take into account the optimization of the distribution of the whole computational process among calculators. The theoretical estimate of the acceleration (graph 4) does not take into account the ratio of the number of computational nodes to the total number of nodes, as opposed to the estimate given in Fig. 8 (graph 1).

The ideal (optimal) distribution of the computational process among calculators is given in the theoretical estimate of Algorithm 1 acceleration (graph 1). Thus, graph 1 is the upper estimate of the accelerations of the algorithms obtained practically (graphs 2 and 3), while graph 4 is the lower estimate of the acceleration of the algorithms.

As we see, the efficiency increases by 10 to 20% when using Algorithm 2 (which is based on the k-means method). Thus, both algorithms, the one based on domain decomposition in two spatial directions and the one based on the k-means method, can be effectively used for solving hydrodynamic problems, provided that the number of computational nodes is sufficiently large.

6   Description of the Program Complex

Various air quality control devices were used to determine the degree of air pollution in the Sea of Azov: GANC-4 gas analyzer, chemical cassettes (for detection

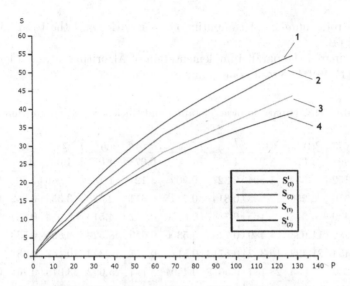

Fig. 8. Acceleration graphs of the parallel algorithms

of NH_3, H_2S, HCl, SO_2, Cl_2, NO_2, NO, soot, inorganic dust $20\% < SiO_2 < 70\%$, suspended dust), absorption devices for air sampling, sensors (for detecting traces of acrolein, gasoline, alkanes C_{12-19}, CO, methane, butyl acetate, kerosene), HPLC-chromatograph, OP431TC aspirator, UCM-1MC analyzer (for measuring the concentration of mercury), and others (see Fig. 9).

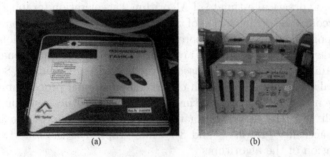

Fig. 9. (a) GANC-4 gas analyzer; (b) OP431TC aspirator

Table 2 contains data on atmospheric pollution and the presence of prevailing hazardous pollutants in Taganrog, on the basis of field measurements performed by Rospotrebnadzor (Rostov region) employees in Taganrog, taking into account the main zones of influence [21]. Figures 10 and 11 show average annual values of nitrites, dissolved oxygen, and basic heavy metals in the Taganrog Bay (Sea of Azov), according to data of the Federal Government Institution (FGI) "Azovmorinformtsentr" [22].

Table 2. Composition and concentration of pollutants in 2018

Main pollutants	Maximum Permissible Concentration (MPC)		
	One-time maximum	Daily average	Hazard class
10/1 Sedova str. (influence zone of "Krasny Kotelshchik" and "Surf" industrial enterprises), mg/m³			
1. Nitrogen dioxide: 0.2	0.085	0.04	II
2. Sulfur dioxide: 0.2	0.5	0.05	III
3. Benzene: 0.2	1.5	0.1	II
4. Benzo[a]pyrene: 0.015	–	0.01	I
Babushkina str.–Shchadenko str. (influence zone of "Tagmet" metallurgical enterprise and transport emissions), mg/l			
1. Manganese: 0.2	0.01	0.001	II
2. Nitrogen dioxide: 0.2	0.085	0.04	II
3. Sulfur dioxide: 0.2	0.5	0.05	III
Shevchenko str.–Obryvnoy lane (influence zone of "TSCP" JSC, (Taganrog Commercial Sea Port), ship repair yard)			
1. Inorganic dust SiO_2 20–70%: 0.051	0.3	0.5	III
2. Nitrogen dioxide: 0.051	0.085	0.04	II
3. Sulfur dioxide: 0.051	0.5	0.05	III
4. Carbon oxide: 0.015	0.15	0.05	IV

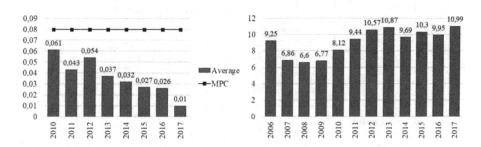

Fig. 10. Dynamics of average annual values of nitrite (a) and dissolved oxygen (b)

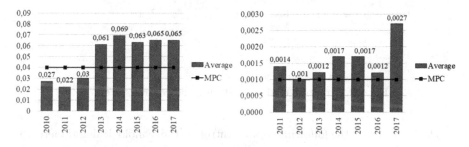

Fig. 11. Dynamics of the concentration of heavy metals in water (mg/dm³): aluminium (a), molybdenum (b)

Our research showed that the most polluted areas of the Sea of Azov are the following: the mouth of the Don (the Kuterma mouth and the Perevoloka branches), river Wet Elanchik, Mius, the estuary of the river Gross Beam (Taganrog), the eastern part of the Taganrog Bay (the ADMC turning buoy, near a sea dump), the port, Central beach, Primorsky beach, Petrushino beach, the release zone.

We developed a monitoring and water quality control system of the Sea of Azov for information support on prevention and mitigation of the effects of catastrophic phenomena and dispersion of contaminants in boundary layers of the atmosphere and bodies of shallow water. The system was implemented on a supercomputer as a software complex (SC) based on the mathematical modeling of the dispersion of pollutants in the in boundary layers of the atmosphere and water bodies. The SC involves the numerical implementation of model problems of aerodynamics, hydrodynamics, and biological kinetics (see Fig. 12).

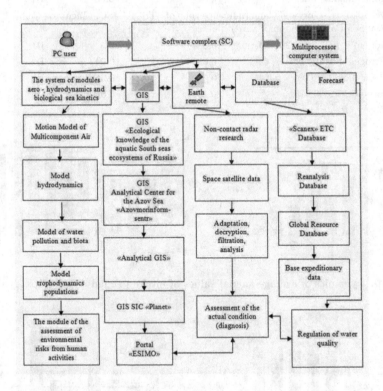

Fig. 12. Structure of the monitoring and control system

The SC includes the following units: control unit, oceanological and meteorological databases, application program library for solving hydrobiology grid problems, integration with various geoinformation systems (GIS), Global Resource Database (GRID) for geotagging and access to satellite data collection systems,

NCEP/NCAR Reanalysis database. The use of GIS provides additional possibilities for more qualitative and complex spatial analysis, and solutions based on it are more accurate. The problems were solved on a high-performance computer system [23] capable of performing a large amount of complex calculations and processing huge amounts of data in limited time. The SC has the following advantages:

- increased efficiency of implementation of mathematical models;
- development of effective methods for numerical solution of problems;
- dynamical change of input data in real time;
- GIS, database sharing with a user-friendly interface;
- integrated monitoring tools for more precise forecasts of the ecological situation in coastal systems;
- promising approaches to parallel implementation.

The hydrophysical models included in the SC were calibrated and verified using data of the ESIMO (see Fig. 13a) and the Analytical GIS portal, developed by the Institute for Information Transmission Problems of the Russian Academy of Sciences (Moscow) [24]. As input data for modeling the hydrophysical processes, we used data provided by the Scientific Research Center (SRC) "Planet" [25] (see Figs. 13b and c), the Azov Fisheries Research Institute ("AzNI-IRH") [22], the FGI "Azovmorinformtsentr" [30], and data from various papers [1, 7–9, 11, 27–29, 31–33].

We used the results of calculations based on the numerical 3D models we constructed to give a complex interpretation of remote sensing data. We also resorted to these results to establish a causal relationship between hydrodynamic and biotic processes and their manifestations on the surface, within the framework of the designed monitoring and control system. As a result, we managed to suggest some approaches to the assessment of the joint use of remote and model data as a possibility. Indeed, remote sensing data give an opportunity to assess the actual manifestation of a certain hydrodynamic or biological phenomenon; at the same time, we can use the set of models to identify and predict in detail intra-aquatic processes in shallow waters [32, 33]. The results of simulations conducted with the monitoring system are portrayed in Fig. 14. By comparing SRC "Planet" data (Fig. 15a, spots of phytoplankton showing the structure of currents) with the results of our SC (Fig. 15b, distribution and concentration of phytoplankton for a time interval of one month), we could detect a qualitative correspondence between simulation results and satellite images.

Error estimates with simultaneous consideration of field data available (n measurements) were taken as criterion of adequacy of the constructed models:

$$\delta = \sqrt{\sum_{k=1}^{n} (S_{k\,\text{nat}} - S_k)^2} \bigg/ \sqrt{\sum_{k=1}^{n} S_{k\,\text{nat}}^2},$$

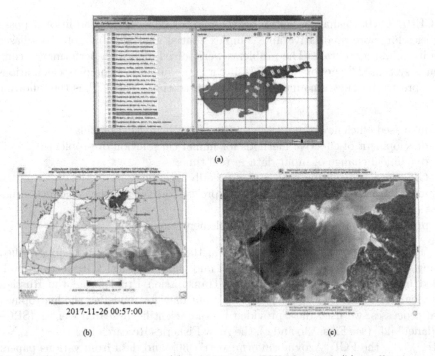

2017-11-26 00:57:00

Fig. 13. Data used for modeling: (a) data from the ESIMO portal, (b) satellite imagery of the Sea of Azov from the SRC "Planet" database, (c) map of water surface temperature of the Azov–Black Sea Basin

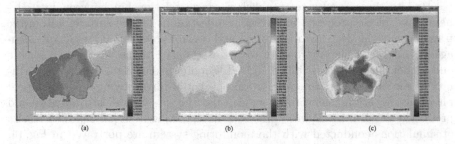

Fig. 14. Distribution and concentration of pollutants and plankton: (a) nitrite (S_7): $\mu_7 = 5 \cdot 10^{-10}, v_7 = 10^{-10}$; (b) phosphates ($S_{15}$): $\mu_{15} = 3 \cdot 10^{-10}, v_{15} = 10^{-8}$; (c) phytoplankton ($S_9$): $\mu_9 = 5 \cdot 10^{-11}, v_9 = 10^{-11}$

where $S_{k\,\mathrm{nat}}$ is the value of the function calculated using field measurements and S_k is the value of the grid function, calculated by simulation. Concentrations of pollutants and plankton calculated under different wind conditions were taken into consideration if the relative error did not exceed 30%.

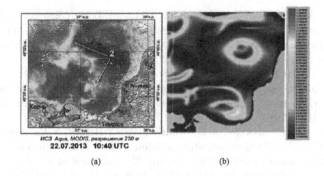

Fig. 15. Comparison of Earth remote sensing data (a—SRC "Planet") and simulation results (b—SC)

7   Conclusions

We studied in this paper stoichiometric ratios of phytoplankton nutrients. Knowing these ratios, we were able to determine a limiting substance. We also considered observation models describing production and destruction processes in "air–water" systems. In addition, we constructed a three-dimensional mathematical model of transformation of phosphorus, nitrogen, silicon, and metal forms in the problem of evolution of multi-species phytoplankton for the Sea of Azov. The numerical implementation of the model was carried out on a multiprocessor computer system with distributed memory. We obtained theoretical estimates for both the acceleration and the efficiency of the parallel algorithms. The algorithm based on the k-means method showed higher efficiency in the solution of the problem. Furthermore, after an analysis of similar SCs, the overall result is that the prediction of changes in pollutants and plankton concentrations in shallow waters obtained by using our SC is 10 to 20% more accurate, depending on the considered model problem from hydrophysics and biological kinetics.

Acknowledgments. The research was partially supported by the Russian Science Foundation (grant No. 17-11-01286).

References

1. GD.52.24.309-2016: Guidance document. Organizing and conducting the routine monitoring of the state and pollution of surface land waters, p. 104. Publishing House of the Federal State Budgetary Institution "GHI", Rostov-on-Don (2016)
2. Unified state system of information on the situation in the world ocean ("ESIMO"). http://esimo.ru/portal/auth/portal/arm-csmonitor/Settlement-Model+Complex
3. Sukhinov, A.I., Khachunts, D.S., Chistyakov, A.E.: Mathematical model of the distribution of impurities in the atmospheric surface layer of the coastal zone and its software implementation. Comput. Math. Math. Phys. **55**(7), 1238–1254 (2015)

4. Sukhinov, A.I., Tishkin, V.F., Ivanov, K.A., Lapin, D.V., Chistyakov, A.E.: Experience in modeling hydrophysical processes in the Azov Sea. In: Supercomputer Technologies in Science, Education and Industry, pp. 156–164. Publishing House of Moscow State University, Moscow (2012)

5. Sukhinov, A.I., Chistyakov, A.E., Semenyakina, A.A., Nikitina, A.V.: Numerical modeling of the ecological state of the Azov Sea with the use of higher order accuracy schemes on a multiprocessor computing system. J. Comput. Res. Model. 8(1), 151–168 (2016)

6. Sukhinov, A.I., Chistyakov, A.E., Nikitina, A.V., Belova, Y.V., Sumbaev, V.V., Semenyakina, A.A.: Supercomputer modeling of hydrochemical condition of shallow waters in summer taking into account the influence of the environment. In: Sokolinsky, L., Zymbler, M. (eds.) PCT 2018. CCIS, vol. 910. Springer, Cham (2018). https://doi.org/10.1007/978-3-319-99673-8

7. Babeshko, O.M., Evdokimova, O.V., Evdokimov, S.M.: On accounting for types of sources and zones of sedimentation of pollutants. Rep. Russian Acad. Sci. 371(1) (2000)

8. Matishov, G.G., Stepanyan, O.V., Grigorenko, K.S., Kharkivsky, V.M., Povazhny, V.V., Sawyer, V.G.: Features of hydrological and hydrochemical regime of the Azov and Black Seas in 2013. Bull. South. Sci. Center 11(2), 36–44 (2015)

9. Menshutkin, V.V.: The Art of Modeling (Ecology, Physiology, Evolution), p. 419. Petrozavodsk, St. Petersburg (2010)

10. Samarski, A.A., Nikolaev, E.S.: Methods for Solving Grid Equations, p. 532. Science, Moscow (1978)

11. Marchuk, G.I., Sarkisyan, A.S.: Mathematical Modeling of Ocean Circulation, p. 297. Science, Moscow (1988)

12. Sukhinov, A., Chistyakov, A., Isayev, A., Nikitina, A., Sumbaev, V., Semenyakina, A.: Complex of Models, High-Resolution Schemes and Programs for the Predictive Modeling of Suffocation in Shallow Waters. J. Commun. Comput. Inf. Sci. 753, 169–185 (2017)

13. Konovalov, A.N.: On the theory of alternating triangular iterative method. Sib. Math. J. 43(3), 552–572 (2002)

14. Vabishchevich, P.N., Zakharov, P.E.: Schemes of the alternating triangular method for convection-diffusion problems. J. Comput. Math. Math. Phys. 56(4), 587–604 (2016)

15. Sukhinov, A.I., Shishenya, A.V.: Improving the efficiency of the alternating triangular method based on refined spectral estimates. J. Math. Model. 24(11), 20–32 (2013)

16. Belotserkovski, O.M.: Numerical modeling in continuum mechanics, 520 p. Nauka, Moscow (1994)

17. Sukhinov, A.I., et al.: Game-theoretic regulations for control mechanisms of sustainable development for shallow water ecosystems. J. Autom. Remote Control 78(6), 1059–1071 (2017)

18. Sukhinov, A.I., Semenyakina, A.A., Chistyakov, A.E., Nikitina, A.V.: Mathematical modeling of transport of pollutants in coastal systems, 200 p. DSTUprint LLC, Rostov-on-Don (2018). ISBN 978-5-6041793-8-3

19. Sukhinov, A.I., Nikitina, A.V., Chistyakov, A.E., Semenyakina, A.A.: Practical aspects of implementation of the parallel algorithm for solving problem of ctenophore population interaction in the Azov Sea. In: Bulletin of the South Ural State University. Series "Computational Mathematics and Computer Science", vol. 7, no. 3, pp. 31–54 (2018). https://doi.org/10.14529/cmse180303

20. Sukhinov, A.I., Nikitina, A.V., Sidoryakina, V.V., Semenyakina, A.A.: Justification of the turbulent method of the formation of the turbulent method. In: Abstracts of the Talks Given at the Property 2nd International Conference on Stochastic Methods, vol. 62, no. 4, pp. 640–674. Society for Industrial and Applied Mathematics (2018). Theory Probab. Appl. c_2018, https://doi.org/10.1137/S0040585X97T988861

21. Territorial department of the Rospotrebnadzor territorial administration of Rostov region in Taganrog. http://rpndon.ru

22. FSBSI "Azov Fisheries Research Institute" ("AzNIIRKH"). http://azniirkh.ru/

23. Sukhinov, A.I., Chistyakov, A.E., Levin, I.I., Semenov, I.S., Nikitina, A.V., Semenyakina, A.A.: Solution of the problem of biological rehabilitation of shallow waters on multiprocessor computer system. In: 5th International Conference on Informatics, Electronics and Vision (ICIEV), pp. 1128–1133 (2016)

24. Analytical GIS. http://geo.iitp.ru/index.php

25. SRC "Planet". http://planet.iitp.ru/index1.html

26. Gushchin, V.A., Sukhinov, A.I., Nikitina, A.V., Chistyakov, A.E., Semenyakina, A.A.: A model of transport and transformation of biogenic elements in the coastal system and its numerical implementation. J. Comput. Math. Math. Phys. **58**(8), 1316–1333 (2018)

27. Winberg, G.G.: Diversity and unity of life phenomena and quantitative methods in biology. J. Total Biol. **61**(5), 549–560 (2000)

28. Abakumov, A.I.: Signs of stability of aquatic ecosystems in mathematical models. In: Proceedings of the Institute for System Analysis of the Russian Academy of Sciences. System Analysis of the Problem of Sustainable Development, vol. 54, pp. 49–60. ISA RAS, Moscow (2010)

29. Gause, G.F.: Verification experimentales de la theorie mathematique de la lutte pour la vie, 87 p. (1935)

30. FSI "Information and analytical center for water use and monitoring of the Azov Sea" ("Azovinformcenter"). http://azovinform.ru/

31. Nikitina, A., Kravchenko, L., Semenov, I., Belova, Y., Semenyakina, A.: Modeling of the production and the supercomputer. In: MATEC Web of Conferences, vol. 226. DTS-2018 (2018). https://doi.org/10.1051/matecconf/201822604025

32. Saminsky, G.A., Debolskaya, E.I., Kuznetsov, I.S., Yakushev, E.V.: Mathematical model of anaerobic contamination process in waters. J. Water Chem. Ecol. **12**(42), 8–17 (2011)

33. Nikitina, A.V., Sukhinov, A.I., Ugolnitsky, G.A., Usov, A.B., Chistyakov, A.E., Puchkin, M.V., Semenov, I.S.: Optimal management of sustainable development in the biological rehabilitation of the Azov Sea. J. Math. Model. **28**(7), 96–106 (2016)

Internal Parallelism of Multi-objective Optimization and Optimal Control Based on a Compact Kinetic Model for the Catalytic Reaction of Dimethyl Carbonate with Alcohols

Kamila F. Koledina[1,2(⊠)], Sergey N. Koledin[2], Liana F. Nurislamova[2], and Irek M. Gubaydullin[1,2]

[1] Institute of Petrochemistry and Catalysis of the UFRC RAS, Ufa, Russia
koledinakamila@mail.ru, irekmars@mail.ru
[2] Ufa State Petroleum Technological University, Ufa, Russia
koledinsrg@gmail.com, nurislamovalf@mail.ru

Abstract. We develop and implement an algorithm for identifying the optimal conditions of catalytic reactions. The algorithm comprises: (i) a mathematical model of the chemical reaction, (ii) an analysis of the sensitivity of kinetic parameters, (iii) the construction of a compact kinetic model, (iv) the identification of optimization criteria for reaction conditions, (v) the determination of variable parameters, and (vi) the setup and solution of multi-objective optimization and optimal control problems. We use the constructed algorithm to model and optimize the catalytic reaction of dimethyl carbonate with alcohols. A compact kinetic model of this reaction is then applied to establish some optimization criteria, such as product yield and profitability criterion. We pose and solve the problems of multi-objective optimization and optimal control for the reaction conditions. For solving the optimal control problem, we suggest reducing the optimization procedure to a nonlinear programming problem. Finally, we determine the optimal conditions for attaining the specified criteria.

Keywords: Optimization · Multi-objective optimization · Optimal control · Dimethyl carbonate · Parameter identification · Kinetic modeling

1 Introduction

The variety of chemical, petrochemical, and oil refinery products constantly grows as their quality becomes better and better. Given the current engineering level in the chemical industry, the development of new catalytic processes

The reported study was funded by the Russian Foundation for Basic Research (projects No. 18-07-00341, 18-37-00015 (paragraph 5)) and by a scholarship awarded by the President of the Russian Federation (SP-669.2018.5).

L. Sokolinsky and M. Zymbler (Eds.): PCT 2019, CCIS 1063, pp. 242–255, 2019.
https://doi.org/10.1007/978-3-030-28163-2_17

and the upgrade of existing ones should require less time. A major problem for the optimal control of the whole engineering system is that of determining the true mechanisms that govern chemical processes. A key phase in the mathematical modeling of chemical engineering processes is the development of adequate kinetic models (KM) for chemical processes [1,2]. A kinetic model comprises kinetic equations, patterns of variation of concentrations over time, and kinetic parameters. Fast, moderate, and slow reactions occur simultaneously in multistep catalytic processes. Kinetic parameters affect the course of a process to different extents. It is necessary to determine the influence exerted by kinetic parameters on the course of a reaction at any point in time. This can help to identify the optimal conditions for complex multistep chemical reactions. For this reason, the efficient use of KMs in studies based on mathematical modeling is associated with the analysis of the sensitivity to kinetic parameters and the development of compact kinetic models which can help in identifying optimal reaction conditions that comply with specified optimization criteria [3,4].

Setting up an optimization problem implies the presence of criteria, variable parameters, and some constraints thereon [5,6]. A problem including several mutually independent optimization criteria is called a multi-objective optimization problem. The study of dynamic (time-dependent) variable parameters gives rise to an optimal control problem. Thus, the determination of the optimal conditions for complex catalytic reactions may lead to single-objective optimization problems, multi-objective optimization problems, and optimal control problems.

2   Algorithm for Determining the Conditions of Catalytic Reactions

The diagram in Fig. 1 describes the interaction between the problems considered in this paper. At the first stage, we develop a mathematical model to describe the process and analyze the sensitivity of the model to kinetic parameters. In this manner, we can identify both the most and the least significant parameters and construct a compact kinetic model for the process. Based on the analysis results, we determine insignificant parameters, such as rate constants for slow and fast steps, which need not be accurately specified. If there are more than one independent optimization criterion, we solve a multi-objective optimization problem. If the process conditions can change over time, we solve an optimal control problem.

3   The Object of Investigation: The Catalytic Reaction of Dimethyl Carbonate with Alcohols

We consider the homogeneous liquid-phase catalytic reaction of dimethyl carbonate (DMC) with alcohols in the presence of a transition metal catalyst, namely $Co_2(CO)_8$. This is a new reaction pertaining to green chemistry. DMC is less reactive than other widely used methylating agents, such as methyl halides

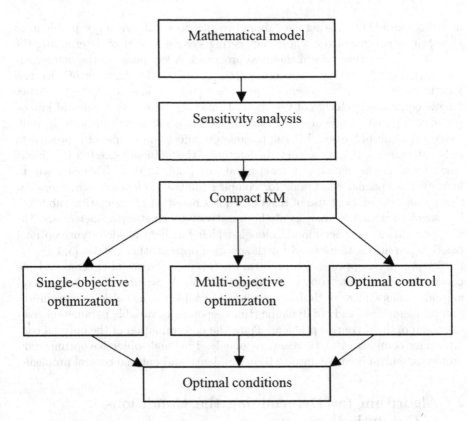

Fig. 1. Stages of optimization of the conditions of a catalytic reaction

(MeX; X = I, Br, Cl) and dimethyl sulfate, and is also less toxic. DMC is preferred for the methylation of alcohols, acids, and amines [7,8]. Nevertheless, for the reaction to proceed successfully, it is necessary to use a large excess of DMC (1 : 40) and conditions must be severe (above 200 °C). This significantly hinders the extensive use of DMC as a methylating agent in synthetic practice. On the other hand, it has been shown [7] that the reactions of alcohols with DMC in the presence of W^-, Mn^-, V^-, and Co^- containing transition metal catalysts yield the expected ethers and esters under milder conditions and with a high selectivity. It was proved that the $Co_2(CO)_8$ complex is the most active transition metal catalyst. The reactions give alkyl methyl ethers and alkyl methyl carbonates. Alkyl methyl ethers are used as engine oil additives and also as intermediates in the preparation of drugs, dyes, and flavors. Carbonates are used as organic solvents, protective groups for alcohols and phenols in organic synthesis, electrolyte solvents in lithium-ion batteries, and monomers for phosgene-free synthesis of polyurethanes and polycarbonates.

It is necessary to develop a kinetic model for the new, industrially valuable, catalytic reaction of DMC with alcohols and apply this model in the identification of the optimal conditions for the reaction.

4   Compact Kinetic Model of the Catalytic Reaction of Dimethyl Carbonate with Alcohols

The mathematical model used in problems of chemical kinetics is a nonlinear system of ordinary differential equations with initial conditions, that is, a Cauchy problem (see [9–13]),

$$\frac{dx_i}{dt} = \sum_{j=1}^{J} \nu_{ij} w_j, \ i = 1, \ldots, I,$$

$$w_j = k_j \prod_{i=1}^{I} x_i^{|\alpha_{ij}|} - k_{-j} \prod_{i=1}^{I} x_i^{\beta_{ij}} \quad (1)$$

$$k_j = k_j^0 \exp\left(-\frac{E_j}{RT}\right),$$

$$t \in [0, t^*],$$

with initial conditions $t = 0, x_i(0) = x_i^0$. Here, ν_{ij} are stoichiometric coefficients; J is the number of steps; x_i are the concentrations of substances that participate in the reaction, [mol/l]; I is the number of substances; w_j is the rate in the j-th step, [1/min]; E_j^+, E_j^- are the activation energies of the forward and backward reactions, [kJ/mol]; R is the universal gas constant, equal to 8.31 J/(mol·K); T is the temperature, [K]; α_{ij} are the negative elements of the matrix (ν_{ij}); β_{ij} are the positive elements of the same matrix; k_j^0, k_{-j}^0 are pre-exponential factors; k_j is the rate constant in the j-th step; [1/min]; and t^* is the reaction time, [min].

Table 1. Kinetic parameters of the reaction of DMC with alcohols catalyzed by $Co_2(CO)_8$

N	Stages	ln k_j^0, 1/(mol · min)	E_j, kJ/mol
k_1	$Co_2(CO)_8 + ROH \rightarrow Co_2^+(ROH) + Co(CO)_4^- + 4CO$	22.00 ± 0.01	100.9 ± 0.5
k_2	$Co(CO)_4^- + (MeO)_2CO \rightarrow Me^+[Co(CO)_4^-] + CO_2 + MeO^-$	11.9 ± 0.8	33.9 ± 3.3
k_3	$Co(CO)_4^- + (MeO)_2CO \rightarrow Co(CO)_4CO_2Me + MeO^-$	20.70 ± 0.01	56.1 ± 0.4
k_4	$Co(CO)_4CO_2Me + ROH \rightarrow HCo(CO)_4 + CO_2 + ROMe$	18.20 ± 0.01	62.8 ± 0.5
k_5	$Co(CO)_4CO_2Me + ROH \rightarrow ROCO_2Me + HCo(CO)_4$	12.0 ± 0.1	23.4 ± 0.5
k_6	$Me^+[Co(CO)_4^-] + ROH \rightarrow ROMe + HCo(CO)_4$	20.1 ± 0.4	90.0 ± 6.3
k_7	$HCo(CO)_4 + MeO^- \rightarrow MeOH + Co(CO)_4^-$	15.90 ± 0.01	44.4 ± 0.1
k_8	$Co(CO)_4CO_2Me + MeO^- \rightarrow Co(CO)_4^- + (MeO)_2CO$	10.90 ± 0.01	37.7 ± 0.4

A mathematical model for this reaction was developed previously (see [14, 15]). We designed the kinetic model considering the proposed reaction scheme.

The pre-exponential factors and the activation energies for single steps are summarized in Table 1.

The uncertainty in the kinetic parameters (see Table 1) is due to the fact that the solution of the inverse kinetic problem is non-unique. The parameters were determined by processing a large number of experimental data obtained at different temperatures and for various initial amounts of the catalyst [7,14,15]. The resulting kinetic parameters vary within the indicated range.

The next stage of the analysis of the reaction mathematical model is concerned with the sensitivity of the kinetic curves to the rate constants, so as to assess the effect of the parameters on the model [16,17]. The study is based on the theory of local sensitivity coefficients applied to mathematical models. The sensitivity analysis determines how the investigated function depends on the variation of parameters as well as on significant and insignificant parameters of the model [18–20].

The local sensitivity coefficients are the partial derivatives of the component concentrations with respect to the rate constants, i.e.

$$S_{ij} = \frac{\partial x_i}{\partial k_j}, \tag{2}$$

where S_{ij} are first-order local sensitivity coefficients.

Sensitivity coefficients may be positive, negative, or zero. A positive (negative) sensitivity coefficient means that, with the assumed values of the rate constants, an increase in a particular rate constant leads to an increasing (decreasing) concentration x_i at a given point in time. If the criterion is equal to zero at time t, then a minor change in k_j does not lead to any changes in x_i at this time point.

Throughout the reaction, the components, including alcohol (ROH), alkyl methyl ether (ROMe), and alkyl methyl carbonate (ROCO$_2$Me), can be monitored experimentally. Figure 2 shows the sensitivity coefficients in particular steps for the experimentally observed compounds. Varying the rate constants k_2 (for the reaction step $Co(CO)_4^- + (MeO)_2CO \rightarrow Me^+[Co(CO)_4^-] + CO_2 + MeO^-$) and k_3 (for the reaction step $Co(CO)_4^- + (MeO)_2CO \rightarrow Co(CO)_4CO_2Me + MeO^-$) in the reaction of the catalytically active species with the initial DMC considerably affects virtually all compounds involved in the reaction. It is interesting that these steps have mutually inverse effects. The coefficients of these steps are equal in magnitude. According to calculations, the variation of the parameter k_8 (for the step $Co(CO)_4CO_2Me + MeO^- \rightarrow Co(CO)_4^- + (MeO)_2CO$) has a minor effect on the kinetic curves. The exclusion of this step from the KM does not affect the description for any observed compound. Thus, we can exclude this step from the KM. The effect of other parameters on the formation/consumption of a substance or a group of substances demonstrates that the scheme cannot be reduced further.

The compact KM constructed for the reaction consisting of seven chemical steps (k_1–k_7) forms the basis for further optimization of the conditions.

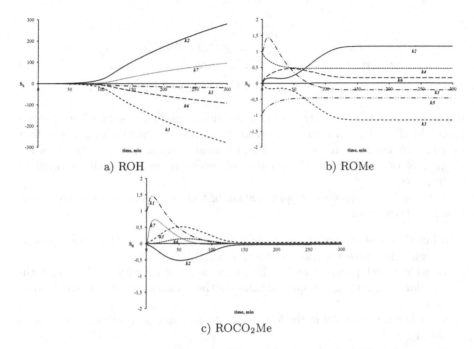

a) ROH

b) ROMe

c) ROCO$_2$Me

Fig. 2. Variation of the sensitivity coefficients for the rate constants of particular steps with respect to the observed compounds

5   Optimization of the Conditions for the Catalytic Reaction

In the general case, optimization criteria based on the KM have the form (see [21–24])

$$Z(x, x^0, t^*, \eta, \mu, T) \rightarrow \max, \tag{3}$$

where η is the vector of specific cost weights of the components, and μ denotes additional expenses.

Both physico-chemical and economic characteristics can be used as optimization criteria for catalytic reactions. We will consider the following criteria based on (3) for the KM-based optimization of the reaction conditions:

(1) Product yield:

$$Z = x_{\mathrm{prod}}(t^*, T, x^0) \rightarrow \max, \tag{4}$$

where x_{prod} is the concentration of reaction products, [mol/l].

(2) Profitability criterion, equal to the sum of incomes divided by capital investment:

$$Z = P = \frac{\sum\limits_{\text{prod}=1}^{\text{Pr}} x_{\text{prod}}(t^*, T, x^0)\eta_{\text{prod}}}{\sum\limits_{\text{source}=1}^{\text{Sr}} x_{\text{source}}(t^*, T, x^0)\eta_{\text{source}} + \psi(t^*, T) + A}, \tag{5}$$

where x_{source} is the concentration of reactants, [mol/l]; ψ are variable expenses (normalized to the sum of component costs and expenses); A are constant expenses (normalized to the sum of component costs and expenses); Pr is the number of products; Sr is the number of reactants; and P is the normalized profitability.

Generally, the problem of optimization of the conditions of a catalytic reaction has the form of:

- a functional of the optimization criteria, $F = (f_1, f_2, \ldots, f_L)$, where f_l is an optimization criterion (one or more), $l = 1, 2, \ldots, L$;
- variable (free) parameters U: the initial amount of catalyst (u_1) and the reaction time $t^*(u_2)$; the optimal values of the variable parameters are denoted by U^*;
- a mathematical model in the form of nonlinear system of ordinary differential Eq. (1);
- direct constraints on the variable parameters (D_U),

$$\max_{U \in D_U} F(U) = F(U^*) = F^*. \tag{6}$$

6   Multi-objective Optimization of Catalytic Reaction Conditions

A solution of a multi-objective optimization problem is a compromise between several independent parameters. It is also called a Pareto solution or Pareto approximation. Solving a multi-objective optimization problem in the domain of optimization criteria is called a Pareto front. Solving the problem in the domain of variable parameters is called a Pareto set.

The numerical determination of the Pareto set and the Pareto front can be described as follows [21]. Let $U = (u_1, u_2, \ldots, u_{|U|})$ be the vector of variable parameters. Let D_U be the set of admissible values of the vector U (see Fig. 3a). Then $F(U) = (f_1(U), f_2(U), \ldots, f_{|F|}(U))$ is the vector of optimization criteria; $F(U)$ maps the set D_U to some set D_F which is called the attainability domain. It is possible to select from D_F the subset D_{F*} of points that are not dominated by other points (see Fig. 3b). The set D_{F*} is called the Pareto front. The subset $D_U^* \subset D_U$ of points of variable parameters that corresponds to the set D_{F*} is called the Pareto set (see Fig. 3a).

Problems with a low-dimensional variable parameter domain can be solved by a grid algorithm. In this case, the set D_U is covered with some grid with the required accuracy of measurements. Then the objective functions are calculated

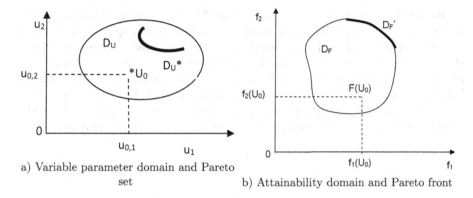

a) Variable parameter domain and Pareto
set

b) Attainability domain and Pareto front

Fig. 3. Multi-objective optimization of the catalytic reaction conditions

at each grid point [21]. The grid points can also be generated randomly. A benefit of this algorithm is an efficient parallelization of the solution.

However, problems with high-dimensional variable parameter domains are encountered more often in practice. The application of the grid algorithm results in large time consumption. The most efficient way to solve the Pareto approximation problem based on ranking is based on the Non-Dominated Sorting Genetic Algorithm (NSGA) [25]. The first non-dominated points (front) are assigned some values of the optimization criteria. Then these points are decomposed into the given values. After each stage, each of the current non-dominated points is assigned a value lower than the minimum decomposed value of the optimization criteria. Further development of multi-objective optimization algorithms was directed towards upgrading the algorithms developed previously. This resulted in the appearance of various algorithms: SPEA2 [26], PESA-II [27], and Non-Dominated Sorting Genetic Algorithm II (NSGA-II) [28,29]. Among these, NSGA-II proved to be the most accurate in determining the Pareto-dominated points, although some drawbacks appear in this approach as the number of criteria increases.

An obvious benefit of NSGA-II is that the algorithm has been implemented in a variety of programming environments, in particular, in Matlab (MATrixLABoratore). Matlab includes packages needed to solve diverse problems that require mathematical calculations and modeling. The multi-objective optimization application can be used for problems of chemical kinetics. It is especially reasonable to optimize the conditions during the analysis of new chemical reactions.

The optimal conditions of the cobalt-carbonyl-catalyzed reaction of DMC with alcohols were found by solving a multi-objective optimization problem. This was done by using the Pareto approximation grid algorithm with determination of the dominated points and application of NSGA-II.

The solution of problems by a genetic algorithm for each population is the solution of a direct problem, i.e. a system of differential equations for the substances in a reaction. The computation of a genetic algorithm may take a few minutes to several hours for complex reactions and requires considerable computational resources. In the calculations, we considered a population of size 100 and 30 generations. The solution of the problem was carried out on an IBM PC-compatible computer using the Matlab software system with the parameter UseParallel set to 'yes'. Figure 4 offers a comparison of the speedup obtained by using a parallel algorithm for p processors against the speedup for serial computations. The speedup drops as the number of substances increases.

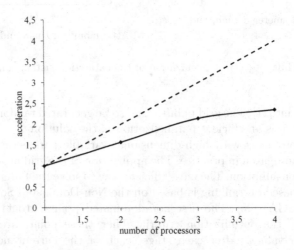

Fig. 4. Speedup of the parallel program for NSGA-II (solid curve: actual; dashed line: theoretically expected)

Figure 5 depicts the effectiveness of using parallel processors for the solution of the problem. As we can see when comparing the graphs for different chemical systems, the effectiveness decreases as the dimension of the task increases. The reduction in effectiveness is apparently connected with an increase in synchronization time between processors.

The Pareto front and Pareto set approximations obtained for optimization criteria (4) and (5) are given in Figs. 6a and b, respectively.

It follows from Figs. 6a and b that the Pareto front and the Pareto set constructed by means of the grid algorithm show a satisfactory agreement with those constructed using NSGA-II. The white symbols in Figs. 6a and b stand for the variable parameters and the corresponding objective functions in real experiments (the triangle corresponds to the Pareto front, the rhombus corresponds to the Pareto set). If the experimental conditions were not taken from the Pareto set, then the values for the objective functions were below the Pareto front. This confirms the adequacy of the determined compromise points of the Pareto approximation.

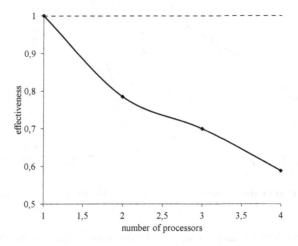

Fig. 5. Effectiveness of parallel program execution for NSGA-II (solid curve: actual; dashed line: theoretically expected)

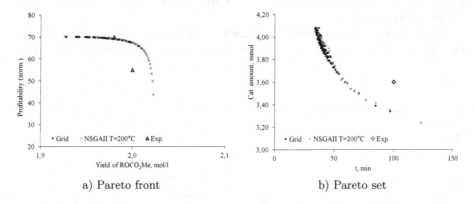

a) Pareto front   b) Pareto set

Fig. 6. Pareto front and Pareto set approximations for the MOO problem associated with the reaction of DMC with alcohols. The approximations were constructed by means of a grid algorithm and NSGA-II.

7   Optimal Control of Catalytic Reaction Conditions

By optimal control of catalytic reaction conditions, we mean changes in conditions throughout the reaction so as to attain a specified criteria. It has been suggested to solve the optimal control problem by reducing it to a nonlinear programming problem [21]. As a control over $u(t)$, we consider the temperature $T(t)$. The interval $[0, t^*]$ is covered by a grid of points t_i, $i \in [0, |U|]$. The optimal control over $u^*(t)$ is defined as a piecewise-constant function. We obtain a nonlinear programming problem with variable parameter $u(t)$.

Figures 7a and b depict the variation of the $ROCO_2Me$ target product concentration over time for the reaction of DMC with alcohols in the presence

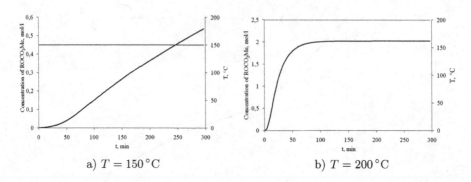

a) $T = 150\,^{\circ}\mathrm{C}$ b) $T = 200\,^{\circ}\mathrm{C}$

Fig. 7. Variation of the concentration of $ROCO_2Me$ under isothermal conditions

of octacarbonyldicobalt and under isothermal conditions, at minimum (150 °C; Fig. 7a) and maximum (200 °C; Fig. 7b) temperatures. The temperatures are plotted along the auxiliary vertical axis. The concentrations determine the possible range of variation of the product yield, from the lowest to the highest value, for 300 min of reaction.

Figure 8 portrays the variation of the $ROCO_2Me$ concentration throughout the solution of the optimal control problem under non-isothermal conditions.

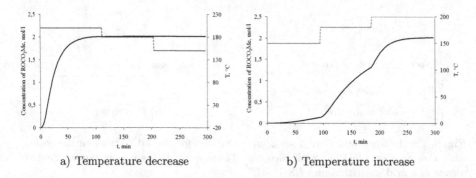

a) Temperature decrease   b) Temperature increase

Fig. 8. Variation of the concentration of $ROCO_2Me$ under non-isothermal conditions

When the reaction temperature decreases stepwise (see Fig. 8a) from 200 °C to 150 °C, the variation of the $ROCO_2Me$ target product concentration is similar to that shown in Fig. 7b. The maximum concentration is attained when the reaction temperature increases in a stepwise manner from 150 °C to 200 °C (see Fig. 8b), but this process takes more time (250 min.). Note that more energy is required to maintain a higher temperature throughout the reaction (Fig. 7b), although the same result is attained under non-isothermal conditions (see Fig. 8a), which requires less energy.

The average temperature for the non-isothermal reaction is 180 °C. For the sake of comparison, let us perform the calculation for the isothermal reaction at $T = 180\,°C$. The results shown in Fig. 9 indicate a change in the concentration of the $ROCO_2Me$ target product, but the concentration maximum is still not attained in 300 min. While the energy consumption is the same as under isothermal conditions, the target-product concentration is lower.

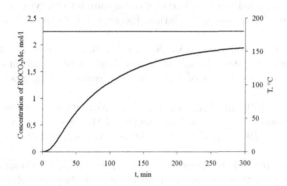

Fig. 9. Variation of the concentration of $ROCO_2Me$ under isothermal conditions, $T = 180\,°C$

Thus, the calculations for isothermal and non-isothermal reactions demonstrate that the conditions of choice suppose a reduction of the temperature from the maximum to the minimum value during the reaction (see Fig. 8a). This is due to the fact that the formation of catalytically active species in the first step (see Table 1) has the highest activation barrier [9,10]. An elevated temperature is needed to overcome the barrier and decrease the induction period.

8   Conclusions

We described in this paper an algorithm for determining the optimal conditions for a catalytic reaction. Moreover, we used the algorithm to construct a mathematical model of complex catalytic reactions of DMC with alcohols. Also, we analyzed the sensitivity of kinetic parameters for particular reaction steps. The compact kinetic model constructed for the reaction of DMC with alcohols was used to design various optimization criteria. In addition, we set up and solved the problems of multi-objective optimization and optimal control of the reaction conditions. We demonstrated the effectiveness of using parallel processors for the solution of these problems.

References

1. Slinko, M.: History of the development of mathematical modeling of catalytic processes and reactors. Theor. Found. Chem. Technol. **41**(1), 16–34 (2007)
2. Boyarinov, A., Kafarov, V.: Methods of optimization in chemical technology. Chemistry, Moscow (1975)

3. Nurislamova, L.F., Gubaydullin, I.M.: Mechanism reduction of chemical reaction based on sensitivity analysis: development and testing of some new procedure. J. Math. Chem. **55**(9), 1779–1792 (2017). https://doi.org/10.1007/s10910-017-0760-x

4. Koledin, S.N., Karpenko, A.P., Koledina, K.F., Gubaydullin, I.M.: Information system for the evaluation of the relationship between objective functions and the study of optimal conditions for carrying out a complex catalytic reaction using the methods of multi-purpose optimization. Electr. Inf. Complexes Syst. **13**(4), 71–81 (2017)

5. Gubaydullin, I.M., Enikeeva, L.V., Naik, L.R.: Software module of mathematical chemistry web-laboratory for studying the kinetics of oxidation of 4-tert-butyl-phenol by aqueous solution of H_2O_2 in the presence of titanosilicates. Eng. J. **20**(5), 263–270 (2016)

6. Koledina, K.F., Koledin, S.N., Gubaydullin, I.M.: Optimization of chemical reactions by economic criteria based on kinetics of the process. In: CEUR Workshop Proceedings, vol. 1966, pp. 5–9 (2017). https://doi.org/10.18287/1613-0073-2017-1966-5-9

7. Khusnutdinov, R.I., Shchedneva, N.A., Mayakova, Y.Y.: Synthesis of alkyl methyl ethers and alkyl methyl carbonates by reaction of alcohols with dimethyl carbonate in the presence of tungsten and cobalt complexes. Russ. J. Org. Chem. **50**(6), 790–795 (2014). https://doi.org/10.1134/S1070428014060050

8. Arico, F., Tundo, P.: Dimethyl carbonate: a modern green reagent and solvent. Russ. Chem. Rev. **79**(6), 479 (2010)

9. Gubaydullin, I.M., Koledina, K.F., Sayfullina, L.V.: Mathematical modeling of induction period of the Olefins Hydroalumination reaction by Diisobutylalumini-umchloride Catalyzed with Cp_2ZrCl_2. Eng. J. **18**(1), 13–24 (2014). https://doi.org/10.4186/ej.2014.18.1.13

10. Nurislamova, L.F., Gubaydullin, I.M., Koledina, K.F.: Kinetic model of isolated reactions of the catalytic hydroalumination of olefins. React. Kinet. Mech. Catal. **116**(1), 79–93 (2015). https://doi.org/10.1007/s11144-015-0876-6

11. Nurislamova, L.F., Gubaydullin, I.M., Koledina, K.F., Safin, R.R.: Kinetic model of the catalytic hydroalumination of olefins with organoaluminum compounds. React. Kinet. Mech. Catal. **117**(1), 1–14 (2016). https://doi.org/10.1007/s11144-015-0927-z

12. Koledina, K.F., Gubaidullin, I.M.: Kinetics and mechanism of olefin catalytic hydroalumination by organoaluminum compounds. Russ. J. Phys. Chem. A **90**(5), 914–921 (2016). https://doi.org/10.1134/S0036024416050186

13. Zainullin, R.Z., Koledina, K.F., Akhmetov, A.F., Gubaidullin, I.M.: Kinetics of the catalytic reforming of gasoline. Kinet. Catal. **58**(3), 279–289 (2017). https://doi.org/10.1134/S0023158417030132

14. Koledina, K.F., Koledin, S.N., Shchadneva, N.A., Gubaidullin, I.M.: Kinetics and mechanism of the catalytic reaction between alcohols and dimethyl carbonate. Russ. J. Phys. Chem. A **91**(3), 444–449 (2017). https://doi.org/10.1134/S003602441703013X

15. Koledina, K.F., Koledin, S.N., Schadneva, N.A., Mayakova, Y.Y., Gubaydullin, I.M.: Kinetic model of the catalytic reaction of dimethylcarbonate with alcohols in the presence $Co_2(CO)_8$ and $W(CO)_6$. React. Kinet. Mech. Catal. **121**(2), 425–428 (2017). https://doi.org/10.1007/s11144-017-1181-3

16. Turany, T.: Sensitivity analysis of complex kinetic systems. Tools and applications. J. Math. Chem. **5**(3), 203–248 (1990)

17. Saltelli, A., Ratto, M., Tarantola, S., Campolongo, F.: Sensitivity analysis practices: strategies for model-based inference. Reliab. Eng. Syst. Saf. **91**(10–11), 1109–1125 (2006)
18. Polak, L.S. (ed.): Application of Computational Mathematics in Chemical and Physical Kinetics. Science, Moscow (1969)
19. Polak, L.S., Goldenber, M.Y., Levitsky, A.A.: Computational Methods in Chemical Kinetics. Science, Moscow (1985)
20. Saltelli, A., Ratto, M., Tarantola, S., Campolong, F.: Sensitivity analysis for chemical models. Chem. Rev. **205**(7), 2811–2828 (2005)
21. Sakharov, M., Karpenko, A.: A new way of decomposing search domain in a global optimization problem. In: Abraham, A., Kovalev, S., Tarassov, V., Snasel, V., Vasileva, M., Sukhanov, A. (eds.) IITI 2017. AISC, vol. 679, pp. 398–407. Springer, Cham (2018). https://doi.org/10.1007/978-3-319-68321-8_41
22. Karpenko, A.P., Mukhlisullina, D.T., Ovchinnikov, V.A.: Multicriteria optimization based on neural network approximation of decision maker's utility function. Opt. Mem. Neural Netw. (Inf. Opt.) **19**(3), 227–236 (2010). https://doi.org/10.3103/S1060992X10030045
23. Vovdenko, M.K., Gubaidulin, I.M., Koledina, K.F., Koledin, S.N.: Isopropylbenzene oxidation reaction computer simulation. In: CEUR Workshop Proceedings, vol. 1966, pp. 20–23 (2017). https://doi.org/10.18287/1613-0073-2017-1966-20-23
24. Baynazarova, N.M., Koledina, K.F., Pichugina, D.A.: Parallelization of calculation the kinetic model of selective hydrogenation of acetylene on a gold clusters. In: CEUR Workshop Proceedings, vol. 1576, pp. 425–431 (2016)
25. Srinivas, N., Deb, K.: Muiltiobjective optimization using nondominated sorting in genetic algorithms. Evol. Comput. **2**(3), 221–248 (1994)
26. Zitzler, E., Laumanns, M., Thiele, L.: SPEA2: improving the strength pareto evolutionary algorithm for multiobjective optimization. In: Evolutionary Methods for Design Optimisation and Control with Application to Industrial Problems, EUROGEN 2001, vol. 3242, no. 103, pp. 95–100 (2002)
27. Corne, D., Jerram, N., Knowles, J., Oates, M.: PESA-II: region-based selection in evolutionary multiobjective optimization. In: Proceedings of the Genetic and Evolutionary Computation Conference, GECCO 2001, pp. 283–290 (2001)
28. Deb, K., Mohan, M., Mishra, S.: Towards a quick computation of well-spread pareto-optimal solutions. In: Fonseca, C.M., Fleming, P.J., Zitzler, E., Thiele, L., Deb, K. (eds.) EMO 2003. LNCS, vol. 2632, pp. 222–236. Springer, Heidelberg (2003). https://doi.org/10.1007/3-540-36970-8_16
29. Kalyanmoy, D., Pratap, A., Agarwal, S., Meyarivan, T.: A fast and elitist multiobjective genetic algorithm: NSGA-II. IEEE Trans. Evol. Comput. **6**(2), 182–197 (2001)

Parallel Numerical Solution
of the Suspension Transport Problem
on the Basis of Explicit-Implicit Schemes

Aleksandr I. Sukhinov[1], Aleksandr E. Chistyakov[1],
Valentina V. Sidoryakina[2(✉)], and Elena A. Protsenko[2]

[1] Don State Technical University, Rostov-on-Don, Russia
sukhinov@gmail.com, cheese_05@mail.ru
[2] A. P. Chekhov University of Taganrog
(Branch of Rostov State University of Economics), Taganrog, Russia
cvv9@mail.ru, eapros@rambler.ru

Abstract. We discuss the construction and study of parallel algorithms for the numerical implementation of a 3D model of suspension transport in coastal marine systems. The corresponding parallel numerical modeling can be helpful in predicting spread of pollutants, as well as changes in the bottom topography which significantly affect the quality of the aquatic environment and the safety of navigation. The numerical method for solving the initial-boundary value problem for the transport of suspensions is based on the idea of constructing explicit-implicit difference schemes involving the approximation of the transfer operator (diffusion-convection) in horizontal directions based on explicit approximation on the previous time layer, and the implicit approximation of the diffusion-convection-gravity operator sedimentation in the vertical direction. Compared with traditional explicit schemes, the constructed scheme requires less time for exchange between processors when implemented on a multiprocessor system. To increase the stability margin and the allowable time step, we add a second-order difference in time to the explicit-implicit approximation. Using this scheme allows you to organize fully parallel computations, when a set of independent one-dimensional three-point problems obtained as a result of approximation with weights at the current and previous time layer of the one-dimensional diffusion-convection-gravity sedimentation operator in the vertical direction is solved in each processor independently of the others. The constructed scheme minimizes the amount of data exchange between adjacent processors when passing from a layer to another in parallel mode at the border nodes of subdomains in the case of a three-dimensional grid domain geometrically decomposed into vertical planes.

Keywords: Coastal zone · Mathematical model ·
Suspension transport · Difference scheme · Explicit-implicit scheme

This paper was partially supported by the Russian Science Foundation (grant No. 17-11-01286).

L. Sokolinsky and M. Zymbler (Eds.): PCT 2019, CCIS 1063, pp. 256–268, 2019.
https://doi.org/10.1007/978-3-030-28163-2_18

1   Introduction

The transport of suspended matter is the most important factor that significantly affects morphological and dynamic regimes of coastal zones in bodies of water, as well as construction and operation zones in port facilities, and others [1–4]. Under the influence of waves and currents, the mass of alluvial material begins to move, while undergoing chemical and mechanical changes. The processes arising at the same time find their expression in the emergence of benthic and coastal formations which can reach a significant size [5–7].

We present in the paper a parallel algorithm for the numerical simulation of suspension transport processes based on a scheme of special type, namely an explicit-implicit scheme. The basic idea of this scheme is to use difference equations containing an explicit approximation of the diffusion-convection operator in the horizontal directions with a regularized addition, concretely a second-order difference derivative with a small factor, as well as an implicit approximation with weights for the diffusion-convection-gravity sedimentation operator in the vertical direction. Restrictions on the time step are in this case less stringent than the use of an explicit scheme which approximates a three-dimensional problem in the traditional way by introducing a derivative of the second order [8]. This is due to the fact that in coastal systems like the Sea of Azov, the north of the Caspian Sea, and others, the typical size of a grid cell is of tens of centimeters vertically, whereas in the horizontal directions, the size is a few meters and tens of hundreds of meters, respectively. This makes it possible to choose in the suggested scheme a time step of the order of the characteristic time of propagation of disturbances in the horizontal directions within one grid cell, which is significantly longer than the time of propagation of disturbances in the vertical direction in a single grid cell. It is precisely this restriction that would have had to be oriented in approximating all the operators of a three-dimensional problem explicitly in the case of regularization. The use of such an explicit-implicit scheme allows to obtain a completely parallel algorithm for solving the problem, which boils down to solving a series of independent three-point difference problems by using the sweep method with exchanges at the border nodes, given that the three-dimensional grid domain is decomposed into a number of equal vertical blocks corresponding to the number of calculators used in the multiprocessor system. In this case, the proposed scheme significantly surpasses traditional implicit schemes and schemes of alternating directions; indeed, it requires a smaller number of exchanges and its parallelization possibilities are superior since it involves the solution of a series of independent three-point problems by an economical sweep method on each computer of a multiprocessor system.

2   Continuous 3D Model of Diffusion-Convection-Aggregation of Suspensions

Let us consider a continuous mathematical model of suspended matter in an aquatic environment, taking into account the diffusion and convection of

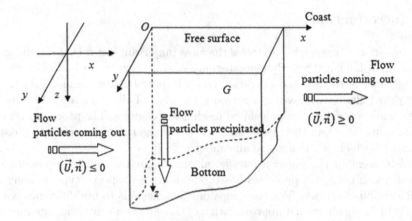

Fig. 1. Domain of the problem of suspension transport

suspensions, the sedimentation of suspended particles under the action of gravity, as well as the presence of the bottom and the free surface [9–11].

We will use a rectangular Cartesian coordinate system $Oxyz$ in which the axis Ox belongs to the undisturbed water surface and is directed towards the sea. We use the following notations: $h = H + \eta$ is the total water depth, [m]; H is the depth of the water body when its surface is undisturbed, [m]; and η is the elevation of the free surface relative to the geoid (sea level), [m].

Let $G \subset \mathbb{R}^3$ be the domain in which the process takes place, namely a parallelepiped beveled towards the shore, with its upper base lying on the free surface $(z = 0)$ and the bottom base a part of the bottom surface $(z = H(x, y))$. Let Γ is the border area G, S, are the lateral boundary surface G; let S be the lateral boundary surface of $\overline{G}$; S_{top} and S_{bottom} are, respectively, the parts of the free surface and the bottom surface limited by the conditions $\{0 \le x \le L_x,\ 0 \le y \le L_y\}$.

We assume that there are particles suspended in the water volume $\overline{G}$ at (x, y, z) and that their concentration at time t is $c = c(x, y, z, t)$, [mg/l], where t is the temporal variable, [sec] (see Fig. 1).

The equation describing the behavior of particles can be written as

$$\frac{\partial c}{\partial t} + \frac{\partial (uc)}{\partial x} + \frac{\partial (vc)}{\partial y} + \frac{\partial ((w + w_g)c)}{\partial z} = \mu \left(\frac{\partial^2 c}{\partial x^2} + \frac{\partial^2 c}{\partial y^2} \right) + \frac{\partial}{\partial z} \left(\nu \frac{\partial c}{\partial z} \right) + F, \quad (1)$$

where u, v, w are the components of the velocity vector U of the fluid, [m/sec]; w_g is the hydraulic particle size or sedimentation rate, [m/sec]; μ, ν are the horizontal and vertical diffusion coefficients of particles, [m²/sec]; and F is the function describing the intensity of distribution of suspended matter sources.

The problem domain of Eq. (1) is $\overline{Q}_T = G \times (0 < t \le T]$, $\overline{G}(x, y, z) = \{0 \le x \le L_x,\ 0 \le y \le L_y,\ -\eta_{\min} \le z \le H_{\max}\}$, $\overline{G} = G \cup \Gamma$, where Γ is the boundary of the domain G. Together with the boundary conditions for the particle-concentration function, the solution of Eq. (1) determines suspended matter fluxes both towards and along the coast.

Let us add to Eq. (1) initial and boundary conditions (assuming that the deposition of particles on the bottom is irreversible):

– initial conditions at $t = 0$:

$$c(x, y, z, 0) \equiv c_0(x, y, z); \tag{2}$$

– boundary conditions on the lateral boundary S at any time, $S \times (0, T]$:

$$\frac{\partial c}{\partial n} = 0, \quad \text{if } (\boldsymbol{U}_\Gamma, \boldsymbol{n}) \leq 0, \tag{3}$$

$$\frac{\partial c}{\partial n} = -\frac{u_\Gamma}{\mu} c, \quad \text{if } (\boldsymbol{U}_\Gamma, \boldsymbol{n}) \geq 0, \tag{4}$$

where $\boldsymbol{n}$ is the velocity vector projection S, $\boldsymbol{U}_\Gamma$ is the fluid velocity vector at the boundary S, u_Γ is the projection of the velocity vector $\boldsymbol{U}_\Gamma$ normal direction $\boldsymbol{n}$ on the border of the region S;

– boundary conditions on the water surface, $S_{\text{top}} \times (0 < t \leq T)$:

$$\frac{\partial c}{\partial z} = 0; \tag{5}$$

– boundary conditions on the bottom, $S_{\text{bottom}} \times (0 < t \leq T)$:

$$\frac{\partial c}{\partial n} = -\frac{w_g}{\nu} c. \tag{6}$$

The conditions for the correctness of the problem of transport of suspensions and, therefore, for the correctness of problem (1)–(6) as its particular case, are investigated in [12] for a multicomponent particle size distribution under the conditions of smoothness

$$c(x, y, z, t) \in C^2(Q_T) \cap C(\overline{Q}_T), \quad \operatorname{grad} c \in C(\overline{Q}_T),$$

imposed upon the solution function, and the required smoothness of the domain boundary.

3 Construction of an Explicit-Implicit Scheme for the Suspension Transport Problem

The terms on the left-hand side (except for the time derivative) of Eq. (1) describe the advective transport of suspended particles under the action of fluid flow and gravity. The terms on the right-hand side describe the diffusion of suspensions. The behavior of the vertical microturbulent diffusion coefficient differs significantly from that of the almost constant diffusion coefficient in the horizontal direction for processes of diffusion-convection of suspensions in coastal systems. The vertical coefficient of microturbulent diffusion can have several extremes depending on the vertical coordinate and varies in magnitude due to physical reasons. An additional complication that substantially affects the coefficient before the second difference derivative in the vertical coordinate is a significant change in depth for coastal systems, which quadratically influences the value of this coefficient. Thus, the one-dimensional discrete operator for diffusion-convection-sedimentation of suspensions in the vertical coordinate as a whole has poor conditioning (a large spread of eigenvalues). If we use an explicit scheme supplemented to increase the stability margin of a second-order differential derivative with a small factor, the admissible time step will be determined by the characteristic time of propagation of perturbations (concentrations of suspended matter) within one grid cell. In coastal systems, the time of propagation of disturbances in the horizontal directions for practically used grids, with cell sizes from many tens to hundreds of meters, is of tens to hundreds of seconds of physical time. For the propagation of disturbances in the vertical direction, when the cell vertical size is tens of centimeters, the characteristic propagation time of the disturbances amounts to several seconds, which will determine the allowable time step in an explicit scheme with the second time derivative. This circumstance is the main motive behind the construction of the explicit-implicit schemes considered below [13,14].

In our presentation, we focus on the use of an explicit-implicit scheme (see Fig. 2) enabling us to build parallel algorithms that are economical in terms of total time costs for arithmetic operations and data exchange operations between processors.

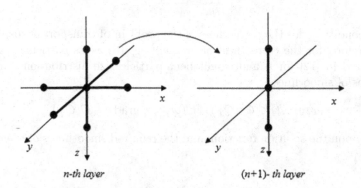

n-th layer   (n+1)- th layer

Fig. 2. Used nodes for the explicit-implicit scheme

Let us write Eq. (1) in the form

$$\frac{\partial c}{\partial t} = Ac + f(x,y,z,t), \quad (x,y,z) \in G, \quad t \in [0,T], \tag{7}$$

where Ac is the following elliptic differential operator with respect to the spatial variables with lower derivatives:

$$Ac = \frac{\partial(uc)}{\partial x} + \frac{\partial(vc)}{\partial y} + \frac{\partial((w + w_g)c)}{\partial z}$$
$$- \mu\left(\frac{\partial^2 c}{\partial x^2} + \frac{\partial^2 c}{\partial y^2}\right) - \frac{\partial}{\partial z}\left(\nu\frac{\partial c}{\partial z}\right) = (A_{12} + A_3)c, \qquad (8)$$

$$A_{12}c = \frac{\partial(uc)}{\partial x} + \frac{\partial(vc)}{\partial y} - \mu\left(\frac{\partial^2 c}{\partial x^2} + \frac{\partial^2 c}{\partial y^2}\right), \quad A_3c = \frac{\partial((w + w_g)c)}{\partial z} - \frac{\partial}{\partial z}\left(\nu\frac{\partial c}{\partial z}\right).$$

Let us construct a uniform grid ω_τ with step τ:

$$\omega_\tau = \{t_n = n\tau, \ n = 0, 1, \ldots, N, \ N\tau = T\}.$$

If we know the solution $c^{n-1}(x, y, z, t_{n-1})$ at some time t_{n-1}, then we can express the solution at time t_n by using the known solution. Thus, at each time step, the solution of problem (7), (2)–(6) can be expressed as

$$\frac{c^{n+1} - c^n}{\tau} + A_{12}c^n + A_3c^{n+1/2} = f, \quad n = 1, \ldots, N, \qquad (9)$$

where $c^{n+1/2} = \frac{1}{2}(c^{n+1} + c^n)$.

To increase the allowable time step in the explicit difference scheme approximation of the two-dimensional problem (7), (8), we add to the left-hand side of Eq. (9) a second-order time derivative with a small regularizing factor (see [15]) that does not exceed the characteristic time distribution of concentration perturbations in the horizontal directions:

$$\frac{\tau^*}{2}\frac{c^{n+1} - 2c^n + c^{n-1}}{\tau^2} + \frac{c^{n+1} - c^n}{\tau} + A_{12}c^n + A_3c^{n+1/2} = f, \quad n = 2, \ldots, N,$$
$$\tau^*\frac{c^{n+1} - c^n}{\tau^2} + \frac{c^{n+1} - c^n}{\tau} + A_{12}c^n + A_3c^{n+1/2} = f, \quad n = 1, \qquad (10)$$

where the coefficient τ^*, $\tau^* \sim \frac{\tau}{\tilde{c}}$, is associated with the characteristic spacing τ of the spatial grid and the characteristic speed of sound $\tilde{c}$ in the aquatic environment.

It has been shown [16] that the solution of problem (10) tends to the solution of problem (9) when $\tau^* \to 0$.

Let us construct a connected mesh $\overline{\omega}_h$ in the domain $\overline{G}$. The set of nodes of this grid consists of internal and boundary nodes. Aggregate ω_h internal nodes are set by a set of points:

$$\omega_h = \{x_i = ih_x, \ y_j = jh_y, \ z_k = -\eta_{min} + kh_z; \ i = \overline{0, N_x}, \ j = \overline{0, N_y}, \ k = \overline{0, N_z};$$
$$N_x h_x = L_x, \ N_y h_y = L_y, \ N_z h_z = \eta_{min} + H_{max}\},$$

where h_x, h_y, h_z are the space steps, and N_x, N_y, N_z are the numbers of nodes along the spatial axes.

By $o_{i,j,k}$, we denote the fullness of the (i, j, k) cell. In addition, we introduce the coefficients q_0, q_1, q_2, q_3, q_4, q_5, and q_6 to describe the occupancy of the domains in the vicinity of the cell [17,18].

Let us use the balance method based on the occupancy ratios of the control regions q_m, $m = 0, \ldots, 6$, to approximate Eq. (10). The discrete analog of the regularized equation of suspension transport takes the following form:

$$
(q_0)_{i,j,k} \frac{\tau^*}{2} \frac{c_{i,j,k}^{n+1} - 2c_{i,j,k}^n + c_{i,j,k}^{n-1}}{\tau^2} + (q_0)_{i,j,k} \frac{c_{i,j,k}^{n+1} - c_{i,j,k}^n}{\tau}
$$

$$
+ (q_1)_{i,j,k} u_{i+1/2,j,k} \frac{c_{i+1,j,k}^n - c_{i,j,k}^n}{2h_x} + (q_2)_{i,j,k} u_{i-1/2,j,k} \frac{c_{i,j,k}^n - c_{i-1,j,k}^n}{2h_x}
$$

$$
+ (q_3)_{i,j,k} v_{i,j+1/2,k} \frac{c_{i,j+1,k}^n - c_{i,j,k}^n}{2h_y} + (q_4)_{i,j,k} v_{i,j-1/2,k} \frac{c_{i,j,k}^n - c_{i,j-1,k}^n}{2h_y}
$$

$$
+ (q_5)_{i,j,k}(w_{i,j,k+1/2} + w_g) \frac{c_{i,j,k+1}^{n+1/2} - c_{i,j,k}^{n+1/2}}{2h_z}
$$

$$
+ (q_6)_{i,j,k}(w_{i,j,k-1/2} + w_g) \frac{c_{i,j,k}^{n+1/2} - c_{i,j,k-1}^{n+1/2}}{2h_z} \tag{11}
$$

$$
= (q_1)_{i,j,k} \mu \frac{c_{i+1,j,k}^n - c_{i,j,k}^n}{h_x^2} - (q_2)_{i,j,k} \mu \frac{c_{i,j,k}^n - c_{i-1,j,k}^n}{h_x^2}
$$

$$
+ (q_3)_{i,j,k} \mu \frac{c_{i,j+1,k}^n - c_{i,j,k}^n}{h_y^2} - (q_4)_{i,j,k} \mu \frac{c_{i,j,k}^n - c_{i,j-1,k}^n}{h_y^2}
$$

$$
+ (q_5)_{i,j,k} \nu_{i,j,k+1/2} \frac{c_{i,j,k+1}^{n+1/2} - c_{i,j,k}^{n+1/2}}{h_z^2}
$$

$$
- (q_6)_{i,j,k} \nu_{i,j,k-1/2} \frac{c_{i,j,k}^{n+1/2} - c_{i,j,k-1}^{n+1/2}}{h_z^2} + f_{i,j,k}^n.
$$

To calculate the components of the velocity vector of the aquatic environment, we used a three-dimensional model of the hydrodynamic flow around the bottom topography with allowance for bottom friction and level rise.

4   The Discrete Equations

Let us write the discrete equations for problem (11) in canonical form [19]:

$$
A_{i,j} c_{i,j,k}^{n+1} - B_{1,i,j} c_{i,j,k+1}^{n+1} - B_{2,i,j} c_{i,j,k-1}^{n+1} = F_{i,j,k}^n, \tag{12}
$$

$$
B_{1,i,j,k} = (q_5)_{i,j,k} \left(-\frac{w_{i,j,k+1/2} + w_g}{4h_z} + \frac{v_{i,j,k+1/2}}{2h_z^2} \right),
$$

$$
B_{2,i,j,k} = (q_6)_{i,j,k} \left(\frac{w_{i,j,k-1/2} + w_g}{4h_z} + \frac{v_{i,j,k-1/2}}{2h_z^2} \right),
$$

$$A_{i,j,k} = (q_0)_{i,j,k} \frac{\tau + \tau^*/2}{\tau^2} + B_{1,i,j} + B_{2,i,j},$$

$$F^n_{i,j,k} = D_{0,i,j,k} c^n_{i,j,k} + D_{1,i,j,k} c^n_{i+1,j,k} + D_{2,i,j,k} c^n_{i-1,j,k} + D_{3,i,j,k} c^n_{i,j+1,k}$$
$$+ D_{4,i,j,k} c^n_{i,j-1,k} + B_{1,i,j,k} c^n_{i,j+1,k} + B_{2,i,j,k} c^n_{i,j-1,k} - E_{i,j,k} c^{n-1}_{i,j,k} + f^n_{i,j,k},$$

$$D_{1,i,j,k} = (q_1)_{i,j,k} \left(-\frac{u_{i+1/2,j,k}}{2h_x} + \frac{\mu}{h^2_x} \right), \quad D_{2,i,j,k} = (q_2)_{i,j,k} \left(\frac{u_{i-1/2,j,k}}{2h_x} + \frac{\mu}{h^2_x} \right),$$

$$D_{3,i,j,k} = (q_3)_{i,j,k} \left(-\frac{v_{i,j+1/2,k}}{2h_y} + \frac{\mu}{h^2_y} \right), \quad D_{4,i,j,k} = (q_4)_{i,j,k} \left(\frac{v_{i,j-1/2,k}}{2h_y} + \frac{\mu}{h^2_y} \right),$$

$$D_{0,i,j,k} = (q_0)_{i,j,k} \frac{\tau + \tau^*}{\tau^2} - \sum_{p=1}^{4} D_{p,i,j,k}, \quad E_{i,j,k} = (q_0)_{i,j,k} \frac{\tau^*}{2\tau^2}.$$

Consequently, $16N$ arithmetic operations are needed to calculate the right-hand side in (12). To solve problem (12), $8N$ arithmetic operations are needed in the first time layer and $5N$ in subsequent layers. Thus, a total of $21N$ operations is required to calculate the new value of the time layer.

The calculation by the explicit scheme has the form

$$c^{n+1}_{i,j,k} = (B_{1,i,j}/A_{i,j}) c^{n+1}_{i,j,k+1} + (B_{2,i,j}/A_{i,j}) c^{n+1}_{i,j,k-1} + (F^n_{i,j,k}/A_{i,j}), \qquad (13)$$

which takes $16N$ arithmetic operations.

5   Description of the Parallel Algorithm

In the parallel implementation, we resorted to methods of decomposition of grid domains for computationally intensive diffusion-convection problems, taking into account the architecture and parameters of the used multiprocessor computing system. We decomposed the computational two-dimensional domain along two spatial axes (see Fig. 3) assuming that the number of vertical blocks of approximately the same number of grid nodes is equal to the number of calculators of the multiprocessor system.

The maximum performance of the multiprocessor computing system is 18.8 teraFLOPS. The system has 128 computing nodes of the same type, namely HP ProLiant BL685c 16-core Blade servers, each equipped with four AMD Opteron 8356 2.3 GHz four-core processors and 32 GB RAM. Table 1 shows the values of the acceleration and efficiency for different numbers of computating cores used for solving the model problem of transport of suspended matter.

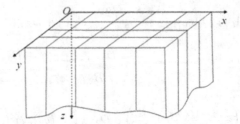

Fig. 3. Decomposition of a two-dimensional grid region

Table 1. Acceleration and performance of the parallel version of the explicit-implicit scheme

Number of cores	Acceleration			Efficiency		
	1000 × 1000	2000 × 2000	5000 × 5000	1000 × 1000	2000 × 2000	5000 × 5000
1	1	1	1	1	1	1
2	1.645	1.716	1.979	0.823	0.858	0.99
4	3.689	3.156	3.064	0.922	0.788	0.766
8	4.843	4.72	8.686	0.605	0.59	1.086
16	5.745	7.184	11.5	0.979	0.449	0.719
32	14.607	13.13	20.936	0.456	0.41	0.654
64	32.8	23.63	37.114	0.513	0.369	0.58
128	75.167	28.454	96.059	0.587	0.222	0.75
256	55.253	42.924	165.434	0.216	0.168	0.646
512	27.883	67.284	228.36	0.054	0.131	0.446

6   The Results of Numerical Experiments

As a model problem, we considered the convective-diffusive advective transport and sedimentation of suspended matter during dumping of extracted bottom material onto the surface of a water body. The following baseline data were used: water depth: 10 m; volume of substance discharged: 741 m^3; flow rate: 0.2 m/s; deposition rate: 2.042 mm/s (according to Stokes); soil density: 1600 kg/m^3; percentage of dust particles (of diameter less than 0.05 mm) in sandy soils: 26.83%. The parameters of the computational domain were as follows: length: 3 km; width: 1.4 km; step along the horizontal axis: 20 m; step along the vertical axis: 1 m; computation interval: 2 h. It should be noted that the initial data are close to those in real conditions for the distribution of suspended matter in landfills of bottom material extracted during dredging works in river mouths.

Figure 4 shows the dynamics of changes in the concentration of suspended particles (mg/l) over time. The values of the suspension concentration field are given in the cross section of the computational domain by a plane passing through the unloading point and formed by two direction vectors: one vertical and one in the direction of the flow. The estimated intervals were 15 min,

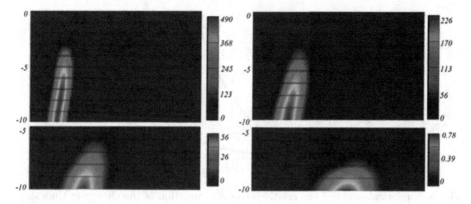

Fig. 4. Field values of suspended particles at 15 min, 30 min, 1 h, and 2 h after the discharge of the suction pump

30 min, 1 h, and 2 h. Flows are directed from left to right. The results of numerical experiments are in good agreement with field data [20].

The problems of transport of suspended particles were solved on the basis of an explicit scheme and an explicit-implicit scheme to obtain the optimal values of the time steps in (11). Figure 5 portrays the plots of the error of the difference schemes (1 is the error function for the explicit scheme; 2 is the error function for the scheme given in (11)). The value of the relative error ψ is indicated along the vertical axis, $\psi = \sqrt{\sum_{i,j,k} (\tilde{c}_{i,j,k} - c_{i,j,k})^2 / \sum_{i,j,k} c_{i,j,k}^2}$, where $c_{i,j,k}$ is the exact value of the solution to the problem of transport of suspended particles at the node (i, j, k), and $\tilde{c}_{i,j,k}$ is the numerical solution, which depends on the time step. The value of the time step referred to $\tau_m = \left(2\mu \left(\frac{1}{h_x^2} + \frac{1}{h_y^2} + \frac{1}{h_z^2} \right) \right)^{-1}$ is given along the horizontal axis. From the condition of stability of an explicit scheme, it follows that the quantity τ_m is an upper bound for the time step [14]. It is convenient to use $\tau_0 = \tau/\tau_m$ [21] to describe the error ψ since the function $\psi = \psi(\tau_0)$ does not practically change when the grid size changes along the spatial axes.

As we can see from Fig. 5, the error achieved with the explicit scheme is such that the restriction on the time step is significantly smaller than in the case of the explicit-implicit scheme given in (11). The relative error of the proposed explicit-implicit scheme (11) is 1% when the value is 0.10087, whereas in the case of the explicit scheme, the corresponding value is 0.01348. Therefore, to achieve an accuracy of 1% with the explicit-implicit scheme (11), it is necessary that the time step be greater by a factor of 7.483 than it is in the explicit scheme. This significantly improves the performance of the programs thanks to a better difference scheme. Scheme (11) is effective if the step along one of the spatial directions is significantly smaller than the steps along the other spatial directions.

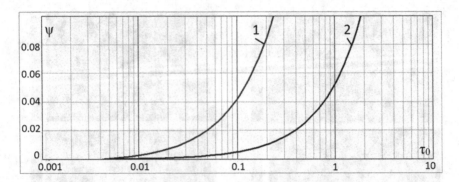

Fig. 5. Numerical study of the relative error as a function of the change in time step, for the explicit (1) and the explicit-implicit (2) schemes

7 Conclusions

We described in the article the construction and study of parallel algorithms for the numerical implementation of a 3D model of transport of suspensions in coastal marine systems. The suggested parallel algorithm is based on an explicit-implicit scheme. This scheme offers an opportunity to fully organize parallel computations by approximating both the two-dimensional diffusion-convection operator explicitly (in the previous time layer) and the diffusion-convection-gravitational precipitation operator by a set of implicit three-point equations with weights (in the upper and previous time layers). We added a second-order difference derivative with a small multiplier to the constructed difference approximation. This made it possible to significantly increase the allowable value of the time step as compared with standard explicit schemes. As a result, we constructed a fully parallel algorithm of independent one-dimensional three-point problems that approximate one-dimensional diffusion-convection problems in the vertical direction by implicit schemes with weights and are solved on separate processors. The constructed scheme minimizes the amount of data exchange between neighboring processors when passing from a layer to another in a parallel way at the border nodes of subdomains in the case of a three-dimensional grid domain geometrically decomposed into vertical planes. In addition, this scheme turned out to be significantly more accurate than the traditional explicit scheme, which was confirmed by the results of numerical experiments.

References

1. Sukhinov, A.I., Chistyakov, A.E., Shishenya, A.V., Timofeeva, E.F.: Predictive modeling of coastal hydrophysical processes on amultiprocessor system using explicit schemes. J. Matem. Model. **30**(3), 83–100 (2018). https://doi.org/10.1134/S2070048218050125
2. Sidoryakina, V.V., Sukhinov, A.I.: Well-posedness analysis and numerical implementation of a linearized two-dimensional bottom sediment transport problem. J. Comp. Math. Math. Phys. **57**(6), 978–994 (2017). https://doi.org/10.7868/S0044466917060138

3. Alekseenko, E., Roux, B., Sukhinov, A., Kotarba, R., Fougere, D.: Nonlinear hydrodynamics in a mediterranean lagoon. J. Nonlinear Process. Geophys. **20**(2), 189–198 (2013). https://doi.org/10.5194/npg-20-189-2013
4. Sukhinov, A.I., Chistyakov, A.E., Protsenko, E.A.: Mathematical modeling of sediment transport in the coastal zone of shallow reservoirs. J. Math. Models Comput. Simul. **6**(4), 351–363 (2014). https://doi.org/10.1134/S2070048214040097
5. Alekseenko, E., Roux, B., Sukhinov, A., Kotarba, R., Fougere, D.: Coastal hydrodynamics in a windy lagoon. J. Comput. Fluids **77**, 24–35 (2013). https://doi.org/10.1016/j.compfluid.2013.02.003
6. Sukhinov, A.I., Chistyakov, A.E., Alekseenko, E.V.: Numerical realization of the three-dimensional model of hydrodynamics for shallow water basins on a high-performance system. J. Math. Models Comput. Simul. **3**(5), 562–574 (2011). https://doi.org/10.1134/S2070048211050115
7. Liu, X., Qi, S., Huang, Y., Chen, Y., Pengfei, D.: Predictive modeling in sediment transportation across multiple spatial scales in the Jialing River Basin of China. Int. J. Sediment Res. **30**(3), 250–255 (2015)
8. D'Ascenzo, N., Saveliev, V.I., Chetverushkin, B.N.: On an algorithm for solving parabolic and elliptic equations. J. Comp. Math. Math. Phys. **55**(8), 1290–1297 (2015). https://doi.org/10.1134/S0965542515080035
9. Sukhinov, A.I., Chistyakov, A.E., Sidoryakina, V.V.: Parallel solution of sediment and suspension transportation problems on the basis of explicit schemes. In: Sokolinsky, L., Zymbler, M. (eds.) PCT 2018. CCIS, vol. 910, pp. 306–321. Springer, Cham (2018). https://doi.org/10.1007/978-3-319-99673-8_22
10. Sukhinov, A.A., Sukhinov, A.I.: 3D model of diffusion-advection-aggregation suspensions in water basins and its parallel realization. In: Parallel Computational Fluid Dynamics, Mutidisciplinary Applications, Proceedings of Parallel CFD 2004 Conference, Las Palmas de Gran Canaria, Spain, pp. 223–230. Elsevier, Amsterdam (2005). https://doi.org/10.1016/B978-044452024-1/50029-4
11. Sukhinov, A.I., Sukhinov, A.A.: Reconstruction of 2001 ecological disaster in the Azov Sea on the basis of precise hydrophysics models. In: Parallel Computational Fluid Dynamics, Mutidisciplinary Applications, Proceedings of Parallel CFD 2004 Conference, Las Palmas de Gran Canaria, Spain, pp. 231–238. Elsevier, Amsterdam (2005). https://doi.org/10.1016/B978-044452024-1/50030-0
12. Sukhinov, A., Sidoryakina, V., Protsenko, S.: Correctness investigation for the suspension transport problem in coastal systems. In: MATEC Web Conferences. XIV International Scientific-Technical Conference on 'Dynamic of Technical Systems' (DTS-2018), vol. 226, p. 04027 (2018). https://doi.org/10.1051/matecconf/201822604027
13. Samarskiy, A.A., Gulin, A.V.: Numerical Methods. Nauka, Moscow (1989). (in Russian)
14. Samarskiy, A.A., Vabishchevich, P.N.: Numerical methods for solving convection-diffusion problems. URSS (2005). (in Russian)
15. Chetverushkin, B.N.: Resolution limits of continuous media models and their mathematical formulations. J. Math. Models Comput. Simul. **5**(3), 266–279 (2013). https://doi.org/10.1134/S2070048213030034
16. Chetverushkin, B.N., D'Ascenzo, N., Saveliev, A.V., Saveliev, V.I.: A kinetic model for magnetogasdynamics. Math. Models Comput. Simul. **9**(5), 544–553 (2017). https://doi.org/10.1134/S2070048217050039
17. Sukhinov, A.I.: Precise fluid dynamics models and their application in prediction and reconstruction of extreme events in the Sea of Azov. J. Izv. Taganrog. Radiotech. Univ. **3**, 228–235 (2006). (in Russian)

18. Sukhinov, A., Chistyakov, A., Sidoryakina, V.: Investigation of nonlinear 2D bottom transportation dynamics in coastal zone on optimal curvilinear boundary adaptive grids. In: MATEC Web of Conferences, XIII International Scientific-Technical Conference on 'Dynamic of Technical Systems' (DTS-2017), Rostov-on-Don, vol. 132 (2017). https://doi.org/10.1051/matecconf/201713204003
19. Samarskiy, A.A.: Theory of Difference Schemes. Nauka, Moscow (1989). (in Russian)
20. Kovtun, I.I., Protsenko, E.A., Sukhinov, A.I., Chistyakov, A.E.: Calculating the impact on aquatic resources dredging in the White Sea. J. Fundamentalnaya i prikladnaya gidrofizika **9**(2), 27–38 (2016). (in Russian)
21. Sukhinov, A.I., Chistyakov, A.E., Shishenya, A.V.: Error estimate for diffusion equations solved by schemes with weights. Math. Models Comput. Simul. **6**(3), 324–331 (2014). https://doi.org/10.1134/S2070048214030120

Supercomputer Stochastic Simulation of Transient Anisotropic Diffusion-Reaction Processes with Application in Cathodoluminescence Imaging

Anastasiya Kireeva$^{(\boxtimes)}$ and Karl K. Sabelfeld

Institute of Computational Mathematics and Mathematical Geophysics,
Novosibirsk, Russia
kireeva@ssd.sscc.ru, karl@osmf.sscc.ru

Abstract. This paper is devoted to supercomputer simulations of transient anisotropic diffusion processes with recombination in GaN semiconductors containing a set of threading dislocations. The random walk on arbitrary parallelepipeds and cubes based on a Monte Carlo algorithm suggested by K. K. Sabelfeld is here applied to a cathodoluminescence imaging problem. The computational time for large diffusion lengths and large numbers of dislocations is of several hours. For this reason, we carried out a parallel implementation of the code by distributing diffusion particle trajectories among several MPI processes and OpenMP threads. The parallel code made it possible to obtain the transient cathodoluminescence intensity, the concentration of survived particles, and the flux to the dislocation surfaces. To verify the algorithm implementation, we compared the simulation results with those obtained in the isotropic case by means of the random-walk-on-spheres algorithm and also with the exact solution of the isotropic diffusion-reaction equation.

Keywords: Cathodoluminescence imaging · Threading dislocations ·
Random-walk-on-cubes algorithm · Anisotropic diffusion ·
Vectorization · MPI with OpenMP programming

1 Introduction

Cathodoluminescence imaging is a powerful technique for exploring fundamental properties of materials, such as the electronic structure, and detecting defect centers and dislocations in crystals [1]. Cathodoluminescence microscopy is effectively applied for research in optics [2,3], materials science, geology [4], and other fields which are directly related to industry. In particular, cathodoluminescence imaging with sub-nanometer spatial resolution is employed for dislocation detection in gallium nitride (GaN) semiconductors used in blue-light-emitting diodes

The work was supported by the ICMMG SB RAS (budget project No. 0315-2019-0002).

L. Sokolinsky and M. Zymbler (Eds.): PCT 2019, CCIS 1063, pp. 269–284, 2019.
https://doi.org/10.1007/978-3-030-28163-2_19

(LEDs) [5]. Cathodoluminescence (CL) is the light emission from a material caused by the impact of an energetic electron beam [6]. In CL images, threading dislocations are visible as dark spots since they act as non-radiative centers [7]. In addition, the density of dislocations has an influence on CL intensity transients. CL intensity can thus be used to evaluate the density of dislocations in semiconductors. Semiconductor properties depending on dislocations have been studied experimentally [1,2,8,9] and by means of computer simulation [2,7,10].

In [7], the exciton diffusion-recombination problem is solved for CL simulation. In [11–13], a random-walk-on-spheres (RWS) method is suggested for calculation of CL and electron-beam induced current (EBIC) maps. RWS is a Monte Carlo method used for solving boundary value problems, such as the isotropic diffusion governed by the Laplace equation with Dirichlet or Robin boundary conditions [13,14]. In many physical problems, it is important to take into account the diffusion anisotropy [15,16]. The random-walk-on-parallelepipeds (RWP) method was suggested in [17] for the solution of the anisotropic diffusion-recombination equation. The RWP method for the isotropic diffusion was described in [18–20]. The exact distributions of the first passage time and exit point on an arbitrary parallelepiped were derived in [17] for the anisotropic diffusion with recombination in a domain.

In the present paper, we use the RWP method [17] to simulate the CL intensity and the flux to the dislocation surfaces for a domain with a set of threading dislocations. According to the Monte Carlo approach [14], many independent exciton trajectories must be simulated to obtain the characteristics with sufficient accuracy. The computational time for large diffusion lengths and large numbers of dislocations is of several hours. For this reason, we carried out a parallel implementation of the code by distributing diffusion particle trajectories among several MPI processes and OpenMP threads. To verify the algorithm implementation, we compared the simulation results with those obtained in the isotropic case by means of the RWS algorithm and also with the exact solution of the isotropic diffusion-reaction equation derived in [21].

The paper is organized as follows. In Sect. 2, we formulate the transient anisotropic diffusion-reaction problem and describe the RWP method. Section 3 is devoted to the description of the parallel implementation of the Monte Carlo RWP algorithm; here we analyze the efficiency of the algorithm. In Sect. 4, by simulation for various densities of threading dislocations, we obtain the transient CL intensity, the concentration of survived particles, and the fluxes to the dislocation surfaces and to the bottom plane. In this section, we also compare the simulation results and the exact solution of the transient diffusion-reaction equation.

2 The Monte Carlo Simulation Algorithm for Solving the Transient Anisotropic Diffusion-Reaction Problem

2.1 The Transient Anisotropic Diffusion-Reaction Problem

Following [7,11,13], we consider the exciton diffusion-recombination problem in the 3D upper half-space, $G = R_+^3 = \{(x,y,z) : z \geq 0\}$, containing a set of dislocations. Each threading dislocation is assumed to be a semi-infinite parallelepiped perpendicular to the plane $z = 0$. The boundary Γ of the domain G consists of the plane $\Gamma_z = \{(x,y,z) : z = 0\}$ and the surfaces Γ_d of the parallelepipeds. The generation of excitons by an incident electron beam [7] is simulated by the generating function $f(\mathbf{r})$ which generates the exciton coordinates in the domain G at the initial time $t = 0$. The excitons can diffuse, recombine in the domain, or annihilate by non-radiative annihilation either on the plane Γ_z or on a dislocation surface Γ_d. The time dependence of the exciton concentration u in the domain G is governed by the anisotropic diffusion-reaction equation:

$$\frac{\partial u(\mathbf{r})}{\partial t} = D_x \frac{\partial^2 u(\mathbf{r})}{\partial x^2} + D_y \frac{\partial^2 u(\mathbf{r})}{\partial y^2} + D_z \frac{\partial^2 u(\mathbf{r})}{\partial z^2} - \frac{1}{\bar{\tau}} u(\mathbf{r}) + f(\mathbf{r})\delta(t), \ \mathbf{r} \in G, \ t \in [0,T]. \quad (1)$$

Here $\mathbf{r}$ is the space coordinate; D_x, D_y, D_z are constant diffusion coefficients in the directions x, y and z, respectively; $\bar{\tau}$ is the exciton mean life time; and $f(\mathbf{r})\delta(t)$ is the instantaneous source of excitons.

We impose the following initial and Dirichlet boundary conditions:

$$u(\mathbf{r},0) = 0, \ \mathbf{r} \in G, \qquad u(\mathbf{r},t) = 0, \ \mathbf{r} \in \Gamma_z, \qquad u(\mathbf{r},t) = 0, \ \mathbf{r} \in \Gamma_d. \quad (2)$$

The Dirichlet boundary conditions correspond to the total absorption of the excitions on the boundaries.

The solution of Eq. (1) with initial and boundary conditions (2) is obtained by the RWP method suggested in [17].

2.2 The Sampling Formulae for the First Passage Time, Exit Point, and Absorption Time

Following [17] and by analogy with the RWS method [11,14], a random-walk-on-parallelepipeds diffusion-reaction process in the domain G, starting from a point $\mathbf{r_0} = (x_0, y_0, z_0) \in G$ at time $t_0 = 0$, considered as a Markov chain of points $\{(\mathbf{r_k}, t_k)\}$, $k = 0, 1, \ldots$, is constructed as follows. The first rectangular parallelepiped $\Pi(\mathbf{r_0}, l_x^0, l_y^0, l_z^0)$ is centered at (x_0, y_0, z_0) and has arbitrary edges such that it lies inside the domain G. The next point, $\mathbf{r_1}$, is randomly distributed on the faces of this rectangular parallelepiped according to a distribution which will be defined below. The same for the next points, namely $\mathbf{r_{k+1}}$ is randomly distributed on the faces of the rectangular parallelepiped centered at $\mathbf{r_k}$, which also has arbitrary edges l_x^k, l_y^k, l_z^k such that it is fully contained inside the domain G. The relevant times t_k are calculated as $t_{k+1} = t_k + \tau_k$, $k = 0, 1, \ldots$, where the random value τ_k is the first passage time, which has a random distribution $p_t(\tau)$

given below. Finally, the Markov chain stops with the break probability P_{abs} of a particle absorption inside the relevant parallelepiped.

The first passage time τ_k has the distribution density $p_t(\tau)$ derived in [17]:

$$p_t(\tau) = \left(\frac{D_x}{l_x^2} F_1(D_x, \tau) F_2(D_y, \tau) F_2(D_z, \tau) + \frac{D_y}{l_y^2} F_1(D_y, \tau) F_2(D_x, \tau) F_2(D_z, \tau) \right.$$

$$\left. + \frac{D_z}{l_z^2} F_1(D_z, \tau) F_2(D_x, \tau) F_2(D_y, \tau) \right) \cdot \frac{64}{\pi} \exp \left\{ -\frac{\tau}{\tilde{\tau}} \right\}, \quad (3)$$

and the functions $F_1(D_i, \tau)$ and $F_2(D_i, \tau)$ are defined by the following formulae:

$$F_1(D_i, \tau) = \sum_{m=1}^{\infty} (-1)^{m+1} (2m-1) \exp \left[-\frac{(2m-1)^2 \pi^2 D_i}{l_i^2} \tau \right],$$

$$F_2(D_i, \tau) = \sum_{n=1}^{\infty} (-1)^{n+1} \frac{1}{2n-1} \exp \left[-\frac{(2n-1)^2 \pi^2 D_i}{l_i^2} \tau \right], \quad i \in \{x, y, z\}. \tag{4}$$

The random point of first exit of a diffusing particle from the center of the parallelepiped with edge lengths l_x, l_y, l_z, assuming that the first passage time is sampled with density $p_t(\tau)$, has the conditional probability density given in [17]:

$$p(x, y, z | \tau) = \frac{1}{Q(\tau)} \left[\frac{D_x}{l_x^2} F_1(D_x, \tau) F_2(D_y, \tau) F_2(D_z, \tau) \, p_{D_y}(y, \tau) p_{D_z}(z, \tau) \right.$$

$$+ \frac{D_y}{l_y^2} F_1(D_y, \tau) F_2(D_x, \tau) F_2(D_z, \tau) \, p_{D_x}(x, \tau) p_{D_z}(z, \tau) \tag{5}$$

$$\left. + \frac{D_z}{l_z^2} F_1(D_z, \tau) F_2(D_x, \tau) F_2(D_y, \tau) \, p_{D_x}(x, \tau) p_{D_y}(y, \tau) \right],$$

where $p_{D_i}(i, \tau)$, $i \in \{x, y, z\}$, is a one-dimensional probability density function,

$$p_{D_i}(i, \tau) = \frac{1}{F_2(D_i, \tau)} \frac{\pi}{2l_i} \sum_{n=1}^{\infty} (-1)^{n+1} \sin \left[\frac{(2n-1)\pi i}{l_i} \right] \exp \left[-\frac{(2n-1)^2 \pi^2 D_i}{l_i^2} \tau \right], (6)$$

and Q is defined by

$$Q(\tau) = \frac{D_x}{l_x^2} F_1(D_x, \tau) F_2(D_y, \tau) F_2(D_z, \tau) + \frac{D_y}{l_y^2} F_1(D_y, \tau) F_2(D_x, \tau) F_2(D_z, \tau)$$

$$+ \frac{D_z}{l_z^2} F_1(D_z, \tau) F_2(D_x, \tau) F_2(D_y, \tau). \tag{7}$$

Efficient sampling methods for the first passage time distributed with density $p_t(\tau)$ and for the random point of exit from the parallelepiped distributed with densities $p_{D_i}(i, \tau)$ were constructed and described in detail in [22]. Below, we give only the main simulating formulae obtained in [22].

The first passage time is sampled by superposition and rejection methods. The time axis is divided into two regions, $[0, \tau^*]$ and $[\tau^*, \infty)$. On each region, the density $p_t(\tau)$ is upper-bounded by an appropriate function.

The value of τ^* is computed by the formula

$$\tau^* = \frac{1}{\pi^2 \Delta} \ln\left(\frac{64}{\pi^3}\right), \quad \Delta = \frac{D_x}{l_x^2} + \frac{D_y}{l_y^2} + \frac{D_z}{l_z^2} + \frac{1}{\pi^2 \bar{\tau}}. \tag{8}$$

The density $p_t(\tau)$ can thus be written as

$$p_t(\tau) = I_{p1} \cdot p_1(\tau) + I_{p2} \cdot p_2(\tau). \tag{9}$$

The weights I_{p1} and I_{p2} are given by

$$I_{p1} = \int_0^{\tau^*} p(\tau)\, d\tau \approx I_1, \quad I_{p2} = \int_{\tau^*}^{\infty} p(\tau)\, d\tau = 1 - I_1, \quad I_1 = \int_0^{\tau^*} g_1(\tau)\, d\tau$$

$$= 2\left[\mathrm{erfc}\left(\frac{l_x}{4\sqrt{D_x \tau^*}}\right) + \mathrm{erfc}\left(\frac{l_y}{4\sqrt{D_y \tau^*}}\right) + \mathrm{erfc}\left(\frac{l_z}{4\sqrt{D_z \tau^*}}\right)\right]. \tag{10}$$

The probabilities $p_1(\tau)$ and $p_2(\tau)$ have the form

$$p_1(\tau) = g_1(\tau)/I_{p1}, \quad \tau < \tau^*, \tag{11}$$
$$p_2(\tau) = g_2(\tau)/I_{p2}, \quad \tau \geq \tau^*. \tag{12}$$

Here

$$g_1(\tau) = 2\left[p_L^{(x)}(\tau) + p_L^{(y)}(\tau) + p_L^{(z)}(\tau)\right], \tag{13}$$

where $p_L^{(i)}(\tau) = \sqrt{\dfrac{l_i^2}{16\pi D_i}} \dfrac{1}{\tau^{3/2}} \exp\left[-\dfrac{l_i^2}{16 D_i \tau}\right]$, $i \in \{x, y, z\}$;

$$\frac{g_1(\tau)}{I_{p1}} = P_x \frac{p_L^{(x)}(\tau)}{I_L^{(x)}} + P_y \frac{p_L^{(y)}(\tau)}{I_L^{(y)}} + P_z \frac{p_L^{(z)}(\tau)}{I_L^{(z)}}, \quad \tau < \tau^*, \tag{14}$$

where $I_L^{(i)} = \mathrm{erfc}\left(\dfrac{l_i}{4\sqrt{D_i \tau^*}}\right)$, $P_i = \dfrac{I_L^{(i)}}{I_L^{(x)} + I_L^{(y)} + I_L^{(z)}}$, $i \in \{x, y, z\}$;

$$g_2(\tau) = \frac{64}{\pi} \Delta \exp[-\pi^2 \Delta \tau], \quad \tau \geq \tau^*. \tag{15}$$

The density $p_t(\tau)$ (see (9)) is simulated by combining the superposition and rejection methods. The algorithm for sampling the first passage time τ with density $p_t(\tau)$ is described below.

Algorithm for Sampling the First Passage Time

1. With probability $P_1 = I_1/(I_1 + 1)$, sample with density $p_1(\tau)$, and with probability $1 - P_1$, sample with density $p_2(\tau)$.

2. Apply the superposition method for sampling a random time τ_s with density $p_1(\tau)$ (see (14)): select with probability P_i, $i \in \{x, y, z\}$, one of the densities $p_L^{(i)}(\tau)/I_L^{(i)}$ and simulate τ_s by the following formula:

$$\tau_s = \frac{l_i^2}{16 D_i \xi_i^2}, \quad i \in \{x, y, z\}, \tag{16}$$

where $\xi_i = \mathrm{erfc}^{-1}\left[rand \cdot \mathrm{erfc}\left(\frac{l_i}{4\sqrt{D_i \tau^*}} \right) \right]$. Apply the rejection method to the value τ_s obtained: if $rand_1 \cdot g_1(\tau_s) \leq p_t(\tau_s)$, then accept the value τ_s; otherwise, go to Step 1. Here $rand \in [0, 1]$ and $rand_1 \in [0, 1]$ are uniformly distributed random numbers sampled independently.

3. Apply the rejection method with upper-bound function $g_2(\tau)$ for sampling a random time τ_s with density $p_2(\tau)$ (see (15)). The function $g_2(\tau)$ is a density; consequently, the following explicit formula generates a random time τ_s

$$\tau_s = \frac{1}{\pi^2 \Delta} \ln\left(\frac{64}{\pi^3 \, rand} \right), \quad rand \in [0, 1]. \tag{17}$$

If $rand_1 \cdot g_2(\tau_s) \leq p_t(\tau_s)$, where $rand_1$ is sampled independently of $rand$, then accept the value τ_s; otherwise, go to the Step 1.

After sampling the first passage time $\tau = \tau_s$, simulate (for this sampled time) with density $p(x, y, z|\tau)$ (see (5)) a spatial random point which is the exit point on the parallelepiped surface, according to [22].

The exit point on the surface of the parallelepiped is sampled by superposition method. Each probability density $p_{D_a}(a, \tau) p_{D_b}(b, \tau)$, $a, b \in \{x, y, z\}$, $a \neq b$, is weighted with probability $P_{ab} = \frac{D_c}{l_c^2} F_1(D_c, \tau) F_2(D_a, \tau) F_2(D_b, \tau)/Q(\tau)$, $c \in \{x, y, z\}$, $c \neq a, c \neq b$. The coordinates a and b are sampled independently with their one-dimensional probability densities $p_{D_a}(a, \tau)$ and $p_{D_b}(b, \tau)$. We take the third coordinate c with probability 0.5 on one of the boundaries of the parallelepiped along the direction of the axis C. For example, we sample with probability P_{yz} the coordinates y and z with densities $p_{D_y}(y, \tau)$ and $p_{D_z}(z, \tau)$, respectively. Then, we take the point (y, z) with probability 0.5 on either the left or the right boundary of the parallelepiped, i.e. on the boundaries lying along the X axis.

We use the Walker algorithm [23] to simulate the one-dimensional probability density $p_{D_i}(i, \tau)$ (see (6)). To this end, we convert the density $p_{D_i}(i, \tau)$ to a dimensionless form using the new variables $v = i/l_i$ and $t = \frac{D_i \tau}{l_i^2}$. Then, we compute the arrays of aliases and reconstructed probabilities for the modified density and store them in files. These arrays are applied for any values of l_i, D_i, and τ through the new variables. In addition, we convert the auxiliary functions $F_1(D_i, \tau)$ and $F_2(D_i, \tau)$ (see (4)) used in (6) to dimensionless forms by means of the change of variable $t = \dfrac{D_i \tau}{l_i^2}$, and then precalculate and store them in files.

For each point of the Markov chain, the probability of absorption of a particle inside the parallelepiped, according to [22], is computed by the formula

$$Q_{\text{abs}} = \frac{64}{\pi^5} \sum_{i=1}^{\infty} \sum_{j=1}^{\infty} \sum_{k=1}^{\infty} \frac{(-1)^{i+j+k+1}}{(2i-1)(2j-1)(2k-1)} \frac{1}{\tilde{\Delta}_{\bar{\tau}}}, \tag{18}$$

where

$$\tilde{\Delta}_{\bar{\tau}} = \frac{(2i-1)^2 D_x}{l_x^2} + \frac{(2j-1)^2 D_y}{l_y^2} + \frac{(2k-1)^2 D_z}{l_z^2} + \frac{1}{\pi^2 \bar{\tau}}. \tag{19}$$

Note that if we use cubes instead of rectangular parallelepipeds in the RWP method, then we can precalculate the absorption probability Q_{abs} for different edge lengths l and tabulate it on the interval $[l_{\min}, l_{\max}]$, where $l_{\min}$ and $l_{\max}$ are, respectively, the minimum and the maximum possible sizes of the cubes. For this reason, in the following computations, we apply the random-walk-on-cubes method, which is the RWP method with $l_x = l_y = l_z$.

If a particle is absorbed inside the parallelepiped, we need to calculate the absorption time. The density of the absorption time is given in [22] and has the form

$$p_{\text{abs}}(\tau) = \frac{1}{Q_{\text{abs}}} \frac{64}{\pi^3} \frac{1}{\bar{\tau}} \exp\left(-\frac{\tau}{\bar{\tau}}\right) F_2(D_x, \tau) F_2(D_y, \tau) F_2(D_z, \tau). \tag{20}$$

The density $p_{\text{abs}}(\tau)$ is simulated by rejection method with the upper-bound function

$$g_3(\tau) = \frac{1}{Q_{\text{abs}}} \frac{64}{\pi^3} \frac{1}{\bar{\tau}} \exp(-\pi^2 \Delta \cdot \tau), \tag{21}$$

where Δ was defined in (8). The random time τ_a is sampled with exponential density by using the well-known formula $\tau_a = -\dfrac{1}{\pi^2 \Delta} \ln(rand)$, in which $rand$ stands for a number uniformly distributed on $[0, 1]$. If $rand_1 \cdot g_3(\tau_a) \leq p_{\text{abs}}(\tau_a)$, where $rand_1$ is sampled independently of $rand$, we accept the value τ_a; otherwise, we sample a new value of τ_a and continue until the random number is sampled.

2.3 The Random-Walk-on-Parallelepiped Algorithm for the Transient Anisotropic Diffusion-Reaction Problem

We will follow [22] to describe the random-walk-on-cubes (RWC) algorithm for the simulation of exciton anisotropic diffusion with recombination in a semiconductor. The algorithm computes the transient CL intensity $I_{\text{cl}}(t)$, the concentration of survived excitons $I_{\text{surv}}(t)$, the flux $F_{z=0}(t)$ to the plain Γ_z, and the flux $F_{\text{dis}}(t)$ to the dislocation surfaces Γ_{d}.

1. Precalculate the auxiliary functions $F_1(\bar{t})$ and $F_2(\bar{t})$ (see (4)), where $\bar{t} = \dfrac{D_i \tau}{l^2}$, $i \in \{x, y, z\}$, $\bar{t} \in [\bar{t}_{\min}, \bar{t}_{\max}]$, $l \in [l_{\min}, l_{\max}]$.

   Precalculate, by using the Walker algorithm [23], the arrays of aliases and reconstructed probabilities for simulation of the one-dimensional probability density $p_{D_i}(v, \bar{t})$ (see (6)), where $v = i/l$.

   Precalculate the absorption probability $Q_{\mathrm{abs}}(l)$ (see (18)) for edge lengths $l \in [l_{\min}, l_{\max}]$.

2. Set all scores to zero: $I_{\mathrm{cl}}(t) := 0$, $I_{\mathrm{surv}}(t) := 0$, $F_{z=0}(t) := 0$, $F_{\mathrm{dis}}(t) := 0$ for $t \in [0, T]$, where T is the maximum simulation time.

3. Set the index of the walking step of the exciton trajectory to one: $k := 1$. Generate the coordinates of the starting position of the exciton, $\mathbf{r}_k = (x_0, y_0, z_0)$, by using the generating function $f(\mathbf{r})$.

   Set the exciton current time to zero: $\tau := 0$.

4. Construct a parallelepiped Π_k with its center at the point $\mathbf{r}_k = (x_k, y_k, z_k)$ and edge lengths equal to twice the minimum of two distances, namely the distance from the center to the plane $z = 0$ and the distance from the center to the closest dislocation surface: $l = 2 \cdot \min(d_{z=0}, d_{\mathrm{dis}})$.

5. Take the absorption probability $Q_{\mathrm{abs}}(l)$ from the precalculated array.

6. If $rand \leq Q_{\mathrm{abs}}(l)$, where $rand$ is a random number uniformly distributed on $[0, 1]$, then the exciton does not survive:

   use the rejection method to simulate the absorption time τ_{abs} inside the cube Π_k with density $p_{\mathrm{abs}}(\tau)$ (see (20));

   increase the exciton current time: $\tau := \tau + \tau_{\mathrm{abs}}$;

   increment by 1 the CL intensity for $\forall \tau_j > \tau$: $I_{\mathrm{cl}}(\tau_j) := I_{\mathrm{cl}}(\tau_j) + 1$;

   the trajectory is terminated; go to Step 3 to start a new trajectory.

7. Otherwise (i.e. if $rand > Q_{\mathrm{abs}}(l)$), the exciton survives:

   simulate the first passage time τ_s with density $p_t(\tau)$ (see (3)) by using the "Algorithm for sampling the first passage time", and the random exit position $\mathbf{r}_{k+1} = (x_{k+1}, y_{k+1}, z_{k+1})$ on the sides of the cube with density $p(x_k, y_k, z_k | \tau_s)$ (see (5)) by using the superposition method;

   increase the exciton current time: $\tau := \tau + \tau_s$;

   set $\mathbf{r}_{k+1}$ as the new coordinate of the exciton.

8. Check whether $\mathbf{r}_{k+1}$ hits the plane $z = 0$. If $z_{k+1} = 0$ or $z_{k+1} < \varepsilon$, where ε is a small positive number of the order of 10^{-5}, the exciton is recombined on the plane $z = 0$;

   increment by 1 the flux $F_{z=0}(\tau_j)$ to the plane $z = 0$ for $\forall \tau_j > \tau$;

   the trajectory is terminated; go to Step 3 to start a new trajectory.

9. Check whether $\mathbf{r}_{k+1}$ hits the surface of one of the dislocations. If the distance d_{dis} from the point $\mathbf{r}_{k+1}$ to the closest dislocation is zero or $d_{\mathrm{dis}} < \varepsilon$, then the exciton recombines on the dislocation surface;

   increment by 1 the flux $F_{\mathrm{dis}}(\tau_j)$ to the dislocation surfaces for $\forall \tau_j > \tau$;

   the trajectory is terminated; go to Step 3 to start a new trajectory.

10. If $\tau \geq T$, i.e. the exciton current time τ exceeds the maximum time T, the trajectory terminates. Go to Step 3 to start a new trajectory.

11. Otherwise (if $\tau < T$), increase the walking step: $k := k + 1$, and go to Step 4.

We obtain the CL intensity and the fluxes to the plane $z = 0$ and to the dislocation surfaces by averaging the scores over N independent trajectories. The concentration of excitons that survived to the time t_k is computed as $I_{surv}(t_k) = 1 - I_{cl}(t_k) - F_{z=0}(t_k) - F_{dis}(t_k)$.

3 Parallel Implementation of the Random-Walk-on-Cubes Algorithm for the Transient Anisotropic Diffusion-Reaction Problem

If we want to compute the CL intensity, the concentration of survived excitons, and the fluxes to the plain $z = 0$ and to the dislocation surfaces with a high degree of accuracy, then we need a quite large number of trajectories. The computational time for large diffusion lengths and a large number of dislocations is of several hours. For this reason, in this paper we considered the parallel implementation of the RWC algorithm for simulation of the transient anisotropic diffusion-reaction problem as the most reasonable choice. The general approach to the parallel implementation of Monte Carlo algorithms consists in distributing independent trajectories among computing nodes [24, 25].

First, we distribute N_{tr} trajectories among n_{mpi} MPI processes. Then, for each MPI process, we distribute (N_{tr}/n_{mpi}) trajectories among n_{omp} OpenMP threads. Each OpenMP thread computes the values of the scores $I_{cl}(t)$, $I_{surv}(t)$, $F_{z=0}(t)$, and $F_{dis}(t)$, using its own arrays for storing the scores. Then, each MPI process sums up the score values obtained by its child OpenMP threads. The root MPI process gathers the values obtained for the scores and averages them. In the code, we use the Mersenne Twister pseudorandom number generator "MT19937", implemented in the Intel MKL. For generating an independent random number sequence for each OpenMP thread, we apply the block-splitting method through the function vslSkipAheadStream. In addition, the most time consuming loops are vectorized by the OpenMP simd directive and by the auto-vectorization supported by the Intel Compiler 17.0.4.196. In the parallel code, we use the OpenMP loop schedule clause "dynamic" with iteration block size $chunk\_size = 1000$. It slightly speeds up the code execution.

The computations were performed on the "Broadwell" partition of the cluster "MVS-10P" at the Joint Supercomputer Center of the RAS [26]. A "Broadwell" node consists of two processors Intel Xeon CPU E5-2697A v4 2.60 GHz, each containing 16 cores with 2 threads per core. Thus, each "Broadwell" node has 32 physical cores or 64 logical CPUs.

To estimate the efficiency of the parallel code, we set the following values for the parameters of the anisotropic diffusion-reaction problem: diffusion coefficients: $D_x = 1$ nm^2/ns, $D_y = 0.5$ nm^2/ns, $D_z = 2$ nm^2/ns; exciton life time: $\bar{\tau} = 1$ ns; domain size along the axes X and Y: $G_{xy} = 100$ nm; size of the neighborhood of the boundaries: $\varepsilon_z = 10^{-4}$ nm for Γ_z and $\varepsilon_{dis} = 0$ nm for Γ_d; number of dislocations: $N_{dis} = 10$. The coordinates x and y are uniformly distributed on $[0, G_{xy}]$, whereas the coordinate z is exponentially distributed, $z \sim \text{Exp}(0.01)$. The number of trajectories $N_{tr} = 10^8$. The functions $F_1(\bar{t})$,

$F_2(\bar{t})$, and $Q_{abs}(l)$ are precalculated for the parameter ranges $\bar{t} \in [10^{-6}, 2]$ with step $d\bar{t} = 10^{-6}$, and $l \in [0.001, 12L^2]$ with step $dl = 0.001$, where the parameter $L^2 = \max(D_x, D_y, D_z) \cdot \bar{\tau} = 2$ nm is the diffusion length.

First, we found the optimal number of MPI processes and OpenMP threads per node. Table 1 contains the computational time for various numbers of MPI processes n_{mpi} and OpenMP threads n_{omp}. The minimum computational time was attained for the parallel code implementation with 2 MPI processes and 32 OpenMP threads. However, the time obtained for $n_{mpi} = 1$ with $n_{omp} = 64$ was close to the one obtained for $n_{mpi} = 2$ with $n_{omp} = 32$. The minimum computational time was reached for $n_{mpi} = 2$ and $n_{omp} = 32$ in several executions of the code, which means that it is not a random result. Therefore, this ratio of MPI processes and OpenMP threads was used for further computations.

Table 1. The computational time of the parallel implementation of the code executed on a single cluster node

n_{mpi}	1	2	4	8	16	32	64
n_{omp}	64	32	16	8	4	2	1
T, s	65.41	**65.38**	65.76	66.28	66.94	67.22	67.56

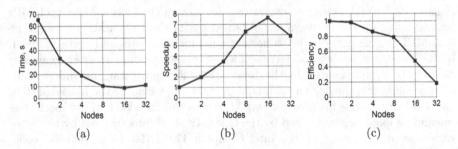

(a)   (b)   (c)

Fig. 1. The characteristics of the code parallel implementation on multiple cluster nodes: (a) computational time, (b) speedup, (c) efficiency

Let us now analyze the parallel code performance on multiple cluster nodes. The following characteristics were calculated for n_{node} cluster nodes: the computational time $T(n_{node})$, the speedup $S(n_{node}) = T(1)/T(n_{node})$, and the efficiency $E(n_{node}) = S(n_{node})/n_{node}$. The calculations speed up when the number of nodes is increased up to 16 (Fig. 1). If the number of nodes is further increased, then the code execution is slowed down: the computational time for 32 nodes is greater than for 16 nodes, speedup drops. Moreover, the efficiency strongly decreases when the number of nodes equals 16. The drop in efficiency and speedup can be explained by an insufficient loading of the cluster nodes. The total number of trajectories is $N_{tr} = 10^8$. Each node computes $N_{tr}/(64 \cdot n_{node})$ trajectories. Experimentally, we found out that a number of trajectories less than

10^6 is not enough to fully load the nodes and leads to a drop in parallelization efficiency. For example, the number of trajectories for 8 nodes is $195\,312.5 > 10^6$; for 16 nodes, $N_{\mathrm{tr}} = 97\,656.3 < 10^6$; for 32 nodes, $N_{\mathrm{tr}} = 48\,828.1 < 10^6$.

On the one hand, the best speedup is equal to 7.6 and is reached when using 16 "Broadwell" nodes. On the other hand, the efficiency on 16 nodes drops to a half of the efficiency on 8 nodes. Thus, the optimal number of nodes for the code execution is eight.

4   Simulation Results for the Transient Anisotropic Diffusion-Reaction Problem

In the computations described below, we used the following values for the modeling parameters: domain size along the X and Y axes: $G_{xy} = 100\,\mathrm{nm}$; size of the neighborhood of the boundaries: $\varepsilon_z = 10^{-4}\,\mathrm{nm}$, $\varepsilon_{\mathrm{dis}} = 0\,\mathrm{nm}$. The source $f(\mathbf{r})$ generates a random exciton position as follows. The coordinates x and y are uniformly distributed on $[0, G_{xy}]$, whereas the z coordinate z is exponentially distributed, $z \sim Exp(0.01)$. The number of trajectories $N_{\mathrm{tr}} = 10^8$. The functions $F_1(\bar{t})$, $F_2(\bar{t})$, and $Q_{\mathrm{abs}}(l)$ are precalculated for the parameter ranges $\bar{t} \in [10^{-6}, 2]$ with step $d\bar{t} = 10^{-6}$, and $l \in [0.001, 12L^2]$ with step $dl = 0.001$, as described above.

Firstly, we verified the parallel implementation of the Monte Carlo RWC algorithm by comparing the simulation results against the exact solution of the isotropic diffusion-reaction equation (see (1)) for a domain without dislocations. The exact solution of Eq. (1) in this case was obtained in explicit form in [21]:

$$u(t)^* = \exp\left(-\frac{t}{\bar{\tau}}\right) \exp\left(\frac{Dt}{z_s^2}\right) \mathrm{erfc}\left(\frac{\sqrt{Dt}}{z_s}\right), \tag{22}$$

where z_s is the size of the exciton source along the axis Z and D is a diffusion coefficient. For this test, we took the following values for the parameters of the diffusion-reaction problem: diffusion coefficients: $D_x = D_y = D_z = D = 2\,\mathrm{nm}^2/\mathrm{ns}$; exciton life time $\bar{\tau} = 10\,\mathrm{ns}$; $z_s = 100\,\mathrm{nm}$. Figure 2 shows that the concentration of survived excitons I_{surv} obtained by simulation is in good agreement with the exact solution $u(t)^*$.

Secondly, to verify the parallel code in the case of a domain with dislocations, we compared the simulation results obtained by two methods: random walk on cubes and random walk on spheres. The RWS method solves the diffusion-reaction problem in the case of isotropic diffusion [13]. Therefore, we took equal diffusion coefficients, $D_x = D_y = D_z = 2\,\mathrm{nm}^2/\mathrm{ns}$, an exciton life time $\bar{\tau} = 10\,\mathrm{ns}$, and assumed that there is a single dislocation in the domain, at the point $(0, 0)$. In this test, we considered a unit point source at $(10, 10, 10)$ as exciton source. Figure 3 portrays the concentration of survived excitons I_{surv}, the CL intensity I_{cl}, the flux $F_{z=0}$ to the plane $z = 0$, and the flux F_{dis} to the dislocation surface, computed by the RWC and the RWS methods. The characteristic transients obtained by both methods coincide with each other.

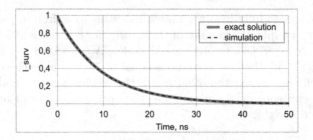

Fig. 2. Comparison of the simulation results with the exact solution of the diffusion-reaction problem in the case of isotropic diffusion and a domain without dislocations

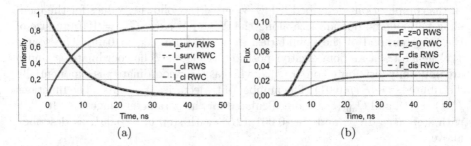

(a)   (b)

Fig. 3. Comparison of the simulation results obtained by the RWC and the RWS methods for a single dislocation at the point $(0, 0)$, a unit source at the point $(10, 10, 10)$, and $D_x = D_y = D_z = 2 \text{ nm}^2/\text{ns}$, $\bar{\tau} = 10 \text{ ns}$. The characteristics obtained by the RWC and the RWS methods are denoted by the characteristic's notation along with the name of the method (for example, "L_cl RWC" is the CL intensity obtained by the RWC method)

Let us now investigate the impact of the diffusion anisotropy on the characteristic transients. The diffusion coefficients along the X and Y axes were taken equal: $D_x = D_y = 2 \text{ nm}^2/\text{ns}$, whereas D_z was varied from 0.1 to 10 nm^2/ns. The exciton life time $\bar{\tau} = 10$ ns. The source $f(\mathbf{r})$ generated the random exciton positions, as described above. The N_{dis} dislocations were uniformly distributed in the domain at random positions. The number of dislocations $N_{\text{dis}} = 50$. As shown in Fig. 4, the increase of D_z from 0.1 to 10 nm^2/ns has a weak impact on the intensities of survived particles and CL, as well as on the flux to the dislocation surfaces. However, a slight decrease in the values of I_{surv}, I_{cl}, and F_{dis} as a whole leads to a noticeable increase in the flux $F_{z=0}$ to the plane $z = 0$.

Next, we wanted to determine how the dislocation density influences the characteristic transients. We took the following diffusion coefficients: $D_x = 1 \text{ nm}^2/\text{ns}$, $D_y = 0.5 \text{ nm}^2/\text{ns}$, $D_z = 2 \text{ nm}^2/\text{ns}$. The life time $\bar{\tau} = 10$ ns. The exciton source $f(\mathbf{r})$ was the same as in the previous test. The number of dislocations N_{dis} varied from 1 to 1000. The dislocation density is the ratio $N_{\text{dis}}/G_{xy}^2 \text{ nm}^{-2}$. When the dislocation density increases, the flux F_{dis} to the dislocation surfaces considerably increases, whereas the other characteristics, i.e. I_{surv}, I_{cl}, and $F_{z=0}$, decrease (Fig. 5).

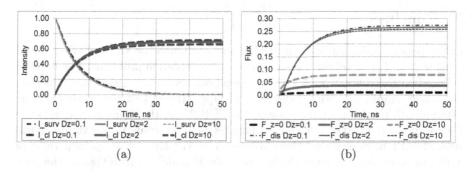

Fig. 4. Influence of the diffusion anisotropy on the intensities of survived excitons and CL (a), as well as on the fluxes to the plane $z = 0$ and to the dislocation surfaces (b), for $N_{\text{dis}} = 50$, $D_x = D_y = 2$ nm^2/ns, $D_z \in \{0.1, 2, 10\}$ nm^2/ns, and $\bar{\tau} = 10$ ns. The characteristics corresponding to different D_z values are denoted by the characteristic's notation along with the value of D_z (for example, "I_cl Dz = 2" is the CL intensity obtained for $D_z = 2$)

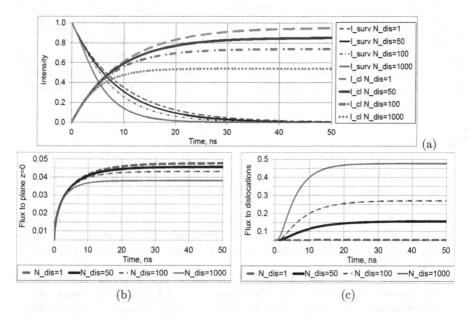

Fig. 5. Influence of the number of dislocations N_{dis} on the intensities of survived excitons and CL (a), the flux to the plane $z = 0$ (b), and the flux to the dislocation surfaces (c), for the following values of the model parameters: $D_x = 1$ nm^2/ns, $D_y = 0.5$ nm^2/ns, $D_z = 2$ nm^2/ns, $\bar{\tau} = 10$ ns. In (a), the characteristics corresponding to different N_{dis} values are denoted by the characteristic's notation along with the value of N_{dis} (for example, "I_cl N_dis = 1" is the CL intensity obtained for $N_{\text{dis}} = 1$)

5   Conclusions

In this paper, we described a parallel implementation of the Monte Carlo method suggested in [17] for solving anisotropic diffusion-reaction problems. The method is based on the random-walk-on-parallelepipeds (or cubes) algorithm. The parallel code, employing MPI standard and OpenMP API, was executed on the "Broadwell" partition of the cluster "MVS-10P" at the Joint Supercomputer Center of the RAS. We analyzed the parallel code performance for various numbers of MPI processes and OpenMP threads. We found the optimal ratio of MPI processes and OpenMP threads on a single node and investigated the code execution speedup and efficiency depending on the number of nodes.

The parallel implementation of the Monte Carlo random-walk-on-cubes (RWC) algorithm was verified for the isotropic diffusion-reaction problem in two cases: for a domain without dislocations and for a domain with a single dislocation. In the first case, the exact solution derived in [21] was used for comparison. In the second case, the Monte Carlo random-walk-on-spheres (RWS) method was applied to solve the problem with a dislocation. The simulation results obtained by the RWC algorithm are in a good agreement with both the theoretical solution and the results of the RWS simulation. We investigated the influence of the diffusion anisotropy on the intensities of survived excitons and CL, as well as on the fluxes to the plane $z = 0$ and to the dislocation surfaces. In addition, we studied the dependence of the characteristic transients on the dislocation density.

References

1. Kalceff, M.A.S., Phillips, M.R.: Cathodoluminescence microcharacterization of the defect structure of quartz. Phys. Rev. B **52**(5), 3122–3135 (1995). https://doi.org/10.1103/PhysRevB.52.3122
2. Liu, W., Carlin, J.F., Grandjean, N., Deveaud, B., Jacopin, G.: Exciton dynamics at a single dislocation in GaN probed by picosecond time-resolved cathodoluminescence. Appl. Phys. Lett. **109**(4), 042101-1–042101-5 (2016). https://doi.org/10.1063/1.4959832
3. Hauser, A.J., et al.: Characterization of electronic structure and defect states of thin epitaxial BiFeO$_3$ films by UV-visible absorption and cathodoluminescence spectroscopies. Appl. Phys. Lett. **92**, 222901 (2008). https://doi.org/10.1063/1.2939101
4. Boggs, S., Krinsley, D.: Application of Cathodoluminescence Imaging to the Study of Sedimentary Rocks. Cambridge University Press, New York (2006). https://doi.org/10.1017/cbo9780511535475.008
5. Weisbuch, C., Piccardo, M., Martinelli, L., Iveland, J., Peretti, J., Speck, J.S.: The efficiency challenge of nitride light-emitting diodes for lighting. Phys. Status Solidi A **212**(5), 885–1176 (2015). https://doi.org/10.1002/pssa.201570427
6. Edwards, P.R., Martin, R.W.: Cathodoluminescence nano-characterization of semiconductors. Semicond. Sci. Technol. **26**, 064005 (8 p.) (2011) https://doi.org/10.1088/0268-1242/26/6/064005

7. Sabelfeld, K.K., Kaganer, V.M., Pfüller, C., Brandt, O.: Dislocation contrast in cathodoluminescence and electron-beam induced current maps on GaN(0001). J. Phys. D **50**, 405101 (2017). https://doi.org/10.1088/1361-6463/aa85c8
8. Rosner, S.J., Carr, E.C., Ludowise, M.J., Girolami, G., Erikson, H.I.: Correlation of cathodoluminescence inhomogeneity with microstructural defects in epitaxial GaN grown by metalorganic chemical-vapor deposition. Appl. Phys. Lett. **70**(4), 420–422 (1997). https://doi.org/10.1063/1.118322
9. Higgs, V., Lightowlers, E.C., Tajbakhsh, S., Wright, P.J.: Cathodoluminescence imaging and spectroscopy of dislocations in Si and $Si_{1-x}Ge_x$ alloys. Appl. Phys. Lett. **61**, 1087–1089 (1992). https://doi.org/10.1063/1.107676
10. Phang, J.C.H., Pey, K.L., Chang, D.S.H.: A simulation model for cathodoluminescence in the scanning electron microscope. IEEE Transact. Electron Devices **39**(4), 782–791 (1992). https://doi.org/10.1109/16.127466
11. Sabelfeld, K.K.: Splitting and survival probabilities in stochastic random walk methods and applications. Monte Carlo Methods Appl. **22**(1), 55–72 (2016). https://doi.org/10.1515/mcma-2016-0103
12. Sabelfeld, K.K.: Random walk on spheres method for solving drift-diffusion problems. Monte Carlo Methods Appl. **22**(4), 265–275 (2016). https://doi.org/10.1515/mcma-2016-0118
13. Sabelfeld, K.K.: Random walk on spheres algorithm for solving transient drift-diffusion-reaction problems. Monte Carlo Methods Appl. **23**(3), 189–212 (2017). https://doi.org/10.1515/mcma-2017-0113
14. Sabelfeld, K.K.: Monte Carlo Methods in Boundary Value Problems. Springer, Berlin (1991)
15. Irkhin, P., Biaggio, I.: Direct imaging of anisotropic exciton diffusion and triplet diffusion length in rubrene single crystals. Phys. Rev. Lett. **107**, 017402 (2011). https://doi.org/10.1103/PhysRevLett.107.017402
16. Lin, J.D.A., et al.: Systematic study of exciton diffusion length in organic semiconductors by six experimental methods. Mater. Horiz. **1**, 280–285 (2014). https://doi.org/10.1039/c3mh00089c
17. Sabelfeld, K.: Stochastic simulation methods for solving systems of isotropic and anisotropic drift-diffusion-reaction equations and applications in cathodoluminescence imaging, submitted to Probabilistic Engineering Mechanics (2018)
18. Milstein, G.N., Tretyakov, M.V.: Simulation of a space-time bounded diffusion. Ann. Appl. Probab. **9**(3), 732–779 (1999)
19. Deaconu, M., Lejay, A.: A random walk on rectangles algorithm. Method. Comput. Appl. Probab. **8**(1), 135–151 (2006). https://doi.org/10.1007/s11009-006-7292-3
20. Campillo, F., Lejay, A.: A Monte Carlo method without grid for a fractured porous domain model. Monte Carlo Methods Appli. De Gruyter **8**(2), 129–147 (2002). https://doi.org/10.1515/mcma.2002.8.2.129
21. Sabelfeld, K.K., Kireeva, A.: Probability distribution of the life time of a drift-diffusion-reaction process inside a sphere with applications to transient cathodoluminescence imaging. Monte Carlo Methods Appl. **24**(2), 79–92 (2018). https://doi.org/10.1515/mcma-2018-0007
22. Sabelfeld, K.K., Kireeva, A.E.: A meshless random walk on parallelepipeds algorithm for solving transient anisotropic diffusion-recombination equations and applications to cathodoluminescence imaging, submitted to Numerische Mathematik, (2018)
23. Walker, A.J.: An efficient method for generating discrete random variables with general distributions. ACM Transact. Math. Softw. **3**(3), 253–256 (1977). https://doi.org/10.1145/355744.355749

24. Rosenthal, J.S.: Parallel computing and Monte Carlo algorithms. Far East J. Theor. Stat. **4**, 207–236 (2000)
25. Esselink, K., Loyens, L.D.J.C., Smit, B.: Parallel Monte Carlo Simulations. Phys. Rev. E **51**(2), 1560–1568 (1995). https://doi.org/10.1103/physreve.51.1560
26. MVS-10P cluster, JSCC RAS. http://www.jscc.ru

Using Multicore and Graphics Processors to Solve The Structural Inverse Gravimetry Problem in a Two-Layer Medium by Means of α-Processes

Elena N. Akimova[1,2]([✉]) [iD], Vladimir E. Misilov[1,2], and Andrey I. Tretyakov[1,2]

[1] Krasovskii Institute of Mathematics and Mechanics, Ural Branch of RAS,
16 S. Kovalevskaya Street, Ekaterinburg, Russia
[2] Ural Federal University, 19 Mira Street, Ekaterinburg, Russia
aen15@yandex.ru, out.mrscreg@gmail.com, fr1z2rt@gmail.com

Abstract. We construct memory-optimized and time-efficient parallel algorithms (and the corresponding programs) taking advantage of regularized modified α-processes, namely the modified steepest descent method and the modified minimal residual method, for solving the nonlinear equation of the structural inverse gravimetry problem. Memory optimization relies on the block-Toeplitz structure of the Jacobian matrix. The algorithms are implemented on multicore CPUs and GPUs through the use of, respectively, OpenMP and NVIDIA CUDA technologies. We analyze the efficiency and speedup of the algorithms. In addition, we solve a model problem of gravimetry and conduct a comparative study regarding the number of iterations and computation time against algorithms based on conjugate gradient-type methods and the componentwise gradient method. The comparison demonstrates that the algorithms based on α-processes perform better, reducing the number of iterations and the computation time by as much as 50%.

Keywords: Nonlinear gradient-type methods · α-processes ·
Parallel algorithms · Gravimetry problems · Toeplitz matrix ·
Multicore CPU · GPU

1 Introduction

An essential importance in the study of the Earth's crust structure is attributed to the solution of the structural inverse gravimetry problem, which consists in finding the interface between layers with different densities by resorting to known density contrasts and a gravitational field [17].

The problem is ill-posed. Thus, it is necessary to apply iterative regularization methods [11,18].

Effective methods to determine structural boundaries were constructed in [1–3,5,7,9,14,21], namely the regularized Newton method, the regularized linearized steepest descent method, the regularized linearized minimum error

© Springer Nature Switzerland AG 2019
L. Sokolinsky and M. Zymbler (Eds.): PCT 2019, CCIS 1063, pp. 285–296, 2019.
https://doi.org/10.1007/978-3-030-28163-2_20

method, the regularized linearized conjugate gradient method, the regularized componentwise gradient method, and the nonlinear analogs of α-processes.

The modified Newton method [11,12,18], the modified steepest descent method [19], the modified analogs of α-processes [20], and the modified conjugate gradient method [8] are based on the idea of computing the derivative of the integral operator in the step operator of the process at the same fixed point throughout the whole iterative process. It was shown that this approach reduces the rate of convergence but expands the domain of convergence. Indeed, we do not need to recalculate the derivative at every iteration; thus, both the number of operations performed at every iteration and the total computation time decrease [8].

Memory-efficient algorithms were constructed in [4,6] and applied to the solution of the inverse problem of finding a variable density in a horizontal or curvilinear layer. It was established that the modification based on the approximation of the SLAE matrix by a Toeplitz-block-Toeplitz matrix significantly reduces memory requirements.

In this paper, we construct parallel algorithms based on modified analogs of both the steepest descent method and the minimal residual method. The algorithms are used for solving the inverse gravimetry problem of finding the density interface. Additionally, we make use of a memory optimization technique based on the Toeplitz-block-Toeplitz structure of the Jacobian matrix calculated at a fixed point.

We implement the parallel algorithms on multicore CPUs and GPUs by using, respectively, OpenMP and NVIDIA CUDA technologies. The parallel programs can be used to solve a model problem of gravimetry with a large grid. The programs were executed on the Uran supercomputer, which employs Intel Xeon E5-2660 CPUs in conjunction with NVIDIA Tesla M2090 GPUs and is installed at the Institute of Mathematics and Mechanics of the Ural Branch of the Russian Academy of Sciences.

We investigate the rate of convergence of the modified and unmodified variants of α-processes. Moreover, we study the efficiency and speedup of the constructed parallel algorithms based on α-processes and compare them regarding the number of iterations and computation time with algorithms based on the conjugate gradient method and the componentwise method.

2   Statement of the Inverse Problem

Let us introduce a Cartesian coordinate system in which the xOy plane coincides with the Earth's surface and the z axis is directed downwards, as shown in Fig. 1. Assume that the lower half-space consists of two layers with constant densities σ_1 and σ_2, separated by the desired surface which is described by a bounded function $\zeta = \zeta(x, y)$ such that $\lim\limits_{|x|+|y|\to\infty} \left(h - \zeta(x, y)\right) = 0$ for some h. The function ζ must satisfy the following equation:

$$f\Delta\sigma \int\limits_{-\infty}^{\infty} \int\limits_{-\infty}^{\infty} \left\{ \frac{1}{\left((x-x')^2 + (y-y')^2 + \zeta^2(x',y')\right)^{1/2}} \right.$$

$$\left. - \frac{1}{\left((x-x')^2 + (y-y')^2 + h^2\right)^{1/2}} \, dx' \, dy' \right\} = \Delta g(x,y,0), \tag{1}$$

where f is the gravitational constant, $\Delta\sigma = \sigma_2 - \sigma_1$ is the density contrast, and $\Delta g(x,y,0)$ is an anomalous gravitational field measured at the surface of the Earth.

Preliminary processing of gravitational data with the aim of extracting the anomalous field is performed according to a technique suggested by Martyshko and Prutkin [15,16].

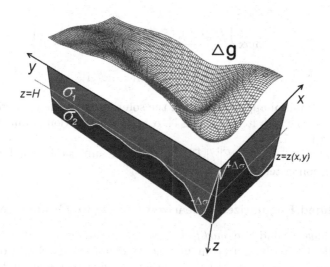

Fig. 1. Model of a two-layer medium for a gravimetry problem

Equation (1) is a nonlinear two-dimensional integral equation of the first kind.

After the discretization of the region $\Pi = \{(x,y) : a \leqslant x \leqslant b, \ c \leqslant y \leqslant d\}$ by means of a grid of size $n = M \times N$ and the approximation of the integral operator by using quadrature rules, we obtain a right-hand part vector F and an approximation of the solution vector z of dimension n. Equation (1) takes the form

$$f\Delta\sigma\Delta x\Delta y \sum_{j=1}^{n} \left[\frac{1}{\sqrt{(x_i - x_j)^2 + (y_i - y_j)^2 + z_j^2}} \right.$$

$$\left. - \frac{1}{\sqrt{(x_i - x_j)^2 + (y_i - y_j)^2 + h^2}} \right] = F_i, \quad i = 1,\ldots,n. \tag{1a}$$

We can rewrite this equation as

$$A(z) = F. \tag{2}$$

3   Numerical Methods for Solving the Problem

3.1   Regularized Linearized Conjugate Gradient Method

The *regularized linearized conjugate gradient method* (RLCGM) [7] has the form

$$
\begin{aligned}
z^{k+1} &= z^k - \psi \frac{\langle p^k, S_\alpha(z^k) \rangle}{\|A'(z^k)p^k\|^2 + \alpha \|p^k\|^2} p^k, \\
p^k &= S_\alpha(z^k) + \beta^k p^{k-1}, \\
p_0 &= S_\alpha(z^0), \\
\beta_k &= \max \left\{ \frac{\langle S_\alpha(z^k), (S_\alpha(z^k) - S_\alpha(z^{k-1})) \rangle}{\|S_\alpha(z^{k-1})\|^2}, 0 \right\}, \\
S_\alpha(z) &= A'(z)^T (A(z) - F) + \alpha(z - z^0),
\end{aligned} \tag{3}
$$

where z^0 is the initial approximation of the solution, z^k is the approximation of the solution at the k-th iteration, $k \in \mathbb{N}$, α is a regularization parameter, and ψ is a damping factor.

The condition $\|A(z^k) - F\|/\|F\| < \varepsilon$ for some sufficiently small ε is chosen as a termination criterion.

3.2   Modified Regularized Linearized Conjugate Gradient Method

The main idea behind the *modified regularized linearized conjugate gradient method* (MRLCGM) [8] is to compute the Jacobian matrix $A'(z^k)$ of the non-linear operator at an initial point z^k without having to update it throughout the entire iterative process. As a result, the method becomes more economical in terms of the number of operations executed at each iteration. Moreover, numerical experiments show that, in many cases, this modification reduces the computation time required for achieving the same accuracy of the desired solution, compared with the unmodified method. The MRLCGM has the form

$$
\begin{aligned}
z^{k+1} &= z^k - \psi \frac{\langle p^k, S_\alpha^0(z^k) \rangle}{\|A'(z^0)p^k\|^2 + \alpha \|p^k\|^2} p^k, \\
p^k &= S_\alpha^0(z^k) + \beta^k p^{k-1}, \\
p^0 &= S_\alpha^0(z^0), \\
\beta^k &= \max \left\{ \frac{\langle S_\alpha^0(z^k), (S_\alpha^0(z^k) - S_\alpha^0(z^{k-1})) \rangle}{\|S_\alpha^0(z)^{k-1}\|^2}, 0 \right\}, \\
S_\alpha^0(z) &= A'(z^0)^T (A(z) - F) + \alpha(z - z^0).
\end{aligned} \tag{4}
$$

3.3 Regularized Componentwise Gradient Method

The *regularized componentwise gradient method* (RCWM) [5] has the form

$$z_i^{k+1} = z_i^k - \psi \frac{A_i(z^k) - F_i + \alpha \left\| z^k - z^0 \right\|^2}{\left\| \nabla A_i(z^k) \right\|^2 + \alpha} \left(\frac{\partial A_i(z^k)}{\partial z_i} \right), \tag{5}$$

where z_i is the i-th component of the approximate solution, $i = 1, \ldots, n$, $k \in \mathbb{N}$, and ψ is a damping factor.

The main idea behind this method is to minimize the residual $A_i(z) - F_i$ at single grid point by changing the value z_i at the same grid point. This idea is founded on the fact that the value of the gravity or magnetic (in the case of a vertically directed magnetization) field depends on $1/r^2$. Thus, the value of z_i has the greatest influence on the field value F_i at the point directly above it.

3.4 Modified α-Processes

A nonlinear analog of the α-processes was proposed by Vasin [20]. It is a class of iterative processes that has the following form:

$$z^{k+1} = z^k - \psi \frac{\left\langle \left(A'(z^k) + \overline{\alpha}I \right)^\varkappa \overline{S_\alpha}(z^k), \overline{S_\alpha}(z^k) \right\rangle}{\left\langle \left(A'(z^k) + \overline{\alpha}I \right)^{\varkappa+1} \overline{S_\alpha}(z^k), \overline{S_\alpha}(z^k) \right\rangle} \overline{S_\alpha}(z^k),$$
$$\overline{S_\alpha}(z) = A(z) - F + \alpha(z - z^0), \tag{6}$$

where α and $\overline{\alpha}$ are positive regularization parameters, $\psi > 0$ is a damping factor, and the parameter $\varkappa$ determines the extremal principle for the gradient-type process. For $\varkappa = -1, 0, 1$, we obtain, respectively, a nonlinear regularized variant of the minimal error method, a nonlinear regularized analog of the steepest descent method, and the minimal residual method.

In this paper, we use a modified variant of the α-processes, namely

$$z^{k+1} = z^k - \psi \frac{\left\langle \left(A'(z^0) + \overline{\alpha}I \right)^\varkappa \overline{S_\alpha}(z^k), \overline{S_\alpha}(z^k) \right\rangle}{\left\langle \left(A'(z^0) + \overline{\alpha}I \right)^{\varkappa+1} \overline{S_\alpha}(z^k), \overline{S_\alpha}(z^k) \right\rangle} \overline{S_\alpha}(z^k). \tag{7}$$

Here, the value of the Jacobian matrix A' is calculated at the initial approximation z^0 and the obtained value is used throughout the entire process.

The *modified steepest descent method* for $\varkappa = 0$ (modSDM), the *modified minimal residual method* for $\varkappa = 1$ (modMRM), and the *modified minimal error method* for $\varkappa = -1$ (modMEM) take, respectively, the forms

$$z^{k+1} = z^k - \psi \frac{\left\| \overline{S_\alpha}(z^k) \right\|^2}{\left\langle \left(A'(z^0) + \overline{\alpha}I \right) \overline{S_\alpha}(z^k), \overline{S_\alpha}(z^k) \right\rangle} \overline{S_\alpha}(z^k); \tag{8}$$

$$z^{k+1} = z^k - \psi \frac{\left\langle \left(A'(z^0) + \overline{\alpha}I \right) \overline{S_\alpha}(z^k), \overline{S_\alpha}(z^k) \right\rangle}{\left\| \left(A'(z^0) + \overline{\alpha}I \right) \overline{S_\alpha}(z^k) \right\|^2} \overline{S_\alpha}(z^k); \tag{9}$$

$$z^{k+1} = z^k - \psi \frac{\left\langle \left(A'(z^0) + \overline{\alpha}I \right)^{-1} \overline{S_\alpha}(z^k), \overline{S_\alpha}(z^k) \right\rangle}{\left\| \overline{S_\alpha}(z^k) \right\|^2} \overline{S_\alpha}(z^k). \tag{10}$$

Remark 1. Apparently, methods (8) and (9) require one matrix-vector product and two dot products per iteration, while conjugate gradient-type methods (3) and (4) require two matrix-vector products and five dot products. In addition, the unmodified RLCGM and RCWM require recalculation of the Jacobian matrix $A'(z^k)$ at each iteration. One iteration of the componentwise gradient method (5) is equivalent to one matrix-vector product and one dot product, but this method requires a larger number of iterations to reach the desired accuracy.

Remark 2. The minimal error method (10) requires inversion of the matrix $A'(z^0)$ or the approximate solution of a system of linear algebraic equations at each iteration. Therefore, the implementation of this method is essentially more time-consuming as compared with the other two.

4   Matrix Structure and Storage Method

Storing the value of the Jacobian matrix A' can be very memory-consuming in the case of large grids; therefore, it is worthwhile investigating the matrix structure in an attempt to optimize the storage method.

Let us assume that A' is a block matrix. Then, its elements can be defined as

$$a_{k,p,l,q} = a_{(k-1)M+p,(l-1)M+q} =$$

$$= f \Delta\sigma \Delta x \Delta y \left(\frac{-z_{(k-1)M+p}}{\left((x_k - x_l)^2 + (y_p - y_q)^2 + z_{(k-1)M+p}^2 \right)^{3/2}} \right),$$

where $k, l = 1, \ldots, M$ are the block indices and $p, q = 1, \ldots, N$ are the indices of the elements inside each block.

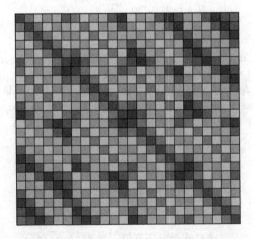

Fig. 2. Matrix structure for a 6×4 grid (Color figure online)

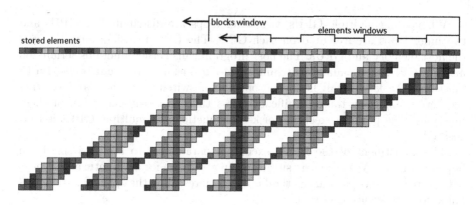

Fig. 3. Matrix storage algorithm

For modified methods (4), (8), and (9), when the Jacobian matrix is computed only once, we can use $z^0 \equiv h$ as the initial approximate solution. Thus, the elements depend only on the terms $(x_k - x_l)^2 + (y_p - y_q)^2$. In this case, A' is a symmetric Toeplitz-block-Toeplitz matrix.

The matrix structure for a 6×4 grid with square grid elements is shown in Fig. 2. The matrix is 24×24 and has 4×4 blocks of dimension 6×6. Equal elements are marked with the same color.

For storing and accessing elements of this matrix, we suggest an algorithm [4] consisting in the following. We store the concatenation of the first and last lines of each block. To get the desired line of the matrix, we use two moving windows on the stored line.

The scheme of this algorithm for a 24×24 matrix is depicted in Fig. 3.

The stored vector has a length of $(2N-1)(2M-1)$ for an $MN \times MN$ matrix.

5   The Parallel Implementation

The parallel algorithms based on the above-described methods were implemented on multicore CPUs and NVIDIA M2090 GPUs taking advantage of OpenMP and CUDA technologies, respectively. The optimized storage algorithm is used only in the GPU implementations.

Note that storing the Jacobian matrix for a $2^9 \times 2^9$ grid takes more than 512 GB, whereas the optimized storage takes only 8 MB.

In the algorithms that do not use the optimized algorithm for storing the Jacobian matrix (OpenMP implementations and GPU implementations of the RLCGM and the RCWM), we must recalculate it at each iteration. In this case, we resort to the on-the-fly calculation of the elements, i.e. the value of the matrix element is computed when calling the element without storing it in memory.

The costliest operation is computing the values of the integral operator and its Jacobian matrix. It consists of four nested loops. In the OpenMP implementation, outer loops are parallelized by means of the #pragma omp parallel directive, while inner loops are vectorized using the #pragma simd directive.

When using multiple GPUs, two outer loops are distributed to GPUs and two inner loops are executed on each GPU. The CPU transfers data between the host memory and GPUs. The host program calls kernel functions serially on multiple GPUs by manually switching devices by means of the `cudaSetDevice()` function. The kernel call is asynchronous; so, switching devices and starting the kernel functions takes significantly less time than the execution of these functions. The parallel execution of kernel functions on multiple GPUs is thus ensured.

The adjustment of the kernel execution parameters to the grid size is an important issue. We previously suggested an algorithm for automatic adjustment of these parameters [8]. This method is based on rescaling the optimal parameters found for a reference grid size.

This imposes some constraints on the input data and GPU configuration:

– the grid size must be divisible by 128 (i.e. $128, 256, 512, 1024, \ldots$);
– the number of GPUs must be a power of 2 (i.e. $1, 2, 4, 8, \ldots$).

6   Numerical Experiments

We considered a model problem of gravimetry that consists in finding the interface between two layers. The model is based on the gravitational field of the area near the city of Ekaterinburg, Russia.

Figure 4 portrays the model gravitational field $\Delta g(x, y, 0)$. This field was obtained by solving the direct gravimetry problem using the known surface z^* with the asymptotic plane $H = 30$ km and the density contrast $\Delta\sigma = 0.2$ g/cm^3. The surface is shown in Fig. 5.

The problems were solved on the Uran supercomputer, with each node having two 8-core Intel E5-2660 CPUs and eight NVIDIA Tesla M2090 GPUs. The solution was found by six methods:

– regularized linearized conjugate gradient method (RLCGM) (see (3));
– modified regularized linearized conjugate gradient method (MRLCGM) (see (4));
– regularized componentwise gradient method (RCWM) (see (5));
– unmodified steepest descent method (SDM) (see (6) for $\varkappa = 0$);
– modified steepest descent method (modCWM) (see (8));
– modified minimal residual method (modCWM) (see (9)).

The reconstructed interface z is shown in Fig. 6.

The termination criterion for all six methods was set for $\varepsilon = 0.01$. The parameters $\psi = 1$ and $\alpha = 0.1$ were used for all six methods.

The relative error of all six solutions found is $\delta = \|z - z^*\|/\|z^*\| < 0.01$.

Table 1 shows the number N of iterations, as well as the average execution time on one core (T_1) and on eight cores (T_8) of the Intel E5-2660 CPU, for a 512×512 grid.

Speedup and efficiency coefficients are used to analyze the scaling of parallel algorithms. The speedup is defined as $S_m = T_1/T_m$, where T_1 is the execution

Table 1. Comparison of methods

Method	N	T_1 [minutes]	T_8 [minutes]	S_8	E_8
LRCGM (3)	30	274	36	7.6	0.95
MLRCGM (4)	40	251	33	7.6	0.95
RCWM (5)	21	140	18	7.8	0.97
SDM (6), $\varkappa = 0$	18	150	19	7.8	0.98
modSDM (8)	15	125	16	7.8	0.98
modMRM (9)	15	125	16	7.8	0.98

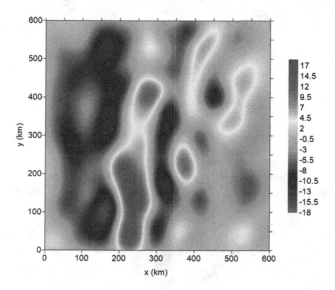

Fig. 4. Model gravitational field

time of the program run on a single CPU and T_m is the execution time on m CPUs. The efficiency is defined as $E_m = S_m/m$. The ideal values are $S_m = m$ and $E_m = 1$; however, real values are lower because of overheads.

Table 2 summarizes the average execution time, speedup, and efficiency for the modSDM method on a 512×512 grid, for various numbers of GPUs. Here, speedup and efficiency are given relative to one GPU.

The experiments showed that the modified algorithms we constructed are very effective. The new algorithms are more economical in terms of number of iterations, number of operations, and time per iteration. For the considered model problem, the modified steepest descent method and the modified minimal residual method showed the same performance in terms of number of iterations and computation time. The parallel algorithms demonstrated a good scaling; their efficiency is near 90% on eight GPUs.

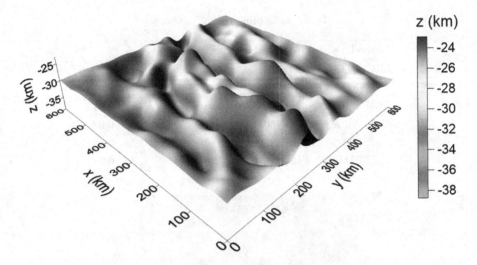

Fig. 5. Original surface z^*

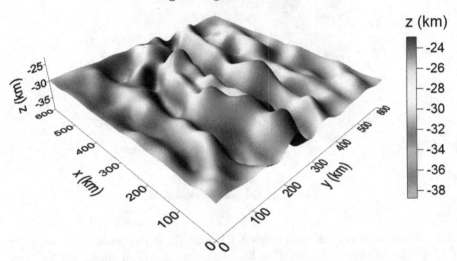

Fig. 6. Reconstructed surface z

Table 2. Speedup and efficiency of the parallel modSDM algorithm

Number m of GPUs	Execution time [minutes]	Speedup S_m	Efficiency E_m
1	5.3	—	—
2	2.8	1.89	0.94
4	1.5	3.6	0.90
8	0.7	7.16	0.89

7   Conclusions

We constructed parallel algorithms based on modified analogs of α-processes, namely the modified steepest descent method and the modified minimal residual method, for solving the structural inverse gravimetry problem. The parallel algorithms were implemented on multicore CPUs and on multiple GPUs taking advantage, respectively, of OpenMP and CUDA technologies. A model problem of gravimetry with a large grid was solved. The parallel algorithms demonstrated excellent scaling and efficiency of nearly 90%.

We compared the constructed algorithms, regarding the number of iterations and computation time, with algorithms based on conjugate gradient-type methods and the componentwise gradient method. The modified α-processes made it possible to reduce the number of iterations and computation time by as much as 50% compared with conjugate gradient-type methods.

References

1. Akimova, E.N., Martyshko, P.S., Misilov, V.E.: Algorithms for solving the structural gravity problem in a multilayer medium. Doklady Earth Sci. **453**(2), 1278–1281 (2013). https://doi.org/10.1134/S1028334X13120180
2. Akimova, E.N., Martyshko, P.S., Misilov, V.E.: Parallel algorithms for solving structural inverse magnetometry problem on multicore and graphics processors. In: Proceedings of 14th International multidisciplinary scientific GeoConference SGEM 2014, vol. 1, no. 2, pp. 713–720 (2014)
3. Akimova, E.N., Martyshko, P.S., Misilov, V.E.: A fast parallel gradient algorithm for solving structural inverse gravity problem. In: AIP Conference Proceedings, vol. 1648, p. 850063 (2015). https://doi.org/10.1063/1.4913118
4. Akimova, E.N., Martyshko, P.S., Misilov, V.E., Kosivets, R.A.: An efficient numerical technique for solving the inverse gravity problem of finding a lateral density. Appl. Math. Inf. Sci. **10**(5), 1681–1688 (2016). https://doi.org/10.18576/amis/100506
5. Akimova, E.N., Misilov, V.E.: A fast componentwise gradient method for solving structural inverse gravity problem. In: Proceedings of 15th International Multidisciplinary Scientific GeoConference SGEM 2015, vol. 3, no. 1, pp. 775–782 (2015)
6. Akimova, E.N., Misilov, V.E. Arguchinsky, M.S.: Memory efficient algorithm for solving the inverse problem of finding a density in a curvilinear layer. In: CEUR Workshop Proceedings, vol. 2076, pp. 1–8 (2018)
7. Akimova, E.N., Misilov, V.E., Tretyakov, A.I.: Regularized methods for solving nonlinear inverse gravity problem. In: EAGE Geoinformatics (2016). https://doi.org/10.3997/2214-4609.201600458
8. Akimova, E.N., Misilov, V.E., Tretyakov, A.I.: Optimized algorithms for solving the structural inverse gravimetry and magnetometry problems on GPUs. Commun. Comput. Inf. Sci. **753**, 144–155 (2017). https://doi.org/10.1007/978-3-319-67035-5_11
9. Akimova, E.N., Vasin, V.V.: Stable parallel algorithms for solving the inverse gravimetry and magnetometry problems. Int. J. Eng. Model. - Univ. Split Croatia **17**(1–2), 13–19 (2004)

10. Akimova, E.N., Vasin, V.V., Misilov, V.E.: Iterative Newton type algorithms and its applications to inverse gravimetry problem. Bull. South Ural State Univ. Ser. Math. Modell. Programm. Comput. Softw. **6**(3), 26–37 (2013)

11. Bakushinskiy, A., Goncharsky, A.: Ill-Posed Problems: Theory and Applications. Springer, Heidelberg (1994). https://doi.org/10.1007/978-94-011-1026-6

12. Kantorovich, L.V., Akilov, G.P.: Functional Analysis in Normed Spaces, International Series of Monographs in Pure and Applied Mathematics. The Macmillan Co., New York (1964)

13. Malkin, N.R.: On solution of inverse magnetic problem for one contact surface (the case of layered masses). In: DAN SSSR, Ser. A, no. 9, pp. 232–235 (1931)

14. Martyshko, P.S., Akimova, E.N., Misilov, V.E.: Solving the structural inverse gravity problem by the modified gradient methods. Izvestiya, Physics of the Solid Earth **52**(5), 704–708 (2016)

15. Martyshko, P.S., Fedorova, N.V., Akimova, E.N., Gemaidinov, D.V.: Studying the structural features of the lithospheric magnetic and gravity fields with the use of parallel algorithms. Izvestiya Phys. Solid Earth **50**(4), 508–513 (2014). https://doi.org/10.1134/S1069351314040090

16. Martyshko, P.S., Prutkin, I.L.: Technology of depth distribution of gravity field sources. Geophys. J **25**(3), 159–168 (2003)

17. Numerov, B.V.: Interpretation of gravitational observations in the case of one contact surface. In: Doklady Akad. Nauk SSSR, pp. 569–574 (1930)

18. Vasin, V.V.: Irregular nonlinear operator equations: Tikhonov's regularization and iterative approximation. J. Inverse Ill-Posed Probl. **21**(1), 109–123 (2013). https://doi.org/10.1515/jip-2012-0084

19. Vasin, V.V.: Modified steepest descent method for nonlinear irregular operator equations. Doklady Math. **91**(3), 300–303 (2015). https://doi.org/10.1134/S1064562415030187

20. Vasin, V.V.: Regularized modified α-processes for nonlinear equations with monotone operators. Dokl. Math. **94**(1), 361–364 (2016). https://doi.org/10.1134/S1064562416040062

21. Vasin, V.V., Skurydina, A.F.: Two-stage method of construction of regularizing algorithms for nonlinear Ill-posed problems. Proc. Steklov Inst. Math. **301**, 173–190 (2018). https://doi.org/10.1134/S0081543818050152

Supercomputer Simulations of the Medical Ultrasound Tomography Problem

Victoria Filatova$^{(\boxtimes)}$ (iD), Alexander Danilin, Vera Nosikova, and Leonid Pestov

Immanuel Kant Baltic Federal University, Kaliningrad, Russia
{ViFilatova,ADanilin,VNosikova,LPestov}@kantiana.ru

Abstract. The inverse problem of medical ultrasound tomography consists in finding small inclusions in the breast tissue by boundary measurements of acoustic waves generated by sources located on the boundary. In this work, we describe some results of simulations of this problem in 2D. The simulations include the solution of direct and inverse problems. We calculate acoustic waves for a specific breast model (direct problem). Furthermore, we solve the inverse problem using method based on the visualization of inclusions and unknown inner boundary between fatty and glandular tissues and resorts to kinematic data for the determination of the speed of sound in the inclusions. The numerical solution of the direct and inverse problems relies on standard libraries of parallel computing (MPI and OpenMP) and is carried out on a cluster system.

Keywords: Medical ultrasound tomography · Breast imaging · Reverse time migration

1 Introduction

Breast cancer is the most common malignant neoplasm and the leading cause of cancer-related deaths in women. The problem of early detection of cancer remains a topical issue despite the existence of a wide spectrum of non-invasive examination methods, such as mammography, computer and magnetic resonance imaging (MRI), and ultrasonography. Diagnosis and detection of pathologies are additionally complicated by the structure of the breast. The informativeness of standard methods decreases in case of a dense background mammary tissue. It is known that mammographic breast density is a breast-cancer risk factor. Women with higher breast density are at an increased risk of developing breast cancer [9].

The greatest complexity in the study is represented by glandular and fibrous tissues since they have approximately the same radiographic densities [13], a factor that hinders the detection of small tumors. The sensitivity of mammography is close to 100% in fatty tissue, whereas examination of women with dense background mammary tissue has a false-negative rate of 10 to 40% of cases [1]. A dense background is caused in many cases by an abundance of glandular tissue, in which the speed of sound propagation (approx. 1.515 m/ms) is close to the

© Springer Nature Switzerland AG 2019
L. Sokolinsky and M. Zymbler (Eds.): PCT 2019, CCIS 1063, pp. 297–308, 2019.
https://doi.org/10.1007/978-3-030-28163-2_21

one in tumors (approx. 1.549 to 1.559 m/ms) [14]; this speed in fatty tissue has a much lower value (approx. 1.470 m/ms).

In connection with these difficulties and the insufficient accuracy of currently existing methods, researchers from various scientific fields are engaged in the search for and development of new methods for detecting breast neoplasms.

One of the most promising approaches to the detection of breast cancer is ultrasound tomography. There are currently several scientific groups involved in the investigation of ultrasound tomography: Ruiter's group in Germany [10], Duric's group in USA [4], and the Russian groups of Rumyantseva [2] and Goncharsky [7]. There are many other scientists working in this field, among which we should mention Andre, Johnson, Greenleaf [15]; Marmarelis, Jeong [11].

At present, ultrasound tomographs are mainly used in testing mode. Existing two-dimensional and three-dimensional prototypes use 256 transducers or more. In the two-dimensional case, the transducers are located around a circle of radius 0.1 m. In the three-dimensional case, transducers are located on a hemisphere. The dominant impulse frequency is in the range from 1 to 8 MHz. In this paper, we describe simulations conducted with a two-dimensional mathematical breast acoustic model which is similar to the model studied by Duric [4] (see Fig. 1). The model consists of a complex and unknown boundary FG between fatty and glandular tissues, as well as six weak inclusions with radiuses varying from 0.8 mm to 5 mm. The values of the speed of sound in the model are close to the real ones in the corresponding tissues: in the fatty tissue, $c_{\text{fat}} = 1.47$ m/ms; in the glandular tissue, $c_{\text{gland}} = 1.515$ m/ms; in the inclusions, the speed of sound is assumed to vary from 1.47 m/ms to 1.559 m/ms (tumors): $c_1 = 1.47$ m/ms, $c_2 = 1.559$ m/ms, $c_3 = 1.549$ m/ms, $c_4 = 1.47$ m/ms, $c_5 = 1.557$ m/ms, and $c_6 = 1.47$ m/ms. We suppose that the speed of sound in fatty and glandular tissues are known.

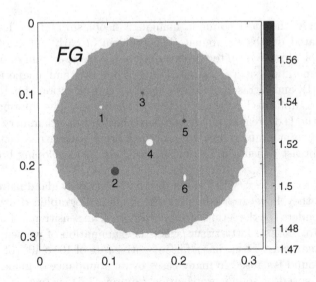

Fig. 1. Breast acoustic model (2D)

The paper is organized as follows. The statement of the inverse problem and the reconstruction method are given in Sect. 2. The parameters of the numerical experiments, the numerical solution of the direct and reversal time problems, the parallelization of the algorithm, as well as two numerical experiments are described in Sect. 3. The first experiment was carried out with exactly known values of the speed of sound in fatty and glandular tissues. The second was conducted with shifted speeds.

2   Ultrasound Tomography Problem

2.1   The Statement of the Inverse Problem

Let us consider linear first-order acoustic equations with zero Cauchy data in $\mathbb{R}^2 \times (0, T)$:

$$\begin{cases} \dfrac{1}{c^2} p_t = \operatorname{div} v + \delta(x - x_s)\, r(t), \\[2mm] v_t = \nabla p, \\[2mm] p|_{t=0} = 0, \ v|_{t=0} = 0. \end{cases} \tag{1}$$

Here $p(x, t; x_s)$ is the acoustic pressure; $v(x, t; x_s)$ is the velocity vector field of a particle; $\delta(x - x_s)$ is Dirac delta function, which models a point source located at $x_s \in \Gamma = \{x : |x - x_0| = R\}$; $r(t)$ is a Ricker impulse, $r|_{t \le 0} = 0$ (see Fig. 2); $c = c(x)$ is the speed of sound. Outside the disc $\Omega = \{x : |x - x_0| \le R\}$, the function c is equal to a constant. The waves generated by the boundary sources are reflected/scattered from the inner boundary FG and inclusions and recorded on the circle Γ.

Let T^* be the "acoustic" radius of the disc Ω, i. e. the minimum time required for filling the disc Ω with waves emitted by all boundary sources (filling time). Throughout the paper, we will assume that the registration time T is greater than T^*.

The **medical ultrasound tomography problem** consists in the reconstruction of the speed of sound on the basis of pressure measurements on the boundary for different positions of the sources (inverse data):

$$p_0(x, t; x_s) = p(x, t; x_s), \ x, x_s \in \Gamma, \ t \in [0, T].$$

2.2   Combined Method

The combined method [6] for solving the medical ultrasound tomography problem consists of two steps: (1) visualization of the acoustic medium; (2) determination of the speed of sound in the inclusions. The first step (visualization) is based on the reverse time migration (RTM) method, which is well-known in geophysics. We use the energy version of the RTM [12]. A brief description of the visualization is given below.

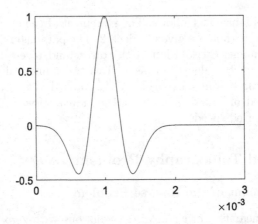

Fig. 2. Ricker impulse (f_0 is the dominant frequency, $f_0 = 1$ MHz)

Visualization

i. We calculate the pair $p^d(x, t; x_s), v^d(x, t; x_s)$ for the known *background speed of sound* $c_0(x)$ (direct problem (1)).
ii. For a given $p_0(x, t; x_s)$ (inverse data), we solve the reversal time problem:

$$\begin{cases} \dfrac{1}{c_0^2} p_t = \text{div } v, \\ v_t = \nabla p + p_0 \delta(|x| - R) \nu_\Gamma, \\ p|_{t=T} = 0, \ v|_{t=T} = 0, \end{cases} \tag{2}$$

where ν_Γ is the outward unit normal vector to Γ, $t \in (0, T)$, and $p^b(x, t; x_s)$ is the solution of problem (2).
iii. For visualization, we use the so-called "energy imaging condition":

$$I_E(x) = \sum_{x_s} \int_0^T [p^d p^b + (v^d, v^b)](x, t)\, dt. \tag{3}$$

Notice that a good speed background plays a crucial role for good visualization. We applied the Energy RTM procedure for a constant background speed, namely $c_0 = c_{\text{gland}}$ [6]. In this case, we can see neither the inner boundary FG nor the inclusions. Obviously, this is due to waves reflected from the boundary FG (see Fig. 1).

We suggest the following two steps to obtain the image of the acoustic media. In the first step, we reconstruct the boundary FG by applying the Energy RTM for a short observation time $T_0 < T$ and $c_0 = c_{\text{fat}}$. As a result, we obtain the boundary FG (see the numerical experiments below). Thus, we have constructed a suitable background breast model. Indeed, we know the speeds of sound c_{fat} and c_{gland}, as well as the inner boundary FG. In the second step, we apply the Energy RTM for the full observation time T, thereby obtaining the final image (see details in [6]).

Determination of the Speed of Sound in the Inclusions

The speed of sound in the inclusions can be determined by using kinematic data, as in [5]. It is important to determine what kind of inclusion is visualized: fatty tissue, glandular tissue, or a tumor. Since the values of the speed of sound in these tissues are quite close, we must compute them with high accuracy. Below, we describe an algorithm for reconstruction of the speed of sound (for more details, see [5]).

For a fixed inclusion, we built all possible rays connecting the transducers on the circle Γ, passing through the center of this inclusion and not intersecting other inclusions (see Fig. 3).

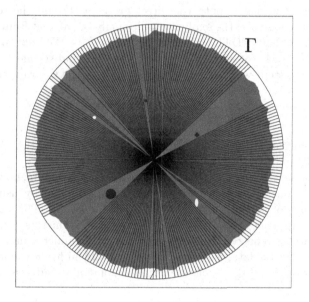

Fig. 3. Scheme of construction of the rays

The actual travel time between transducers is obtained from boundary pressure measurements. We know the speed of sound in fatty and glandular tissues, the sizes of the inclusions, and their locations. Therefore, we can determine the residual function for the travel time through inclusions. By minimizing the residual function with respect to the values of the speed of sound in the inclusions, we obtain the final result (see below the descriptions of the numerical experiments).

3 The Numerical Solution

3.1 Parameters of the Numerical Experiments

The numerical experiments were performed on a rectangular grid with the following parameters:

- The computational domain is a square of size 0.32 m × 0.32 m.
- The transducers are uniformly distributed along a circle of radius $r = 0.15$ m.
- Number of grid nodes: $N = 10\,240\,000$.
- Number of transducers: $N_{\mathrm{tr}} = 256$ (both types of emitters/receivers are called "transducers").
- Grid step: $\Delta x = 0.0001$ m.
- Time step: $\Delta t = 0.00001$ ms.
- Dominant frequency of the Ricker wavelet: 1 MHz.

3.2   Numerical Solution of the Direct and Reversal Time Problems

The numerical solutions of both the direct problem (1) and the reversal time problem (2) were based on the explicit conditionally convergent Finite-Difference Time-Domain method (FDTD) with a staggered grid [6]. We used an approximation of the 12-th order of accuracy in space variables and second order in time. Spatial and time steps were determined according to the Courant condition,

$$\Delta t < \frac{\Delta x}{k c_{\max} \sqrt{2}},$$

where the coefficient k depends on the order of the approximation (for the 12-th order, $k \approx 1.34$).

The function c is a constant outside the disc Ω and, therefore, no waves reflect from $\mathbb{R}^2 \setminus \Omega$. Thus, we restrict the computational domain by using the well-known technique of perfectly matched layers (PML; see [3]). The computational domain (a square) contains an absorption layer with parameters specially chosen to attenuate waves reflected from the boundary of the inner square (see Fig. 4). To obtain the inverse data, we solve the direct problem for the model shown in Fig. 1. The pressure on the circle Γ induced by a single source is illustrated in Fig. 5.

The algorithm of visualization consists in solving the direct and reversal time problems for 256 transducers. Given that medical ultrasound tomography requires a high resolution, we use a small-scale grid. Thus, we need to parallelize this part of the solution to the inverse problem. The numerical solution of the problem of visualization uses standard libraries of parallel computing, namely MPI and OpenMP. The solution is carried out on a cluster system.

3.3   Parallelization of the Algorithm

Parallelization was performed at two levels (see Fig. 6):

(1) Distribution of computations by means of the MPI library: each cluster node solves the direct or the reversal time problem for a fixed transducer.
(2) Multithreading within a single node by using OpenMP: the algorithm for the solution of the direct or reversal time problem is parallelized.

More precisely, we parallelize the algorithm of visualization as follows:

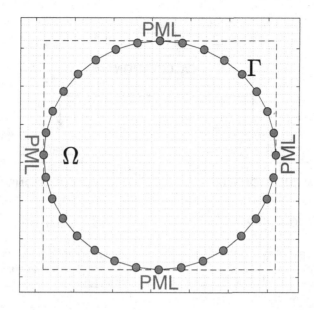

Fig. 4. Scheme of the experiment

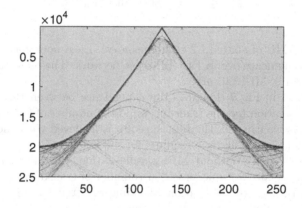

Fig. 5. Pressure "measured" on the circle Γ

1. A transducer number (as a source) is assigned to each computational node and the corresponding direct or reversal time problem is computed.
2. Once a node has been assigned a number, the processing is carried out in parallel using the CPU cores available on the node. Parallel execution is carried out through the OpenMP interface.
3. Data estimated for each transducer are written to disk.

The simulations and program tests were conducted on the HPC "RocsCluster", which consists of 128 nodes, each built on the basis of an Intel Server Board S5400SF and equipped with two 4-core Intel Quad-Core Xeon E5472 processors

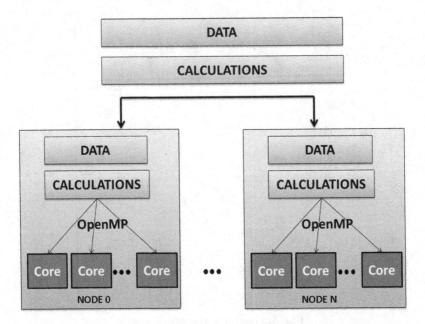

Fig. 6. Scheme of parallelization

(3.00 GHz, 2×6 MB of shared L2 cache memory). Each node has 32 GB RAM. The nodes communicate over a Fast Ethernet network. The operating system is OS Linux x86-64 + MPI Library.

The bar chart in Fig. 7 describes the dependence between the speed of the direct problem solution for one transducer and the number of CPU cores on one cluster node. As we can see, the best speedup is achieved with the eight CPU cores of a cluster node. The time required for solving the direct problem for one transducer is 143 min. Without MPI parallelization, the time for solving the direct problem for all transducers is 256 (number of transducers) $\times$ 143 min = 36 608 min. With MPI parallelization, the time required for solving the direct problem on all 128 nodes is equal to 351 min.

3.4 Numerical Experiment #1

The result of the visualization by the above-described method is illustrated in Figs. 8 (visualization of the boundary FG) and 9 (visualization of the inclusions) [6]. The computed values of the speed of sound in the inclusions are shown in Fig. 10. The left and right columns correspond, respectively, to the real and the reconstructed values of the speed of sound in the inclusions. The relative error of the reconstruction does not exceed 2.43% [5].

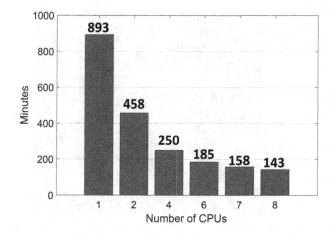

Fig. 7. Bar chart showing the dependence between the speedup of the direct problem solution for one transducer and the number of CPU cores used on one cluster node (with OpenMP parallelization)

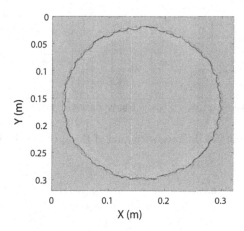

Fig. 8. Visualization of the boundary FG (Experiment #1)

3.5 Numerical Experiment #2

It is impossible to set up accurately the background speed for the RTM. The speed of sound in fatty tissues ranges from 1.430 to 1.470 m/ms approximately [8]. For the speed of sound in fatty and glandular tissues, we took $c_{fat} = 1.4553$ m/ms and $c_{gland} = 1.5301$ m/ms, respectively. Furthermore, we conducted an experiment with shifted background speed of sound. The results of the visualization are given in Figs. 11 (visualization of the boundary FG) and 12 (visualization of the inclusions).

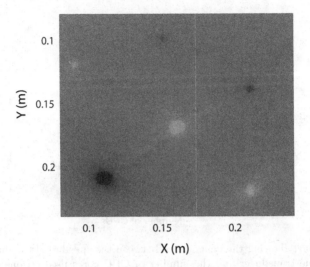

Fig. 9. Visualization of the inclusions (Experiment #1)

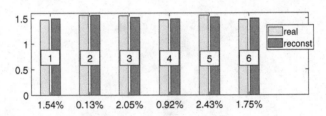

Fig. 10. Results of the reconstruction of the speed in the inclusions

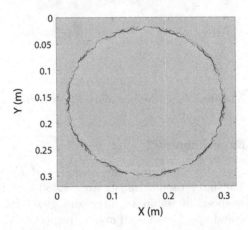

Fig. 11. Visualization of the boundary FG (Experiment #2)

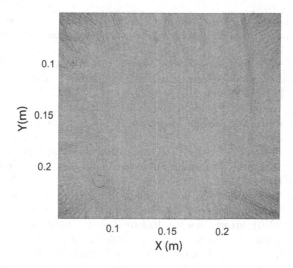

Fig. 12. Visualization of the inclusions (Experiment #2)

4 Conclusions

Medical ultrasound tomography is an important and complex inverse problem. The approach suggested in this paper is based on both the visualization of acoustic media and the determination of the speed of sound in inclusions, which is achieved relying on kinetic data. We described the results of tests conducted on a complex two-dimensional acoustic model. The efficiency of the proposed approach was demonstrated by the results of numerical simulations. There are two factors essential to medical tomography: high resolution and quick results. If one needs a high-resolution result, then the problem must be solved on a small-scale grid. This, however, significantly increases the time required for solving the problem. The advisability of using parallel technologies (OpenMP and MPI) for the solution of the considered task is, therefore, confirmed.

Acknowledgments. This work was supported by the Russian Science Foundation (grant No. 16-11-10027).

References

1. Borisova, M.S., Martynova, N.V., Bogdanov, S.N.: X-ray mammography in the diagnosis of breast cancer. Vestnik Rossiyskogo nauchnogo tsentra rentgenoradiologii Minzdrava Rossii **3**(13), (2013)
2. Burov, V., Zotov, D., Rumyantseva, O.: Reconstruction of the soundvelocity and absorption spatial distributions in soft biologicaltissue phantoms from experimental ultrasound tomography data. Acoust. Phys. **61**(2), 231–248 (2015). https://doi.org/10.1134/S1063771015020013

3. Colino, F., Tsogka, C.: Application of PML absorbing layer model to the linear elastodynamic problem in anisotropic heterogeneous media. Geophysics **66**(1), 294–307 (2001)

4. Duric, N., et al.: Detection of breast cancer with ultrasound tomography: First results with the computed ultrasound risk evaluation (CURE) prototype. Med. Phys. **34**(2), 773–785 (2007). https://doi.org/10.1118/1.2432161

5. Filatova, V., Nosikova, V.: Determining sound speed in weak inclusions in the ultrasound tomography problem. Eurasian J. Math. Comput. Appl. **6**(1), 11–20 (2018)

6. Filatova, V., Nosikova, V., Pestov, L.: Application of reverse time migration (RTM) procedure in ultrasound tomography, numerical modeling. Eurasian J. Math. Comput. Appl. **4**(4), 5–13 (2016)

7. Goncharsky, A., Romanov, S., Seryozhnikov, S.: A computer simulation study of soft tissue characterization using low-frequency ultrasonic tomography. Ultrasonics **67**, 136–150 (2016). https://doi.org/10.1016/j.ultras.2016.01.008

8. Goss, S., Johnston, R., Dunn, F.: Comprehensive compilation of empiricalultrasonic properties of mammalian tissue. J. Acoust. Soc. Am. **54**(2), 423–457 (1978)

9. Huo, C., et al.: Mammographic density-a review on the currentunderstanding of its association with breast cancer. Breast Cancer Res. Treat. **144**(3), 479–502 (2014). https://doi.org/10.1007/s10549-014-2901-2

10. Jirik, R., et al.: Sound-speed image reconstruction in sparse-aperture 3-D ultrasound transmission tomography. IEEE Transact. Ultrason. Ferroelectr. Freq. Controly **59**(2), 254–264 (2012). https://doi.org/10.1109/TUFFC.2012.2185

11. Marmarelis, V., Jeong, J., Shin, D., Do, S.: High-resolution 3-D imaging and tissue differentiation with transmission tomography. In: André, M.P., et al. (eds.) Acoustical Imaging, vol. 28, pp. 195–206. Springer, Dordrecht (2007). https://doi.org/10.1007/1-4020-5721-0_21

12. Rocha, D., Tanushev, N., Sava, P.: Acoustic wavefield imaging using the energy norm. In: 2015 SEG Annual Meeting, pp. 49–68. Society of Exploration Geophysicists, New Orleans, Louisiana (2015). https://doi.org/10.1190/segam2015-5878824.1

13. Sadykov, S.S., Bulanova, Y.A., Zakharova, E.A.: Computer diagnosis of tumors in mammograms. Comput. Opt. **38**(1), 131–138 (2014)

14. Sandhu, G., Li, C., Roy, O., Schmidt, S., Duric, N.: Frequency domain ultrasound waveform tomography: breast imaging using a ring transducer. Phys. Med. Biol. **60**(14), 5381–5398 (2015). https://doi.org/10.1088/0031-9155/60/14/5381

15. Wiskin, J., et al.: Three-dimensional nonlinear inverse scattering: quantitative transmission algorithms, refraction corrected reflection, scanner design and clinical results. In: Proceedings of Meetings on Acoustics ICA 2013, vol. 19, no. 1, p. 075001. Acoustical Society of America, San Diego (2013). https://doi.org/10.1121/1.4800267

Numerical Modeling of Hydrodynamic Turbulence with Self-gravity on Intel Xeon Phi KNL

Igor Kulikov[1]([✉])(iD), Igor Chernykh[1], Evgeny Berendeev[1], Victor Protasov[1,2],
Alexander Serenko[1], Vladimir Prigarin[1,2], Ivan Ulyanichev[1,2],
Dmitry Karavaev[1], Eduard Vorobyov[3], and Alexander Tutukov[4]

[1] Institute of Computational Mathematics and Mathematical Geophysics SB RAS,
Novosibirsk, Russia
kulikov@ssd.sscc.ru, chernykh@parbz.sscc.ru, evgeny.berendeev@gmail.com,
inc_13@mail.ru, fafnur@yandex.ru, vovkaprigarin@gmail.com,
wmzonacomvn@mail.ru, kda@opg.sscc.ru
[2] Novosibirsk State Technical University, Novosibirsk, Russia
[3] University of Vienna, Vienna, Austria
eduard.vorobiev@univie.ac.at
[4] Institute of Astronomy RAS, Moskva, Russia
atutukov@inasan.ru

Abstract. In this paper, we present the results of numerical simulations of hydrodynamic turbulence with self-gravity, employing the latest Intel Xeon Phi accelerators with KNL architecture. A new vectorized numerical method with a high order of accuracy on a local stencil is described in details. We outline the main features of the program implementation of the method for massively parallel architectures and study the code parallel implementation. We achieved a performance of 173 gigaFLOPS and an acceleration factor of 48 using a single Intel Xeon Phi KNL. Using 16 accelerators, we were able to achieve a scalability of 97%.

Keywords: Computational astrophysics · Intel Xeon Phi ·
Numerical methods

1 Introduction

The study of physical processes in the Universe, their influence on the self-organization and evolution of astronomical objects, as well as on their further dynamics and interaction constitute the subject of modern astrophysics. The importance of considering gravitational and magnetic fields and the difficulty of reproducing cosmic conditions in the laboratory impose significant restrictions on the experimental study of astronomical objects. Thus, mathematical modeling is the main, and often the only, approach to the theoretical study of astrophysical processes and astronomical objects.

© Springer Nature Switzerland AG 2019
L. Sokolinsky and M. Zymbler (Eds.): PCT 2019, CCIS 1063, pp. 309–322, 2019.
https://doi.org/10.1007/978-3-030-28163-2_22

The evolution of hydrodynamic turbulence and the formation of compact objects as a result of gravitational collapse are among the important processes occurring in astrophysical objects at various spatial scales [1,2]. Magnetohydrodynamic (MHD) turbulence was simulated at the scales of clusters of galaxies in [3]. Problems of gravitational and magneto-gravitational instability [4], dynamics of clouds falling into a black hole [5], and cloud collapse and its fragmentation [6] have been considered in the context of modeling the dynamics of molecular clouds.

An important role is given to the influence of magnetic fields on the evolution of interstellar turbulent flows, in which the magnetic fields are quite strong [7–9]. The energy spectrum [10], the subalfvenian flows [11], and the star formation rate [12] have been studied in the context of the evolution of MHD turbulence. A comparison of various codes for simulation of supersonic turbulence was made in [13]. Turbulence in the solar wind was investigated in [14]. It has been noted that turbulence is the main mechanism for the transition of the deflagration process into detonation in supernova explosion problems [15]. It is important to realize that significant computational high-performance resources are required if one wants to simulate the evolution of hydrodynamic turbulence with self-gravity taken into account.

A trend for using hybrid supercomputers equipped with graphics accelerators and Intel Xeon Phi or Sunway accelerators has become obvious. There are a variety of codes adapted for hybrid supercomputers to simulate hydrodynamic flows in astrophysics [16–23]. However, the main potential for improving the performance in hydrodynamic computing on Intel Xeon Phi accelerators using low-level vectorization of computations has not been sufficiently explored.

In this paper, we shall consider the model problem of turbulence evolution using a new vectorized code developed for supercomputers equipped with Intel Xeon Phi KNL accelerators. The peak performance of Intel Xeon Phi dual accelerators is about three teraFLOPS. Of course, such a value is unreachable in real-world applications but a value of the order of one teraFLOPS can be achieved on synthetic tests. We will be guided by this value when designing our computational model. At present, some program codes (based on publications in the Computer Physics Communications journal) using Intel Xeon Phi accelerators have been implemented in the fields of plasma physics [24], molecular dynamics [25,26], statistical mechanics [27], and hydrodynamics [28].

In 2015, we developed the AstroPhi code [18], based on the implementation of an original numerical method by using the offload programming model of the Intel Xeon Phi. The used accelerator architecture did not allow us to implement vector instructions, although switching to the native mode made it possible to achieve a code performance of 28 gigaFLOPS [29]. The use of low-level vectorization of cycles in the AstroPhi code allowed us to increase the performance to a value of the order of 100 gigaFLOPS [30]. There became evident the necessity to use low-level vectorizing tools to achieve a maximum performance. The new version of the code was based on the HLL method and used a single accelerator

[31,32]. With this implementation, we achieved performances of 245 gigaFLOPS on Intel Xeon Phi 7250 and 302 gigaFLOPS on Intel Xeon Phi 7290.

The computational model and the numerical method will be briefly described in Sect. 2. Section 3 is devoted to the development and investigation of the parallel implementation. In Sect. 4, we formulate the main problems of vectorization. Section 5 is devoted to the simulation of hydrodynamic turbulence taking self-gravity into consideration. Finally, we summarize the conclusions of our research in Sect. 6.

2   The Computational Model

The mathematical model is based on the equations of multicomponent gravitational hydrodynamics. An important condition for the subsequent construction of a vectorized numerical method is to write the equations in vector form. We will use an overdetermined system of hydrodynamic equations with an entropy equation. This will enable us to write the system of hydrodynamic equations in a divergent form, making it possible to formulate a vector numerical method:

$$
\frac{\partial}{\partial t}
\begin{pmatrix} \rho \\ \rho_i \\ \rho \boldsymbol{u} \\ \rho S \\ \rho E \end{pmatrix}
+ \nabla \cdot
\begin{pmatrix} \rho \boldsymbol{u} \\ \rho_i \boldsymbol{u} \\ \rho \boldsymbol{u} \otimes \boldsymbol{u} + p \\ \rho S \boldsymbol{u} \\ (\rho E + p)\, \boldsymbol{u} \end{pmatrix}
=
\begin{pmatrix} 0 \\ s_i \\ \rho \nabla \Phi \\ (\gamma - 1)\, \rho^{1-\gamma}\, (\Lambda - \Gamma) \\ \Lambda - \Gamma \end{pmatrix} ,
\tag{1}
$$

where ρ_i is the density of the species, $\rho = \sum_i \rho_i$ denotes the density of the gas mixture, $\boldsymbol{u} = (u_x, u_y, u_z)$ is the velocity vector, S stands for the entropy, $p = p(\rho, S, T)$ denotes the pressure, γ is the adiabatic index, $\rho E = \rho \varepsilon + \frac{1}{2} \rho u^2$ is the total mechanical energy, T is the temperature, s_I represents the rate of formation of the corresponding species and, finally, Φ is the gravitational potential satisfying the Poisson equation

$$
\triangle \Phi = 4\pi G \rho,
\tag{2}
$$

in which G is the gravitational constant, Λ is the cooling function and Γ is the heating function. In this article, we restrict ourselves to considering the equation of state based on a combination of the isothermal and adiabatic regimes:

$$
p = c_s^2 \rho + c_s^2 \rho_{\text{crit}} \left(\rho / \rho_{\text{crit}} \right)^\gamma ,
\tag{3}
$$

where c_s^2 is the isothermal velocity of sound and ρ_{crit} is the critical density of the gas during the transition from isothermal to adiabatic mode, which can be expressed as

$$
\rho_{\text{crit}} = \mu m_H n_{\text{crit}},
\tag{4}
$$

with μ the average molecular weight of gas, m_H the mass of a hydrogen atom, and n_{crit} the critical gas concentration. In this work, we assume $n_{\text{crit}} = 10^{10}$ cm^{-3}.

We will consider neither cooling/heating processes nor chemical kinetics processes. Consequently, to simulate hydrodynamic turbulence, we will use the following simplified form of the equations:

$$\frac{\partial}{\partial t}\begin{pmatrix} \rho \\ \rho \boldsymbol{u} \end{pmatrix} + \nabla \cdot \begin{pmatrix} \rho \boldsymbol{u} \\ \rho \boldsymbol{u} \otimes \boldsymbol{u} + p \end{pmatrix} = \begin{pmatrix} 0 \\ \rho \nabla \Phi \end{pmatrix}. \tag{5}$$

However, we will describe all the calculations and the structure of the code for the entire system given in (1).

The equations of hydrodynamics can be written in vector form:

$$\frac{\partial U}{\partial t} + \frac{\partial F(U)}{\partial x} = 0. \tag{6}$$

To solve the equations, one can use a numerical method based on a combination of the operator splitting approach, the Godunov method, the HLL scheme, and the piecewise-parabolic method on a local stencil. The flow through the boundary between the left (L) and the right (R) cells is calculated with the help of the equation

$$F = \frac{F\left(-\lambda_{\mathrm{L}}\tau\right) + F\left(\lambda_{\mathrm{R}}\tau\right)}{2} + \frac{c + \|\boldsymbol{u}\|}{2}\left(U\left(-\lambda_{\mathrm{L}}\tau\right) - U\left(\lambda_{\mathrm{R}}\tau\right)\right), \tag{7}$$

where

$$\lambda_{\mathrm{L}} = c - \|\boldsymbol{u}\|, \qquad \lambda_{\mathrm{R}} = c + \|\boldsymbol{u}\|, \tag{8}$$

with $c = \sqrt{\frac{\gamma p}{\rho}}$ the speed of sound. The modification of the parabolas construct given in [33] is based on the reduction of the order of the first element in the parabola.

The application of the procedure suggested in [33] for the construction of a local parabola to increase the order of accuracy would have made more difficult the transition to an adaptive nested mesh, due to the difference in size of the cells. Therefore, we set two features: to take the original PPML approach using a compact template and the ability to integrate parabolas along the characteristics in each cell. To this end, we save the solver notation and, therefore, the parallel computing algorithms. To solve the problems posed, we will rewrite the parabola construction algorithm from [33] and integrate the parabolas within each cell.

The blocks are the parabolas constructed for the numerical scheme. We construct a piecewise-parabolic function $q(x)$ on a regular mesh with step size h on the interval $[x_{i-1/2}, x_{i+1/2}]$. The general equation of the parabola can be written as

$$q(x) = q_i^{\mathrm{L}} + \xi\left(\triangle q_i + q_i^{(6)}(1 - \xi)\right),$$

where q_i is the value at the center of the cell, $\xi = (x - x_{i-1/2})h^{-1}$, $\triangle q_i = q_i^{\mathrm{L}} - q_i^{\mathrm{R}}$, and $q_i^{(6)} = 6\left(q_i - 1/2(q_i^{\mathrm{L}} + q_i^{\mathrm{R}})\right)$, according to conservation laws:

$$q_i = h^{-1}\int_{x_{i-1/2}}^{x_{i+1/2}} q(x)\,dx.$$

To construct $q_i^R = q_{i+1}^L = q_{i+1/2}$, we use an interpolation function of second order of accuracy:

$$q_{i+1/2} = 1/2(q_i + q_{i+1}),$$

where $\delta q_i = 1/2(q_{i+1} - q_{i-1})$. The input value for the construction of the parabola is q_i. The output procedure involves all parameters of the parabola on each interval $[x_{i-1/2}, x_{i+1/2}]$.

1. Construct $\delta q_i = 1/2(q_{i+1} - q_{i-1})$ without extreme regularization:

$$\delta_m q_i = \begin{cases} \min(|\delta q_i|, 2|q_{i+1} - q_i|, 2|q_i - q_{i-1}|)\,\text{sgn}(\delta q_i), \\ \qquad\qquad \text{if } (q_{i+1} - q_i)(q_i - q_{i-1}) > 0, \\ 0, \quad \text{if } (q_{i+1} - q_i)(q_i - q_{i-1}) \leq 0. \end{cases}$$

2. Compute the boundary values for the parabola:

$$q_i^R = q_{i+1}^L = q_{i+1/2} = 1/2(q_i + q_{i+1}).$$

3. Reconstruct the parabola according to the following equations:

$$\triangle q_i = q_i^L - q_i^R, q_i^{(6)} = 6(q_i - 1/2(q_i^L + q_i^R)).$$

To obtain a monotone parabola, we use the following equations for the boundary values q_i^L, q_i^R:

$$q_i^L = q_i, \ q_i^R = q_i, \ (q_i^L - q_i)(q_i - q_i^R) \leq 0,$$
$$q_i^L = 3q_i - 2q_i^R, \ \triangle q_i q_i^{(6)} > (\triangle q_i)^2,$$
$$q_i^R = 3q_i - 2q_i^L, \ \triangle q_i q_i^{(6)} < -(\triangle q_i)^2.$$

4. Make a final upgrade of the parabola parameters:

$$\triangle q_i = q_i^L - q_i^R,$$
$$q_i^{(6)} = 6(q_i - 1/2(q_i^L + q_i^R)).$$

At the final stage of the solution of the hydrodynamic equations, we execute an adjustment procedure. In the case of a gas vacuum border, we have

$$\|u\| = \sqrt{2(E - \epsilon)}, \ (E - u^2/2)/E < 10^{-3}. \qquad (9)$$

In other regions, we apply an adjustment to ensure a nondecreasing entropy:

$$\rho\epsilon = \left(\rho E - \frac{\rho u^2}{2}\right), \ (E - u^2/2)/E \geq 10^{-3}. \qquad (10)$$

This modification provides a detailed balance of energy and ensures a nondecreasing entropy.

After solving the hydrodynamic equations, it is necessary to restore the gravitational potential with respect to the gas density. To this end, we will use a 27-point template to approximate the Poisson equation. The algorithm for solving the Poisson equation consists of three stages:

1. Setting the boundary conditions for the gravitational potential at the boundary of the region.
2. Transforming the density function to the harmonics space. A fast Fourier transform is used for this.
3. Solving the Poisson equation in the harmonics space. Next, it is necessary to perform the inverse fast Fourier transformation of the potential of the harmonics into the functional space of the harmonics.

The details of the method are given in [33].

3   Parallel Implementation

The parallel implementation is based on a multi-level decomposition of the computations:

1. One-dimensional decomposition of the computational domain by means of MPI, which, for consistency with the solution of the Poisson equation, is specified by the FFTW library.
2. One-dimensional decomposition of the computations by means of OpenMP as part of a single process running on a single Intel Xeon Phi accelerator.
3. Vectorization of computations within a single cell.

The geometric decomposition of the computational domain is carried out by means of MPI processes and by means of OpenMP threads. In the case of a decomposition of the computations by means of MPI, it is necessary to take into account overlapping subregions. The compact calculation template allows for the use of only one overlapping layer.

Next, we describe the basic instructions used to implement the method. We will dwell only on the declarative description:

– **_mm512_set1_pd** – Formation of a vector with each element being a scalar.
– **_mm512_load_pd** – Loading the addresses of the eight double elements of the vector.
– **_mm512_mul_pd** – Multiplication of vectors.
– **_mm512_add_pd** – Addition of vectors.
– **_mm512_sub_pd** – Subtraction of vectors.
– **_mm512_stream_pd** – Writing the vector to memory.
– **_mm512_abs_pd** – Getting the absolute value of the vector elements.

The instructions given here are sufficient to implement a numerical method for the solution of the hydrodynamic equations. We used the following line to compile the code:

icc -xMIC-AVX512 -qopenmp -O3 -no-prec-div

-o gooPhi.mic gooPhi.cpp -lm

It is worth noting only the acceleration of the division through the option -no-prec-div, which is recommended when using SSE extensions.

Table 1. Speedup and real performance of the code on a single Intel Xeon Phi

Cores	GFLOPS	Speedup
1	3.63	1.000
2	7.62	2.099
4	14.67	4.041
8	31.49	8.675
16	62.27	17.154
32	109.57	30.184
64	157.66	43.432
72	173.35	47.755
128	131.68	36.275
256	111.19	30.631

We studied the acceleration of the gooPhi code on a 512^3 grid. We measured the time of the numerical method (Total) in seconds on different numbers of logical cores (Cores). The acceleration P (Speedup) was calculated with the formula

$$P = \frac{\text{Total}_1}{\text{Total}_K}, \tag{11}$$

where Total_1 is the computation time on one logical core and Total_K is the computation time on K logical cores. We also assessed the actual performance. Table 1 contains the results on acceleration and performance on a mesh of size 512^3. We achieved a performance of 173 gigaFLOPS and a speedup factor of 48 using a single Intel Xeon Phi KNL.

In addition, we studied the scalability of the gooPhi code on a mesh of size $512 \times 512 \times 512$ points using all logical cores of each accelerator. Thus, each accelerator has a subdomain size of 512^3. For scalability assessment purposes, we measured the time of the numerical method (Total) in seconds while varying the number of Intel Xeon Phi (KNL) accelerators. The scalability T was computed using the formula

$$T = \frac{\text{Total}_1}{\text{Total}_p}, \tag{12}$$

where Total_1 is the computation time for one accelerator when using a single accelerator and Total_p is the computing time for one accelerator when using p accelerators. The results on acceleration are given in Table 2. Using 16 accelerators, we achieved a 97% scalability. Note that this is a fairly high result.

Table 2. Scalability of the code for various numbers of Intel Xeon Phi accelerators

KNL	Scalability
1	1.000
2	0.999
3	0.998
4	0.994
8	0.988
12	0.972
16	0.968

4   Discussion

In this section, we will discuss several important issues related to the organization of computations, constraints, and new features.

1. In the study, we used the eight elements of the vector (four density functions, three components of the velocity, and the entropy). This is connected with the use of all elements of a 512-bit double-precision vector. We hope that the size of the vector in future versions of the processors will be increased. This would allow us to take into account a greater number of species. At the same time, the multiplicity of eight requires in some cases the use of dummy elements for the organization of computations.
2. When writing the first version of the AstroPhi code and performing subsequent studies, an interesting fact emerged: a greater performance is achieved when using separate arrays to describe hydrodynamic quantities (density, angular momentum, pressure, etc.) than when using an array of C/C++ language structures in which each object contains all the information about the cell. Apparently, this is due to the use of a larger cache. This means that, when accessing multiple arrays, the corresponding cache lines are filled. Thus, we efficiently used as many cache lines as arrays. In the case of structures (or 4D arrays as in the present paper), only one or two cache lines were used.
3. In our implementation, we did not use combined instructions of FMA type. Performance tests, especially in linear algebra applications, where the main operation is a daxpy instruction, show that using FMA instructions improves performance. However, this trend was not observed. Moreover, there was a slowdown of the code, after which we decided to reject such instructions.

5   Modeling of Hydrodynamic Turbulence with Self-Gravity

For the simulation, we considered the test problem in the cubic region $[-1; 1]^3$ with $c_s = 0.1$. The initial density was assumed to be 1. The initial velocity perturbations followed a Gaussian distribution [34].

The main analysis of turbulent flows with gravity consists in estimating the Jeans criterion and the free-fall time, during which a local collapse occurs. To estimate of Jeans criterion, let us write the equations of gravitational hydrodynamics in 1D form using the isothermal equation of state:

$$\frac{\partial \rho}{\partial t} + \frac{\partial}{\partial x}(\rho u) = 0,$$

$$\frac{\partial \rho u}{\partial t} + \frac{\partial}{\partial x}(\rho u u) = -\frac{\partial p}{\partial x} - \rho\frac{\partial \Phi}{\partial x},$$

$$\frac{\partial^2 \Phi}{\partial x^2} = 4\pi G\rho,$$

$$p = c_s^2\rho. \tag{13}$$

The adiabatic term of the equation of state (3) starts working when the critical density is reached. This density is attained during the development of instability. For the analysis, we need the Jeans criterion, which is achieved at the initial stage by using the isothermal equation of state.

We will consider a linear perturbation of the physical variables:

$$\rho = \rho_0 + \rho_1, \quad p = p_0 + p_1, \quad u = u_1, \quad \Phi = \Phi_0 + \Phi_1. \tag{14}$$

Let us rewrite the equations of gravitational hydrodynamics for the considered perturbation of the physical variables:

$$\frac{\partial \rho_1}{\partial t} + \rho_0\frac{\partial u_1}{\partial x} = 0,$$

$$\frac{\partial u_1}{\partial t} = -\frac{c_s^2}{\rho_0}\frac{\partial \rho_1}{\partial x} - \frac{\partial \Phi_1}{\partial x},$$

$$\frac{\partial^2 \Phi_1}{\partial x^2} = 4\pi G\rho_1. \tag{15}$$

We seek a nontrivial solution proportional to $\exp[i(kx + \omega t)]$. Consequently,

$$\frac{\partial}{\partial t} = i\omega, \quad \frac{\partial}{\partial x} = ik.$$

Let us write the equations for (ρ_1, u_1, Φ_1) in the following form:

$$\omega\rho_1 + k\rho_0 u_1 = 0,$$

$$\frac{kc_s^2}{\rho_0}\rho_1 + \omega u_1 + k\Phi_1 = 0,$$

$$4\pi G\rho_1 + k^2\Phi_1 = 0. \tag{16}$$

By equating to zero the determinant of the system,

$$\begin{vmatrix} \omega & k\rho_0 & 0 \\ \frac{kc_s^2}{\rho_0} & \omega & k \\ 4\pi G & 0 & k^2 \end{vmatrix},$$

we obtain the condition

$$\omega^2 = k^2 c_s^2 - 4\pi G \rho_0. \tag{17}$$

We should write the critical wavenumber of the Jeans criterion in the form

$$k_J = \left(\frac{4\pi G \rho_0}{c_s^2}\right)^{1/2}, \tag{18}$$

and the critical wavelength of the Jeans criterion in the form

$$\lambda_J = \frac{2\pi}{k_J} = \left(\frac{\pi}{G \rho_0}\right)^{1/2} c_s. \tag{19}$$

By applying a perturbation of the wavelength $\lambda > \lambda_J$, we trigger the gravitational instability.

To estimate the free-fall time, we consider the collapse of a homogeneous sphere of mass M and radius R. We need to estimate the time it takes the sphere radius to decrease from R to zero. Let us write the equation for the moment of impulse in the following form:

$$\frac{d^2 r}{dt^2} = -\frac{Gm}{r^2}, \tag{20}$$

where $m = 4\pi \int_0^r r^2 \rho_0 \, dr$ and $M = \frac{4\pi R^3 \rho_0}{3}$. Here we omit the cumbersome but rather trivial computations. It follows from Eq. (20) that

$$dt = -\left(\frac{8\pi G \rho_0}{3}\right)^{-1/2} \left(\frac{r}{R-r}\right)^{1/2} \frac{dr}{R}. \tag{21}$$

By integrating the last equation from the initial state of the sphere $r = R$ to the final stage $r = 0$, when it collapses, we obtain the equation for the free-fall time t_{ff}:

$$t_{ff} = \left(\frac{3\pi}{32 G \rho_0}\right)^{1/2}. \tag{22}$$

We will use the last equation to find the characteristic time for the local collapse. Obviously, a collapse is not achievable in a hydrodynamic model in that time. However, since the computational cells have finite size we can consider the process of local collapse in various subdomains of the computational domain. That is especially important in the context of the process of star formation and supernovae explosions.

The results of the computational experiments on the evolution of hydrodynamic turbulence are portrayed in Fig. 1. As we can see, density fragmentation occurs throughout the evolution of turbulence. It would be interesting to consider each individual density wave since in the context of star formation these waves can potentially correspond to young stars. It would also be interesting from the point of view of nuclear reactions to consider the high density regions in the case of turbulent combustion of carbon in white dwarfs.

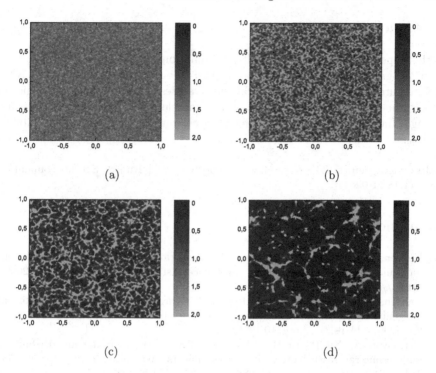

Fig. 1. The density distribution during the evolution of the turbulence process for a model time equal to one quarter (a), two quarters (b), three quarters (c), and four quarters (d) of the free-fall time for cold matter

The problem of hydrodynamic turbulence is one of interest in various astrophysical applications. Our main interest is related to the organization of parallel and distributed computations of supernova explosions. Despite the variety of mechanisms involved in supernova explosions, the distributed computations in these problems are used to correctly reproduce the nuclear combustion of chemical elements and, therefore, correctly compute the injected energy in each computational cell of the domain.

The distributed run of such problems is a very expensive and complicated procedure, and a detailed elaboration is not always required. This is a consequence of the fact that perturbations in the computational cell do not always lead to instabilities. The main criterion for running a hydrodynamic problem should be the analysis of the Jeans criterion λ_J. If it is attained, then it is enough to carry out the simulation for a time less than free-fall time t_{ff}, rather than for the characteristic time step of the main task. All density waves are formed in that time, and this allows one to fully take into account all nuclear reactions in supernovae of all types.

6 Conclusions

In this paper, we presented the results of simulations of hydrodynamic turbulence with self-gravity, employing the latest Intel Xeon Phi accelerators with KNL architecture. A new vector numerical code was described in detail. We achieved a performance of 173 gigaFLOPS and an acceleration factor of 48 by using a single Intel Xeon Phi KNL. Using 16 accelerators, we reached a scalability of 97%.

Acknowledgments. The research was supported by the Russian Science Foundation (project 18-11-00044).

References

1. Klessen, R., Heitsch, F., Mac Low, M.-M.: Gravitational collapse in turbulent molecular clouds I. Gasdynamical turbulence. Astrophys. J. **535**, 887–906 (2000). https://doi.org/10.1086/308891
2. Heitsch, F., Mac Low, M.-M., Klessen, R.: Gravitational Collapse in turbulent molecular clouds II. Magnetohydrodynamical turbulence. Astrophys. J. **547**, 280–291 (2001). https://doi.org/10.1086/318335
3. Beresnyak, A., Xu, H., Li, H., Schlickeiser, R.: Magnetohydrodynamic turbulence and cosmic-ray reacceleration in galaxy clusters. Astrophys. J. Suppl. Ser. **771**, 131 (2013). https://doi.org/10.1088/0004-637X/771/2/131
4. Kim, W., Ostriker, E.: Amplification, saturation, and Q Thresholds for runaway: growth of self-gravitating structures in models of magnetized galactic gas disks. Astrophys. J. **559**, 70–95 (2001). https://doi.org/10.1086/322330
5. Alig, C., Burkert, A., Johansson, P., Schartmann, M.: Simulations of direct collisions of gas clouds with the central black hole. Mon. Not. Roy. Astron. Soc. **412**(1), 469–486 (2011). https://doi.org/10.1111/j.1365-2966.2010.17915.x
6. Petrov, M., Berczik, P.: Simulation of the gravitational collapse and fragmentation of rotating molecular clouds. Astron. Nachr. **326**(7), 505–513 (2005)
7. Beresnyak, A.: Basic properties of magnetohydrodynamic turbulence in the inertial range. Mon. Not. Roy. Astron. Soc. **422**(4), 3495–3502 (2012). https://doi.org/10.1111/j.1365-2966.2012.20859.x
8. Mason, J., Perez, J.C., Cattaneo, F., Boldyrev, S.: Extended scaling laws in numerical simulations of magnetohydrodynamic turbulence. Astrophys. J. Lett. **735**, L26 (2011). https://doi.org/10.1088/2041-8205/735/2/L26
9. Perez, J.C., Boldyrev, S.: Numerical simulations of imbalanced strong magnetohydrodynamic turbulence. Astrophys. J. Lett. **710**, L63–L66 (2010). https://doi.org/10.1088/2041-8205/710/1/L63
10. Beresnyak, A.: Spectra of strong magnetohydrodynamic turbulence from high-resolution simulations. Astrophys. J. Lett. **784**, L20 (2014). https://doi.org/10.1088/2041-8205/784/2/L20
11. McKee, C.F., Li, P.S., Klein, R.: Sub-alfvenic non-ideal MHD turbulence simulations with ambipolar diffusion II. Comparison with observation, clump properties, and scaling to physical units. Astrophys. J. **720**, 1612–1634 (2010). https://doi.org/10.1088/0004-637X/720/2/1612

12. Federrath, C., Klessen, R.: The star formation rate of turbulent magnetized clouds: comparing theory, simulations, and observations. Astrophys. J. **761**, 156 (2012). https://doi.org/10.1088/0004-637X/761/2/156

13. Kritsuk, A., et al.: Comparing numerical methods for isothermal magnetized supersonic turbulence. Astrophys. J. **737**, 13 (2011). https://doi.org/10.1088/0004-637X/737/1/13

14. Galtier, S., Buchlin, E.: Multiscale hall-magnetohydrodynamic turbulence in the solar wind. Astrophys. J. **656**, 560–566 (2007). https://doi.org/10.1086/510423

15. Willcox, D., Townsley, D., Calder, A., Denissenkov, P., Herwig, F.: Type Ia supernova explosions from hybrid carbon-oxygen-neon white dwarf progenitors. Astrophys. J. **832**, 13 (2016). https://doi.org/10.3847/0004-637X/832/1/13

16. Schive, H., Tsai, Y., Chiueh, T.: GAMER: a GPU-accelerated adaptive-mesh-refinement code for astrophysics. Astrophys. J. **186**, 457–484 (2010). https://doi.org/10.1088/0067-0049/186/2/457

17. Kulikov, I.: GPUPEGAS: a new GPU-accelerated hydrodynamic code for numerical simulations of interacting galaxies. Astrophys. J. Supp. Ser. **214**, 1–12 (2014). https://doi.org/10.1088/0067-0049/214/1/12

18. Kulikov, I.M., Chernykh, I.G., Snytnikov, A.V., Glinskiy, B.M., Tutukov, A.V.: AstroPhi: a code for complex simulation of dynamics of astrophysical objects using hybrid supercomputers. Comput. Phys. Commun. **186**, 71–80 (2015). https://doi.org/10.1016/j.cpc.2014.09.004

19. Schneider, E., Robertson, B.: Cholla: a new massively parallel hydrodynamics code for astrophysical simulation. Astrophys. J. Suppl. Ser. **217**, 2–24 (2015). https://doi.org/10.1088/0067-0049/217/2/24

20. Benitez-Llambay, P., Masset, F.: FARGO3D: a new GPU-oriented MHD code. Astrophys. J. Suppl. Ser. **223**, 1–11 (2016). https://doi.org/10.3847/0067-0049/223/1/11

21. Pekkilaa, J., Vaisalab, M., Kapylac, M., Kapylad, P., Anjum, O.: Methods for compressible fluid simulation on GPUs using high-order finite differences. Comput. Phys. Commun. **217**, 11–22 (2017). https://doi.org/10.1016/j.cpc.2017.03.011

22. Griffiths, M., Fedun, V., Erdelyi, R.: A fast MHD code for gravitationally stratified media using graphical processing units: SMAUG. J. Astrophys. Astron. **36**(1), 197–223 (2015). https://doi.org/10.1007/s12036-015-9328-y

23. Mendygral, P., et al.: WOMBAT: a scalable and high-performance astrophysical magnetohydrodynamics code. Astrophys. J. Suppl. Ser. **228**, 2–23 (2017). https://doi.org/10.3847/1538-4365/aa5b9c

24. Surmin, I., et al.: Particle-in-cell laser-plasma simulation on Xeon Phi coprocessors. Comput. Phys. Commun. **202**, 204–210 (2016). https://doi.org/10.1016/j.cpc.2016.02.004

25. Needham, P., Bhuiyan, A., Walker, R.: Extension of the AMBER molecular dynamics software to Intel's Many Integrated Core (MIC) architecture. Comput. Phys. Commun. **201**, 95–105 (2016). https://doi.org/10.1016/j.cpc.2015.12.025

26. Brown, W.M., Carrillo, J.-M.Y., Gavhane, N., Thakkar, F.M.: Optimizing legacy molecular dynamics software with directive-based offload. Comput. Phys. Commun. **195**, 95–101 (2015). https://doi.org/10.1016/j.cpc.2015.05.004

27. Bernaschia, M., Bissona, M., Salvadore, F.: Multi-Kepler GPU vs. multi-Intel MIC for spin systems simulations. Comput. Phys. Commun. **185**, 2495–2503 (2014). https://doi.org/10.1016/j.cpc.2014.05.026

28. Nishiura, D., Furuichi, M., Sakaguchi, H.: Computational performance of a smoothed particle hydrodynamics simulation for shared-memory parallel computing. Comput. Phys. Commun. **194**, 18–32 (2015). https://doi.org/10.1016/j.cpc.2015.04.006

29. Kulikov, I., Chernykh, I., Tutukov, A.: A new hydrodynamic model for numerical simulation of interacting galaxies on Intel Xeon Phi supercomputers. J. Phys: Conf. Ser. **719**, 012006 (2016). https://doi.org/10.1088/1742-6596/719/1/012006

30. Glinsky, B., Kulikov, I., Chernykh, I., et al.: The co-design of astrophysical code for massively parallel supercomputers. Lect. Notes Comput. Sci. **10049**, 342–353 (2017). https://doi.org/10.1007/978-3-319-49956-7_27

31. Kulikov, I.M., Chernykh, I.G., Glinskiy, B.M., Protasov, V.A.: An efficient optimization of HLL method for the second generation of Intel Xeon Phi processor. Lobachevskii J. Math. **39**(4), 543–550 (2018). https://doi.org/10.1134/S1995080218040091

32. Kulikov, I.M., Chernykh, I.G., Tutukov, A.V.: A new parallel Intel Xeon Phi hydrodynamics code for massively parallel supercomputers. Lobachevskii J. Math. **39**(9), 1207–1216 (2018). https://doi.org/10.1134/S1995080218090135

33. Kulikov, I., Vorobyov, E.: Using the PPML approach for constructing a low-dissipation, operator-splitting scheme for numerical simulations of hydrodynamic flows. J. Comput. Phys. **317**, 318–346 (2016). https://doi.org/10.1016/j.jcp.2016.04.057

34. Kulikov, I., Chernykh, I., Protasov, V.: Mathematical modeling of formation, evolution and interaction of galaxies in cosmological context. J. Phys: Conf. Ser. **722**, 012023 (2016). https://doi.org/10.1088/1742-6596/722/1/012023

Parallel Computational Algorithm of a Cartesian Grid Method for Simulating the Interaction of a Shock Wave and Colliding Bodies

Dmitry A. Sidorenko and Pavel S. Utkin$^{(\boxtimes)}$ (iD)

Institute for Computer Aided Design, Russian Academy of Sciences,
Moscow, Russia
sidr1234@mail.ru, pavel_utk@mail.ru

Abstract. We conduct direct numerical simulations for clarifying the mechanics of high-speed flows in two-phase media with shock waves (e.g., shock wave interaction with a cloud or a layer of particles). We describe the interaction between the flow and particles by solving the Euler equations in a domain with variable boundaries. Cartesian grid methods (also known as cut-cell methods, immersed boundary methods) are the most appropriate class of numerical methods in this case. The work describes the development of an effective parallel computational algorithm for a Cartesian grid method and the application of the algorithm for simulating the relaxation of several interacting particles behind a passing shock wave corresponding to natural experiments.

Keywords: Numerical simulation · Cartesian grid method ·
Parallel computations · Shock wave · Particles

1 Introduction

The interfacial interaction between particles and gas in high-speed two-phase media flows with shock waves (SW) is associated with the supersonic aerodynamics of closely spaced bodies and the collective effects of their mutual influence on each other. In a real flow with a large ensemble of particles, any spatial configuration of particles is possible. Owing to this, it is expedient to reduce the problem to two model situations for the interaction of the flow with a pair of particles located either along or across the flow. Some results achieved in studies done in this direction in the case of supersonic flows are given in [1,2]. It should be noted that the concept of the study of pair interactions has certainly evolved, and current technical capabilities as well as the development of numerical methods allow for modeling the interaction of the flow with a set of irregularly spaced bodies [3].

Multidimensional gas dynamics simulations are conducted to elucidate the features of the interaction between SW and ensembles of particles. In [2], the

© Springer Nature Switzerland AG 2019
L. Sokolinsky and M. Zymbler (Eds.): PCT 2019, CCIS 1063, pp. 323–334, 2019.
https://doi.org/10.1007/978-3-030-28163-2_23

authors considered the dynamics of the motion of two and three spheres behind a passing SW (two-dimensional computations in an axially symmetric setup). The simulations resorted to the Navier–Stokes equations to specify the mechanical models of heterogeneous media. It was shown that the relaxation model for a single particle is unsuitable in this situation and it is necessary to take into account collective effects. Two-dimensional gas dynamics numerical simulations were performed in [3] for qualitative determination of flow features of a three-dimensional cloud. The solution obtained was used as the "exact" solution for comparison with one-dimensional volume-averaged models. The simulation results highlighted the importance of multidimensional effects associated with Reynolds stress terms inside a particle cloud as well as in its turbulent wake.

An important matter that significantly affects the capability of such multidimensional gas dynamics modeling is the numerical algorithm used, since the computational domain is multiply connected and cannot be covered by a structured computational grid. In [2], the simulations were carried out using ANSYS Fluent. The computational cells in two-dimensional simulations were quadrilateral; the details of the grid structure and the numerical algorithm used are unknown. In [3], the authors used a Cartesian grid and a general Riemann-type solver. The cylinders had stepwise boundaries, which can lead to artificial effects during the reflection of shocks. Apparently, the immersed boundary methods (see, for example, [4] and references in [5]) are the most appropriate class of numerical methods for the kind of problems under consideration. In this case, all cells have a square or rectangular shape and it is necessary to solve somehow the problem of calculation of the parameters in the cells intersected by the curved boundaries of the computational domain. It is also worth noting the approach associated with the use of several overlapping grids [6,7]. The first grid is in the core of the flow and is not consistent with the boundaries of the bodies. The second grid is in a vicinity of the curvilinear boundaries and is consistent with them. The matching of solutions on overlapping grid sections is achieved through interpolation procedures.

Immersed boundary methods can be divided into two classes. In the methods of the first class, dummy terms simulating the effect of a wall are included in the system of defining equations. Computations are carried out consistently throughout the extended computational domain, including both the cells intersected by the boundaries of the solid bodies and the cells totally covered by the bodies. Such a new method for the calculation of compressible gas flows was proposed in [8]. The algorithmic uniformity of the method is one of its main advantages. Among its disadvantages, we have the absence of an unambiguous physical sense for the parameters in the cut cells, especially in problems from physical and chemical hydrodynamics. Another disadvantage is a need to control the conservative property of the algorithm in the context of the existence of source terms. The second class of methods comprises Cartesian grid methods in which the system of defining equations is approximated in a special manner in the cells that are intersected by the computational domain boundaries. The approach is characterized by the possibility of the "small cell" problem, when the size of a cut cell is much smaller than the size of a regular one.

Thanks to the simplicity of the topology involved, Cartesian grid methods allow for an obvious decomposition of the computational domain to implement the algorithm on multiprocessor computing systems. In commercial codes, parallelization is often used in conjunction with adaptive mesh refinement (see, for example, [9]). In the early work [10], parallelism is implemented by decomposing the computational domain that is common in space-time problems of mathematical physics but not in functional parallelism which is used in computational fluid dynamics [11,12]. One of the most important aims of program parallelization is to minimize the time required for data exchange between processes.

In this work, we develop a parallel algorithm for a Cartesian grid method for solving the Euler equations and conduct a numerical study of the mechanisms governing the motion of two particles in a supersonic gas flow behind the front of a passing SW, in a setup corresponding to the full-scale experiment in [13].

2 Parallel Numerical Algorithm of the Cartesian Grid Method

2.1 Mathematical Model and Numerical Algorithm of the Cartesian Grid Method

Consider the rectangular computational domain $[0; R] \times [0; Z]$ with possibly several movable bodies inside. The computational domain is covered by a fixed Cartesian grid with quadrilateral cells. The computational grid is uniform, with N_z cells along the axial direction and N_r cells along the radial direction. The mathematical model is based on the two-dimensional Euler equations in a computational domain with internal moving boundaries, under the assumption of axial symmetry of the flow, which is written in the cylindrical frame (r, z):

$$\mathbf{U}_t + \mathbf{F}_r + \mathbf{G}_z = -\frac{\mathbf{H}}{r},$$

$$\mathbf{U} = \begin{bmatrix} \rho \\ \rho u \\ \rho v \\ E \end{bmatrix}, \ \mathbf{F} = \begin{bmatrix} \rho u \\ \rho u^2 + p \\ \rho u v \\ u(E + p) \end{bmatrix}, \ \mathbf{G} = \begin{bmatrix} \rho v \\ \rho u v \\ \rho v^2 + p \\ v(E + p) \end{bmatrix}, \ \mathbf{H} = \begin{bmatrix} \rho u \\ \rho u^2 \\ \rho u v \\ u(E + p) \end{bmatrix},$$

$$E = \frac{\rho}{2}(u^2 + v^2) + \rho e(p, \rho), \quad e(p, \rho) = \frac{p}{\rho(\gamma - 1)}.$$

Here and below, we use standard notations: t is the time, ρ is the gas density, u and v are the radial and axial velocity components, p is the gas pressure, and γ is the specific heat ratio. At each time step, each computational cell is assigned a status. The following statuses are possible: outer cell (the cell is completely inside a body or is intersected by the boundary of a body) and inner cell (the whole cell is inside the computational domain). The solution is built only in inner cells.

The numerical algorithm is based on a Godunov-type finite volume scheme. In addition, we use an explicit Euler scheme for time integration:

$$\mathbf{U}_{i,j}^{n+1} = \mathbf{U}_{i,j}^n - \frac{\Delta t^n}{\Delta x}\left(\mathbf{F}_{i+1/2,j} - \mathbf{F}_{i-1/2,j} + \mathbf{G}_{i,j+1/2} - \mathbf{G}_{i,j-1/2}\right) - \frac{\mathbf{H}_{i,j}^n}{r_{i,j}}, \quad (1)$$

where i and j are the spatial indexes of the inner cells, $r_{i,j}$ is the radius of the center of the cell, measured from the symmetry axis of the cylindrical frame (r, z), n is the time index, Δt^n is the dynamic time step, and Δx is the size of the computational grid cell. The numerical fluxes $\mathbf{F}$ and $\mathbf{G}$ through the edges common to inner cells are calculated using the Steger–Warming approach [14] since it is free from the numerical effect of SW instability, called the "carbuncle" effect. The "carbuncle" effect can occur in our problem near the symmetry axis in case of a classical Godunov scheme. To compute the numerical fluxes through the edges that are common to both an inner cell and an outer one intersected by the boundary of a body, we follow a special procedure allowing for the definition of the parameters on the side of the intersected cell (see [15, 16]).

By way of example, we will describe the algorithm of calculation of the numerical flux $\mathbf{F}_{i-1/2,j}$ through a vertical edge that is common to both the outer cell $(i-1, j)$ and the inner cell (i, j). Let us assume that the outer cell corresponds to a cell intersected by the boundary of a body that moves with velocity $\mathbf{v}_b$. Denote the vector of conservative variables in the cell (i, j) by

$$\mathbf{U}_{i,j} = \begin{bmatrix} \rho \\ \rho u \\ \rho v \\ E(u, v, p) \end{bmatrix}.$$

To compute the flux $\mathbf{F}_{i-1/2,j}$, we should solve the Riemann problem $\mathbf{U}_R(.,.)$ with left state

$$\mathbf{U}_{\text{ghost}} = \begin{bmatrix} \rho \\ \rho(2\mathbf{v}_{br} - u) \\ \rho v \\ E(2\mathbf{v}_{br} - u, v, p) \end{bmatrix}$$

and right state $\mathbf{U}_{i,j}$. The solution of this problem in the local one-dimensional case along the r axis contains two SWs or two rarefaction waves connected by a contact surface without density gaps and moving with gas velocity $\mathbf{v}_{br}$. Denote the gas parameters and the speed of sound at the contact surface by $\mathbf{U}_{i-1/2,j}^*$ and $c_{i-1/2,j}^*$, respectively. The numerical flux is then determined as

$$\mathbf{F}_{i-1/2,j} = \begin{cases} \mathbf{F}[\mathbf{U}_{i,j}], & \text{if } \mathbf{v}_{br} < -c_{i-1/2,j}^*, \\ \mathbf{F}[\mathbf{U}_R(\mathbf{U}_{i-1/2,j}^*, \mathbf{U}_{i,j}], & \text{if } -c_{i-1/2,j}^* < \mathbf{v}_{br} < c_{i-/2,j}^*, \\ \mathbf{F}[\mathbf{U}_R(\mathbf{U}_{\text{ghost}}, \mathbf{U}_{i,j}], & \text{if } \mathbf{v}_{br} > c_{i-1/2,j}^*. \end{cases}$$

The other fluxes are calculated in the same manner.

After determining the fluxes, the equations of movement of the bodies are integrated. Then, both the status of each cell and its vector of conservative variables are updated. If, as a consequence of the movement of the bodies, a cell becomes outer, then we do not build the solution in it. If the cell is still inner, then the solution is found in accordance with (1). If the cell becomes inner, but it was previously outer, then the vector of conservative variables is determined by the formula

$$
\mathbf{U}_{i,j}^{n+1} = \frac{\alpha_z}{\alpha_z + \alpha_r} \frac{\alpha_{i+1,j}\mathbf{U}_{i+1/2,j}^* + \alpha_{i-1,j}\mathbf{U}_{i-1/2,j}^*}{\alpha_{i+1,j} + \alpha_{i-1,j}} +
$$
$$
+ \frac{\alpha_r}{\alpha_z + \alpha_r} \frac{\alpha_{i,j+1}\mathbf{U}_{i,j+1/2}^* + \alpha_{i,j-1}\mathbf{U}_{i,j-1/2}^*}{\alpha_{i,j+1} + \alpha_{i,j-1}},
$$

$$
\alpha_{i\pm1,j} = \begin{cases} 1, & \text{if the cell } (i \pm 1, j) \text{ was inner in the previous time step,} \\ 0, & \text{otherwise,} \end{cases}
$$

$$
\alpha_{i,j\pm1} = \begin{cases} 1, & \text{if the cell } (i, j \pm 1) \text{ was inner in the previous time step,} \\ 0, & \text{otherwise,} \end{cases}
$$

$$
\alpha_z = \max(\alpha_{i+1,j}, \alpha_{i-1,j}), \quad \alpha_r = \max(\alpha_{i,j+1}, \alpha_{i,j-1}).
$$

The algorithm we have described is relatively easy to implement and, unlike the methodology suggested in [19], does not require the determination of the shape or the area of the intersection of a body boundary with a regular computational cell.

Modeling the motion of multiple bodies in a flow leads to the necessity of taking into account not only the interaction of the bodies with a gas but also with each other. As model of interaction, we chose the theory of kinematic impact [17]. This model considers the inelastic collision of two bodies of arbitrary shape taking into account the restitution coefficient e. Assuming conservation laws under the instantaneous impact approximation as well as the absence of deformations, we can express the velocities of two particles of masses m_1 and m_2 after the impact as

$$
v_1 = v_1^0 + \frac{(1+e)(v_2^0 - v_1^0)}{1 + m_1/m_2}, \quad v_2 = v_2^0 - \frac{(1+e)(v_2^0 - v_1^0)}{1 + m_2/m_1}.
$$

Here the superscript 0 denotes the velocities before the impact.

The numerical algorithm and its program realization were tested using one- and two-dimensional problems from [15], concerned with defined bodies and their motion under pressure forces.

We also simulated the liftoff of a cylinder behind a SW [16], which is frequently considered in the literature. In this test, the algorithm was compared with a number of other known similar methods (see [20, 21]). The results showed that the SW patterns and the integral dynamics of motion of the cylinder are quite similar in different approaches. The most remarkable differences are in the structure of vortexes near the cylinder.

We also made a comparison with simulations that use boundary-fitted computational grids and the corresponding methods. Additionally, we considered the problem of interaction of a SW with a static cylinder. We made a quantitative comparison of the distribution of pressure on the cylinder surface against simulation results obtained in [22] using the Navier–Stokes equations and an unstructured adaptive computational grid; this comparison showed a match with a 2% error.

The model of interaction of bodies was tested on the problem of interaction of a pair of initially static cylinders and a SW with Mach number 5.0 [18]. The test showed that the cylinders start moving, later collide and scatter from each other. The predicted density isolines correlate well with data from [18].

Finally, we considered the motion of one spherical bronze particle ($170\,\mu$m of diameter) in a supersonic flow behind a passing wave with Mach number $M = 2.6$, in the setup from [23]. The simulation time was $500\,\mu$s. The error in the coordinate of the particle did not exceed 2% relative to the experimental value (see Fig. 1).

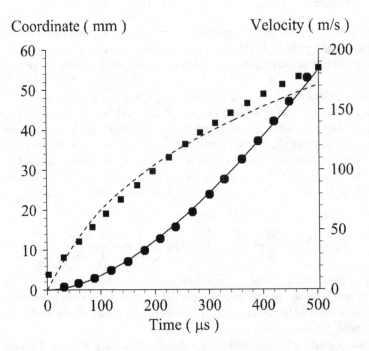

Fig. 1. Dynamics of the motion of a particle behind a SW. Lines correspond to the simulation; dots (coordinates) and squares (velocities) correspond to the experiment (see [23])

2.2 Parallelization

The parallelization is based on the method of static domain decomposition using MPI. The rectangular computational domain is subdivided into K_z equal parts along the axial direction and K_r along the radial direction. Each rectangular part is delegated to a separate processor. Thus, the total number of processors in use is equal to $K_z \times K_r + 1$, since one processor is used as the control one ("master–slave" scheme). A uniform grid is generated independently by each processor. So the first critical part which is common to most CFD methods, namely the generation of a huge computational grid, is avoided. Since we use the explicit time integration scheme (1), communication between processors is necessary only to exchange the conservative variable vectors at the boundary cells for each subdomain and to find the next time step. The second important infrastructure problem is the visualization of large output files containing the spatial distributions of the fields of primitive variables. We do not collect the whole output file containing the distributions. The maximum number of output files for one moment of recording time is equal to the number of processors. Given that the problems we are considering are nonstationary, we introduce matrices of output containing information about which parts of the whole computational domain are of interest at a concrete moment of time. This way, we reduce the volume of output data by an order of magnitude and avoid large files by replacing them with a certain number of smaller ones.

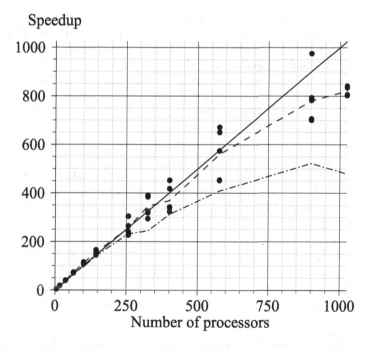

Fig. 2. Speedup estimation. Solid line: ideal speedup. Dashed line and dots: speedup in the case of 1000 time steps. Dashed-dotted line: speedup in the case of 100 time steps

To estimate the speedup of the parallel algorithm, we considered a test problem of interaction between a rarefaction wave and two spherical particles. The statement involving a rarefaction wave ensures the existence of nontrivial gas dynamics simulations in each computational cell. The computational domain is a square with 7200×7200 cells. Computations of 100 and 1000 time steps were carried out on the MVS-10P supercomputer at the Joint Supercomputer Center of the RAS (JSCC RAS). A repetitive series of computations was performed for each number of processors, and then the results were averaged. Owing to the two-dimensional partition of the computational domain and the presence of a single control processor, the minimum number of processors throughout the test was five. The maximum number of processors was 1025. Even in this case, there were about 50 000 calculation cells per processor. Figure 2 illustrates the estimation of the speedup of the algorithm. We measured the speedup relative to the simulation time on five processors and then scaled it to one processor, so the plot should be considered as an upper-bound estimate of the speedup. Results demonstrate the good scalability of the numerical algorithm.

3   Numerical Simulation of the Dynamics of Two Particles in a Supersonic Flow

3.1   Problem Statement

The statement of the problem corresponds to the full-scale experiment in [13]. Two spherical particles are situated on the axis of symmetry of the channel. The width of the channel is 52 mm; the length of the window of the measuring section of the channel is 250 mm. The channel is filled with quiescent air under initial pressure $p_0 = 0.226$ atm and initial temperature 293 K. Thus, the initial air density equals 0.269 kg/m^3. The sound speed is $c_0 = 343$ m/s. The diameters of the particles are $d_1 = 4.78$ mm (the left one in Fig. 3) and $d_2 = 4.98$ mm (the right particle in the figure). Their masses are $m_1 = 24.7$ mg and $m_2 = 22.5$ mg, respectively. Initially, the centers of the first and the second particles are located at (0 mm; 10 mm) and (0 mm; 25.9 mm), respectively. Thus, the initial distance between the centers of the particles is $l_0 = 3.2d_2$. The Mach number of the incident SW is $M = 3.0$; the SW speed is 1029 m/s. Initially, the SW is at $z_0 = 7.465$ mm, i.e. immediately before the first particle. The parameters behind the SW front are as follows:

$$p_M = p_0 \left(\frac{2\gamma}{\gamma+1} M^2 - \frac{\gamma-1}{\gamma+1} \right) = 2.34 \text{ atm}, \quad \rho_M = \frac{p_0(\gamma+1)M^2}{2+(\gamma-1)M^2} = 1.04 \text{ } \frac{\text{kg}}{\text{m}^3},$$
$$v_M = \frac{2c_0}{\gamma+1} \frac{M^2-1}{M} = 762.2 \text{ } \frac{\text{m}}{\text{s}}.$$

The possible interaction of the particles is considered to be perfectly elastic. The boundary conditions are as follows:

– for $r = 0$ mm: inflow with parameters p_M, ρ_M, and v_M. The compression phase of the SW in the experiments was long enough to keep the parameters constant behind the SW.

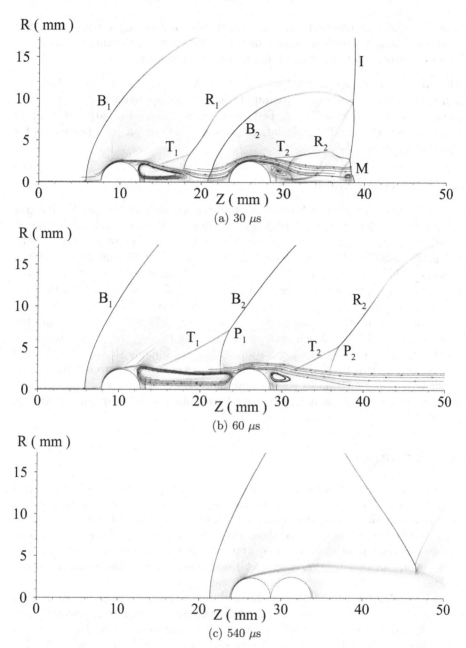

Fig. 3. Numerical schlieren visualization (distributions of the modulus of the gas density gradient) and several instant streamlines at successive moments

- for $r = 250$ mm: free outflow. The condition is valid as long as the gas velocity at the boundary is supersonic, which is indeed true for the times considered.
- for $z = 0$ mm and $z = 26$ mm: non-penetrating conditions.

The size of the computational grid is $N_r \times N_z = 4160 \times 39\,936 \approx 166$ million cells. There are about 750 computational cells per particle diameter. The computational time was 570 μs. The results were recorded in 30 μs increments as in the experiment. The calculations were carried out on 1000 processors for approximately one day.

3.2   Simulation Results

Figure 3 portrays the spatial distribution of the modulus of the gas density gradient (the so-called numerical schlieren visualization) and several instantaneous streamlines. In Figs. 3a and b, which correspond to 30 μs and 60 μs, respectively, the main elements of the flow structure are indicated at the initial stage of the process. When the first particle is pushed by the supersonic flow, there arise a bow shock B_1, a wave R_1, reflected from the axis of symmetry behind the particle, and a trailing wave T_1 (see Fig. 3a). A similar system of waves is observed in the flow around the second particle. The leading SW is denoted by I and the Mach wave by M (the Mach configuration forms behind the second particle). At 60 μs, there appears a configuration with three oblique shocks, B_1, B_2, and R_2, as well as two trailing waves, T_1 and T_2, which are connected to B_1 and B_2 at triple points P_1 and P_2. The bow shock B_2 near the second particle disappears at about 120 μs. All observations are consistent with experimental data given in [13].

4   Conclusions

A parallel computational algorithm of a Cartesian grid method was developed for numerical simulations of flows with shock waves in domains of varying shape. The mathematical model was based on the two-dimensional Euler equations in an axially symmetric statement. We measured the speedup of the algorithm. A distinctive feature of the suggested approach is the absence of a stage of grid generation and visualization of spatial distributions for the whole computational domain.

We conducted a numerical study of the mechanisms governing the motion of two particles in a supersonic gas flow behind the front of a passing shock wave, in a setup corresponding to a full-scale experiment. In addition, we gave a description of the main stages of the process from the point of view of the simulated shock wave configurations. It must be noted that we achieved the effect of interchange of the two main possible modes of supersonic flow around two bodies. As a result of the movement of both particles, the front particle overtakes the rear one, and then their interaction and flying off follow. The results we obtained correspond qualitatively to experimental data. The maximum computational grid size for the problem contained about 700 million cells during grid refinement studies; the simulations were carried out for approximately two days on 2000 processors.

Acknowledgments. The reported study was funded by the RFBR and the Moscow government (research project No. 19-38-70002). The authors express their deep gratitude to Professor V. M. Boyko (Khristianovish Institute of Theoretical and Applied Mechanics SB RAS) for the detailed explanation of the statement of the problem. Simulations were performed on the MVS-10P supercomputer at the JSCC RAS.

References

1. Boiko, V.M., Klinkov, K.V., Poplavski, S.V.: Collective bow ahead of a transverse system of spheres in a supersonic flow behind a moving shock wave. Fluid Dyn. **39**(2), 330–338 (2004). https://doi.org/10.1023/B:FLUI.0000030316.35579.73
2. Bedarev, I., Fedorov, A.: Direct simulation of the relaxation of several particles behind transmitted shock wave. J. Eng. Phys. Thermophys. **90**(2), 423–429 (2017). https://doi.org/10.1007/s10891-017-1581-2
3. Regele, J.D., Rabinovitch, J., Colonius, T., Blanquart, G.: Unsteady effects in dense, high speed, particle laden flows. Int. J. Multiph. Flow **61**, 1–13 (2014). https://doi.org/10.1016/j.ijmultiphaseflow.2013.12.007
4. Mittal, R., Iaccarino, G.: Immersed boundary methods. Annu. Rev. Fluid Mech. **37**, 239–261 (2005). https://doi.org/10.1146/annurev.fluid.37.061903.175743
5. Gorsse, Y., Iollo, A., Telib, H., Weynans, L.: A simple second order cartesian scheme for compressible Euler flows. J. Comput. Phys. **231**, 7780–7794 (2012). https://doi.org/10.1016/j.jcp.2012.07.014
6. Henshaw, W.D., Schwendeman, D.W.: Moving overlapping grids with adaptive mesh refinement for high-speed reactive and non-reactive flow. J. Comput. Phys. **216**, 744–779 (2006). https://doi.org/10.1016/j.jcp.2006.01.005
7. Maksimov, F.A., Churakov, D.A., Shevelev, Y.D.: Development of mathematical models and numerical methods for aerodynamic design on multiprocessor computers. Comput. Math. Math. Phys. **51**(2), 284–307 (2011). https://doi.org/10.1134/S0965542511020126
8. Menshov, I.S., Kornev, M.A.: Free-boundary method for the numerical solution of gas-dynamic equations in domains with varying geometry. Math. Modelsand Comput. Simul. **6**(6), 612–621 (2014). https://doi.org/10.1134/S207004821406009X
9. FlowVision. https://fv-tech.com/en/
10. Aftosmis, M.J., Berger, M.J., Adomavicius, G.: A domain-decomposed multilevel method for adaptively refined Cartesian grids with embedded boundaries. In: Proceedings of 38th Aerospace Sciences Meeting and Exhibit. Reno NV. AIAA Paper 2000–0808 (2000)
11. Agbaglah, G., et al.: Parallel simulation of multiphase flows using octree adaptivity and the volume-of-fluid method. C. R. Mec. **339**(2–3), 194–207 (2011). https://doi.org/10.1016/j.crme.2010.12.006
12. Dong, S., Karniadakis, G.E.: Dual-level parallelism for high-order CFD methods. Parallel Comput. **30**(1), 1–20 (2004). https://doi.org/10.1016/j.parco.2003.05.020
13. Boiko, V.M., Klinkov, K.V., Poplavski, S.V.: On a mechanism of interphase interaction in non-relaxing two-phase flow. In: Proceedings of 11th International Conference on the Methods of Aerophysical Research, Novosibirsk, pp. 24–27 (2002)

14. Steger, J.L., Warming, R.F.: Flux vector splitting of the inviscid gasdynamic equations with application to finite-difference methods. J. Comput. Phys. **40**, 263–293 (1981). https://doi.org/10.1016/0021-9991(81)90210-2
15. Chertock, A., Kurganov, A.: A simple Eulerian finite-volume method for compressible fluids in domains with moving boundaries. Commun. Math. Sci. **6**(3), 531–556 (2008)
16. Sidorenko, D.A., Utkin, P.S.: Numerical modeling of the relaxation of a body behind the transmitted shock wave. Math. Models Comput. Simul. **11**(4), 509–517 (2018). https://doi.org/10.1134/S2070048219040136
17. Goldsmith, W.: Impact: The Theory and Physical Behaviour of Colliding Solids. Published by Edward Arnold, London (1960)
18. Nourgaliev, R.R., Dinh, T.N., Theofanous, T.G., Koning, J.M., Greenman, R.M., Nakafuji, G.T.: Direct numerical simulation of disperse multiphase high-speed flows. In: Proceedings of 42nd Aerospace Sciences Meeting and Exhibit. Reno NV. AIAA Paper 2004–1284 (2004)
19. Sidorenko, D.A., Utkin, P.S.: A Cartesian grid method for the numerical modeling of shock wave propagation in domains of complex shape. Numer. Methods Program. **17**(4), 353–364 (2016). https://doi.org/10.26089/NumMet.v17r433
20. Arienti, M., Hung, P., Morano, E., Shepherd, J.E.: A level set approach to Eulerian-Lagrangian coupling. J. Comput. Phys. **185**(1), 213–251 (2003). https://doi.org/10.1016/S0021-9991(02)00055-4
21. Tan, S., Shu, C.-W.: A high order moving boundary treatment for compressible inviscid flows. J. Comput. Phys. **230**(15), 6023–6036 (2011). https://doi.org/10.1016/j.jcp.2011.04.011
22. Drikakis, D., Ofengeim, D., Timofeev, E., Voionovich, P.: Computation of non-stationary shock wave/cylinder interaction using adaptive-grid methods. J. Fluids Struct. **11**(6), 665–692 (1997). https://doi.org/10.1006/jfls.1997.0101
23. Boiko, V.M., Fedorov, A.V., Fomin, V.M., Papyrin, A.N., Soloukhin, R.I.: Ignition of small particles behind shock waves. In: Shock Waves, Explosions and Detonations. Progress in Astronautics and Aeronautics, vol. 87, pp. 71–87. AIAA (1983)

Accelerated Boundary Element Method for 3D Simulations of Bubble Cluster Dynamics in an Acoustic Field

Yulia A. Pityuk[1]([✉])(iD), Nail A. Gumerov[1,2], Olga A. Abramova[1],
Ilnur A. Zarafutdinov[1](iD), and Iskander S. Akhatov[3]

[1] Center for Micro and Nanoscale Dynamics of Dispersed Systems,
Bashkir State University, Ufa, Russia
`PityukYulia@gmail.com`
[2] Institute for Advanced Computer Studies, University of Maryland,
College Park, USA
[3] Skolkovo Institute of Science and Engineering (Skoltech), Moscow, Russia

Abstract. In this study, we develop a numerical approach based on the fast multipole method (FMM) to accelerate the iterative solution of the boundary element method (BEM) for bubble dynamics in the presence of an acoustic field. The FMM for 3D Laplace equation is accelerated by applying heterogeneous hardware, including multi-core CPUs and graphics processors. Problems of mesh stabilization are resolved by using a shape filter based on the spherical harmonic expansions of the bubble surface. We discuss the accuracy and performance of the algorithm. We demonstrate that the approach enables the simulation of the dynamics of regular monodisperse bubble clusters with thousands of bubbles and millions of boundary elements on modern personal workstations. The algorithm is scalable and can be extended to larger systems.

Keywords: Bubble cluster · Acoustic field · Potential flow ·
Boundary element method · Fast multipole method ·
Graphics processors · Heterogeneous hardware

1 Introduction

Bubbles can be found in natural conditions and also used in many technological processes, including surface cleaning by ultrasound. Moreover, biomedicine and microelectronics manifest particular interest in bubbly liquids. To understand the problems associated with the use of gas–liquid systems, a detailed study of bubble cluster dynamics is necessary.

The study of the dynamics of a single spherical bubble in an infinite liquid is given much attention; a considerable number of theoretical works were written on this subject in the 20th century (e.g., [1]). The main advantage of using a spherically symmetric model of a bubble is the possibility to extend

© Springer Nature Switzerland AG 2019
L. Sokolinsky and M. Zymbler (Eds.): PCT 2019, CCIS 1063, pp. 335–349, 2019.
https://doi.org/10.1007/978-3-030-28163-2_24

the model taking into account various effects, such as compressibility, viscosity, inertia, diffusion, temperature (e.g., [2]). It is almost impossible in the case of direct numerical simulation. Several works (e.g., [3,4]) have been devoted to describing the collective behavior of bubbles in acoustic fields, including bubble self-organization. However, bubbles may lose their spherical shape while interacting with each other, due to the excitation of shape modes. Obviously, it is necessary to develop other methods to study deformable bubble dynamics. We note that theoretical approaches have many limitations, so the use of numerical methods is the most appropriate for a more detailed study of the dynamics of non-spherical bubbles. There are various methods of direct numerical simulation, such as the finite difference method (FDM), the control volume method (CVM), and the finite element method (FEM). All these methods are based on the solution of the classical equations of fluid mechanics with boundary conditions. These methods allow one to adequately take into account the viscosity and the inertia of the liquid, the heat exchange with the bubble, and many other effects. However, the weakness of these methods is that they require the discretization of the entire domain which inevitably leads to a considerable number of computational points, especially in three-dimensional problems. At the same time, the use of the boundary element method (BEM), which requires only the surface discretization, allows one to reduce the practical dimension of the problem by a unit.

The BEM for potential flows is described in [5]. The BEM for two-dimensional dynamics of a single bubble near a solid wall and a free surface was successfully used in [6–8]. The three-dimensional boundary element method was applied to study the dynamics of bubbles arising from an underwater explosion or induced by a laser or a spark [9–11]. Lee et al. [12] extended the boundary integral formulation for the study of energy dissipation during bubble collapse.

Note that large-scale three-dimensional problems are computationally very complex and resource-intensive. So the essential aspect is the development and application of methods to accelerate such computing. For many years, the progress in studying the behavior of large bubble systems was mainly empirical with just several examples of large-scale direct simulations (e.g., [11]). Today, modern computational methods and powerful computer resources allow one to implement codes for fast large-scale simulations of bubble dynamics.

The primary goal of this work is the development and improvement of efficient research tools for the study of the complex behavior of three-dimensional two-phase potential flows. It is worth noting that the authors achieved substantial improvements in the performance of the algorithm reported earlier [13,14]. The approach is based on the BEM accelerated both via advanced scalable algorithms, particularly, the fast multipole method (FMM), and via utilization of advanced hardware, particularly, graphics processors (GPUs) and multicore CPUs.

The application of the BEM to direct simulation of large non-periodic systems is very limited. The major computational challenge is related to the solution of a large dense system of N algebraic equations for each time step, where

$N = M \times \Delta N$ is the number of computational points, M is the number of bubbles, ΔN is the number of discretization points on the bubble surface. In this case, the cost of the direct solution is $O\left(N^3\right)$. This cost can be reduced to $O\left(N_{\text{iter}}N^2\right)$, where $N_{\text{iter}} \ll N$ is the number of iterations and $O\left(N^2\right)$ is the cost of a single matrix–vector product (MVP).

Moreover, the application of the FMM allows one to reduce the complexity of MVP calculation to $O\left(N\right)$. The efficient implementation of the FMM for the Laplace equation was presented in [18], while in [19] the considered FMM was applied to the BEM. Bui et al. [11] suggested a numerical strategy that combines the BEM, the single level FMM, and the fast Fourier transform (FFT) to study the physics of multiple bubble dynamics. However, the $O\left(N\right)$ scalability of the FMM can be achieved only on hierarchical (multilevel) data structures. Moreover, the FMM can be efficiently parallelized [20].

In the present study, the developed approach was applied to study the dynamics of regular bubble clusters.

2   Problem Statement

Consider the dynamics of a single gas bubble in an incompressible inviscid liquid of density ρ:

$$\rho\frac{d\mathbf{v}}{dt} = -\nabla p + \rho\mathbf{g}, \quad \nabla \cdot \mathbf{v} = 0, \quad \frac{d}{dt} = \frac{\partial}{\partial t} + \mathbf{v} \cdot \nabla, \tag{1}$$

where $\mathbf{v}$ is the liquid velocity, p is the pressure, and $\mathbf{g}$ is the gravitational acceleration. Equation (1) allows one to find a solution in the form of a potential flow, $\mathbf{v} = \nabla\phi$, where ϕ is the velocity potential which is a harmonic function: $\nabla^2\phi = 0$. Thus, Eq. (1) can be represented in the form of a Cauchy–Lagrange integral:

$$\frac{\partial\phi}{\partial t} + \frac{1}{2}|\nabla\phi|^2 + \frac{p}{\rho} = \mathbf{g} \cdot \mathbf{x} + F(t), \tag{2}$$

where $\mathbf{x}$ is the coordinate of the point in space to which this equation applies, while $F(t)$ is an integration constant which should be determined from boundary conditions far from the bubble. To the liquid resting far from the bubble, the following conditions are imposed:

$$\phi|_{|\mathbf{x}|\to\infty} = 0, \quad p|_{|\mathbf{x}|\to\infty} = p_\infty\left(t\right) + \rho\mathbf{g} \cdot \mathbf{x}. \tag{3}$$

Then $F(t)$ is determined as

$$F\left(t\right) = p_\infty\left(t\right)/\rho, \quad p_\infty(t) = p_0 + p_a(t), \quad p_a(t) = P_a\sin(\omega t + \varphi), \tag{4}$$

where $p_\infty(t)$ is the liquid pressure far from the bubble; p_0 is the liquid pressure in the absence of acoustic field; $p_a(t)$ is the acoustic pressure; P_a, ω, and φ are the amplitude, the frequency, and the phase shift of the acoustic field, respectively.

The liquid pressure $p(\mathbf{x}, t)$ and the gas pressure $p_g(t)$ on the bubble surface S are related by

$$p(\mathbf{x},t) = p_g(t) - 2\gamma k(\mathbf{x},t), \tag{5}$$

where γ is the surface tension and $k(\mathbf{x},t)$ is the mean surface curvature. The gas pressure is related to the bubble volume according to the polytropic law:

$$p_g(t) = p_{g0}\left(\frac{V_0}{V}\right)^\kappa, \quad p_{g0} = p_0 + \frac{2\gamma}{a_0}, \tag{6}$$

where κ is the polytropic exponent, the subscript "0" refers to the initial value at $t = 0$, V is the bubble volume, and a is the equivalent bubble radius.

The evolution of the velocity potential and that of the interface are determined by the dynamic and kinematic conditions:

$$\frac{d\phi}{dt} = \frac{1}{2}|\mathbf{v}(\mathbf{x})|^2 - \frac{p_g(t) - 2\gamma k(\mathbf{x},t)}{\rho} + \mathbf{g}\cdot\mathbf{x} + F(t), \quad \mathbf{x} \in S, \tag{7}$$

$$\frac{d\mathbf{x}}{dt} = \mathbf{v}(\mathbf{x}), \quad \mathbf{x} \in S. \tag{8}$$

where $\mathbf{v}(\mathbf{x})$ is the interface velocity which is determined by the solution of the elliptic boundary value problem stated above. Despite the fact that the governing equations and the boundary conditions are linear with respect to $\mathbf{v}$, the dynamics of the interface is a non-linear problem since $\mathbf{v}(\mathbf{x})$ depends on the bubble shape.

3   The Accelerated Boundary Element Method

3.1   Boundary Integral Equations

The BEM uses a formulation in terms of boundary integral equations (BIE) whose solution with boundary conditions produces $\phi(\mathbf{x})$ and $q(\mathbf{x}) = \partial\phi(\mathbf{x})/\partial\mathbf{n}(\mathbf{x})$ on the boundary and subsequently determines $\phi(\mathbf{y})$ for external and boundary domain points $\mathbf{y}$. Using Green's identity, the boundary integral equations for $\phi|_{|\mathbf{y}|\to\infty} = 0$ can be written as

$$L[q](\mathbf{y}) - M[\phi](\mathbf{y}) = \begin{cases} -\phi(\mathbf{y}), & \mathbf{y} \notin S, \ \mathbf{y} \notin V, \\ -\frac{1}{2}\phi(\mathbf{y}), & \mathbf{y} \in S, \\ 0, & \mathbf{y} \in V, \end{cases} \tag{9}$$

Here $L[q]$ and $M[\phi]$ are, respectively, single and double layer potentials which denote the following operators:

$$L[q](\mathbf{y}) = \int_S q(\mathbf{x}) G(\mathbf{y}, \mathbf{x}) dS(\mathbf{x}),$$
$$M[\phi](\mathbf{y}) = \int_S \phi(\mathbf{x}) \frac{\partial G(\mathbf{y}, \mathbf{x})}{\partial n(\mathbf{x})} dS(\mathbf{x}), \tag{10}$$

where $G(\mathbf{y}, \mathbf{x})$ is the free-space Green's function for the Laplace equation and $\partial G(\mathbf{y}, \mathbf{x})/\partial n(\mathbf{x})$ is its normal derivative

$$G(\mathbf{y}, \mathbf{x}) = \frac{1}{4\pi r}, \quad \frac{\partial G(\mathbf{y}, \mathbf{x})}{\partial n(\mathbf{x})} = \frac{\mathbf{n}\cdot\mathbf{r}}{4\pi r^3}, \quad \mathbf{r} = \mathbf{y} - \mathbf{x}, \quad r = |\mathbf{y} - \mathbf{x}|. \tag{11}$$

3.2   The Discretization

The numerical method is based on the discretization of the bubble surface by means of triangular meshes. For accurate computation of boundary integrals (10) and surface properties, the mesh should be of good quality. The regular integrals over the patches are computed using the trapezoidal rules of second-order accuracy. Collocation points for the bubble surface are located at the mesh vertices. Using the vertex collocation approach, the second equation in BIE (9) can be rewritten at $\mathbf{y} = \mathbf{x}_j$ in discrete form,

$$\sum_{i=1}^{N} I_{ji}^{(G)} q_i = -\frac{1}{2}\phi_j + \sum_{i=1}^{N} I_{ji}^{(\partial G/\partial n)}\phi_i, \quad j = 1,\dots,N, \tag{12}$$

where N is the number of computational points; I_{ji} is the unit dyadic, represented by the identity matrix; $I_{ji}^{(G)}$ and $I_{ji}^{(\partial G/\partial n)}$ are the dyadics corresponding to kernels G and $\partial G/\partial n$, respectively; the subscripts placed to the right of the spatial functions denote the index of the $\mathbf{x}_i$ collocation point.

Obviously, Eq. (12) is a linear system of size $N \times N$ which can be rewritten in matrix–vector form as

$$\mathbf{A} \cdot \mathbf{X} = \mathbf{b}, \tag{13}$$

where $\mathbf{A}$ is the system matrix, $\mathbf{X}$ is the solution vector, and $\mathbf{b}$ is the right-hand-side vector.

To calculate surface properties, such as the area, the normal vector, and the curvature, each plane triangle is divided by its medians. Thus, the area corresponding to a collocation point can be defined as a sum of partial areas of triangles, while the normal vectors at the collocation point can be defined by averaging the normal vectors to the triangles containing the point. The mean curvature is calculated by the fitted paraboloid method. Singular integrals can be computed based on integral identities that allow one to express those integrals as sums of regular integrals over the surface.

To solve system (13), we use the unpreconditioned general minimal residual method (GMRES) [15].

It is also well-known from practice that the evolution of the mesh must be stabilized by some techniques. For this purpose, we implemented a parametric spherical filter whose idea is based on the representation of the mapping of each bubble surface on a topologically equivalent object (a unit sphere in our case) and the expansion of each surface function (Cartesian coordinates of the surface and the velocity potential) over an orthogonal basis (spherical harmonics in our case) up to harmonics of a certain degree using the least square fitting (low-pass filter) [14].

Note that only a solver for the Dirichlet problem is needed for the bubble evolution. A solver for the Neumann problem is needed only if the potential is unknown at $t = 0$, which is the case of prescribed nonzero initial conditions for the normal velocity.

3.3   Fast Multipole Method

In the conventional BEM the full system matrix $\mathbf{A}$ (Eq. (13)) must be computed
to solve the linear system either directly or iteratively. The memory needed to
store the matrix is fixed and is not affected by the accuracy imposed on the
computation of the surface integrals. Since the system matrix has a special type,
it is not necessary to store the elements of the matrix: it is sufficient to calculate
the matrix–vector product. The MVP in the GMRES can be computed using
the FMM, which represents a substantial shift in the computational strategy.
Moreover, additional runs of the FMM for computation of both the singular
elements and the right-hand-side vector before the start of the GMRES can
reduce the algorithm complexity as well.

The main idea of using the FMM for the solution of the discretized BIE (9)
is based on a formal decomposition of the $\mathbf{A}$ matrix in the $\mathbf{A} \cdot \mathbf{X}$ MVP in the
form

$$\mathbf{A} = \mathbf{A}^{\mathrm{sparse}} + \mathbf{A}^{\mathrm{dense}}, \tag{14}$$

where $\mathbf{A}^{\mathrm{sparse}}$ is the sparse matrix accounting for nonzero entries of interaction
of mesh vertices $\mathbf{x}_i$ and $\mathbf{y}_j$ such that $|\mathbf{y}_j - \mathbf{x}_i| < r_c$ (near field), where r_c is
a certain distance, usually of the same order as the distance between vertices,
whose selection can be based on some estimates or error bounds, while $\mathbf{A}^{\mathrm{dense}}$
is the dense matrix accounting for nonzero entries of all other interactions for
which $|\mathbf{y}_j - \mathbf{x}_i| \geq r_c$ (far field). The $\mathbf{A}^{sparse}\mathbf{X}$ MVP is computed directly, whereas
$\mathbf{A}^{\mathrm{dense}}\mathbf{X}$ is computed approximately.

To calculate the dense matrix, the factored approximate representation is
built in the considered domain of the space with center at $\mathbf{x}_*$:

$$A_{ji}^{\mathrm{dense}} = \sum_{l=0}^{P-1} S_l(\mathbf{y}_j - \mathbf{x}_*)R_l(\mathbf{x}_i - \mathbf{x}_*) + O(\epsilon), \tag{15}$$

where S_l is a singular set of multipole solutions and R_l is a regular set of local
solutions. Expansion (15) comes from analytical series representations and is
truncated at some number $P = p_{\mathrm{fmm}}^2$ of coefficients, which determines the accuracy of the expansion, $\epsilon = \epsilon(P)$.

We use 2^d-tree data-structures to split the near and far fields and implement
the multilevel FMM (see [16]). Moreover, the translation operator is needed to
converse a representation in a coordinate system with a particular center to
another representation in another coordinate system with another center. The
action of the translation operators for functions expandable over the multipole
basis, $\mathbf{S}$, and the local basis, $\mathbf{R}$, can be represented by the action of linear transforms on the space of coefficients. Depending on the expansion, there are three
main types of translation operators: local-to-local, i.e. $R|R$, far-to-far, i.e. $S|S$,
and far-to-local, i.e. $S|R$ [16].

The FMM in BIE (9) is used to calculate the MVP of the Laplace kernel
(monopole), $G(\mathbf{y}, \mathbf{x})$, and the normal derivative of the velocity potential, $\partial\phi/\partial n$,
and the MVP of the normal derivative of the Laplace kernel (dipole with moment
$\mathbf{n}$), $\partial G(\mathbf{y}, \mathbf{x})/\partial n(\mathbf{x}) = \nabla_{\mathbf{x}}G(\mathbf{y}, \mathbf{x}) \cdot \mathbf{n}$, and the velocity potential, ϕ. Then the

total potential, ϕ_{Laplace}, can be represented as the sum of the monopole (with the intensity $f = \partial\phi/\partial n$) and the dipole (with the intensity $g = \phi$):

$$\phi_{\text{Laplace}} = f\frac{1}{r} + g\frac{(\mathbf{n}\cdot\mathbf{r})}{r^3}, \quad \mathbf{r} = \mathbf{y} - \mathbf{x}, \quad r = |\mathbf{y} - \mathbf{x}|. \tag{16}$$

Thus, the integrals in BIE (9) can be computed as summations of monopoles and dipoles of the Laplace equation. In order to calculate the first integral in Eq. (9), we should use $f = \partial\phi/\partial n$ while $g = 0$. To calculate the second integral in Eq. (9), we should determine $f = 0$ while $g = \phi$. Based on the decomposition given above for the solution of the Laplace equation, the FMM for three-dimensional Laplace equation is employed for the MVP calculation as described in [18].

3.4   GPU Acceleration

GPU acceleration is used in the implementation of the FMM. In [20], it was shown that the GPU architecture is more efficient to calculate the sparse matrix–vector product $\mathbf{A}^{\text{sparse}}\mathbf{X}$ (up to 100 times of effective acceleration), whereas the GPU, when used for the $\mathbf{A}^{\text{dense}}\mathbf{X}$ multiplication, can be accelerated just a few times compared to a single-core CPU. This effect can be explained by the fact that a relatively complicated data structure requires extensive random access to the GPU global memory (costly operations) and the limited local GPU memory is not sufficient for efficient storage of the translation operators. Moreover, the use of CPUs with K cores allows one to implement the relatively easy parallelization of the MVP $\mathbf{A}^{\text{dense}}\mathbf{X}$ via OpenMP. The efficiency of such parallelization is close to 100%, which means about K-fold acceleration. Moreover, additional acceleration can be achieved by parallelization of the independent operations $\mathbf{A}^{\text{sparse}}\mathbf{X}$ and $\mathbf{A}^{\text{dense}}\mathbf{X}$ on GPU and CPU. Finally, the actual acceleration of the entire algorithm can be much more than K-fold, owing to the reduction of the octree depth and additional accelerations [20]. Careful tuning of the algorithm and the octree depth is based on the workload balance between the CPU and the GPU. Thus, the overall acceleration using such approach for a system with 8 to 12 CPU cores and one GPU can be about 100 times compared to a single-core CPU implementation.

3.5   Time Marching

The time step used in explicit schemes (7) and (8) should be sufficiently small to satisfy a Courant-type stability condition,

$$\Delta t = C \min(\triangle_{d0}) \left(\frac{\rho}{P_a}\right)^{1/2}, \tag{17}$$

where $\min(\triangle_d(t))$ is the minimum spatial discretization length (the length of the edge of the mesh) at moment t , $(P_a/\rho)^{1/2}$ is the characteristic velocity of bubble growth/collapse, and C is a constant of the order of 0.1.

For integration of Eqs. (7) and (8), we used the Adams–Bashforth scheme of the sixth order, which requires one call of the right-hand-side function per time step. It also requires initialization, which was provided by a fourth order Runge–Kutta scheme. The initialization requires more calls of the right-hand side; however, we only needed a few initial steps, and the overall performance was not affected substantially.

4   Numerical Experiments

Calculations were performed on a workstation equipped with Intel Xeon 5660 2.8 GHz CPU (12 physical + 12 virtual cores), 12 GB RAM, and one GPU NVIDIA Tesla K20 (5 GB of global memory).

4.1   Tests for the MVP

Tests were conducted for homogeneous and uniformly distributed sources and receivers in a cube. For the comparative analysis of the performance of the FMM, four modules were implemented:

- Module 1: the direct algorithm of MVP calculation, CPU parallel computing using OpenMP;
- Module 2: the direct algorithm of MVP calculation, GPU parallel computing using CUDA;
- Module 3: the FMM algorithm of MVP calculation, CPU parallel computing using OpenMP;
- Module 4: the heterogeneous FMM algorithm of MVP calculation, CPU/GPU parallel computing.

All GPU calculations were performed in single and double precision. Thus, six cases were considered:

- Case 1: MVP calculation in Module 1;
- Case 2: MVP calculation in Module 2 for single-precision GPU computing;
- Case 3: MVP calculation in Module 2 for double-precision GPU computing;
- Case 4: MVP calculation in Module 3;
- Case 5: MVP calculation in Module 4 for single-precision GPU computing;
- Case 6: MVP calculation in Module 4 for double-precision GPU computing.

The wall-clock time of MVP calculation using different modules is portrayed in Fig. 1 in a logarithmic coordinate system. We can see that the direct algorithm has quadratic complexity, whereas the FMM is a linear one. Moreover, the direct MVP calculation on GPU is faster for up to 32 768 sources. Table 1 shows the acceleration of calculations using Modules 2, 3, and 4 compared to calculations by Module 1, and a maximum L_2-norm error. It is seen that the acceleration of GPU calculations does not change starting at 65 536 sources. Note that the relative error does not depend substantially on N and the accuracy of the MVP in Cases 5 and 6 is not practically affected by the precision of GPU computing. Single precision computations are 2.6 times faster compared to double precision ones. However, the FMM gives acceleration when the number of sources is increased. For example, for 1 048 576 sources, Case 5 runs 2896 times faster than Case 1.

Table 1. Acceleration and relative error of the MVP using different modules ($p_{fmm} = 12$ for the FMM)

N	Case 2	Case 3	Case 4	Case 5	Case 6
4096	40	31	3	8	7
16 384	111	44	13	66	54
65 536	86	33	31	157	125
262 144	87	33	125	395	339
1 048 576	87	33	667	2896	2309
Relative error	$7 \cdot 10^{-5}$	$5 \cdot 10^{-9}$	$4 \cdot 10^{-7}$	$5 \cdot 10^{-7}$	$4 \cdot 10^{-7}$

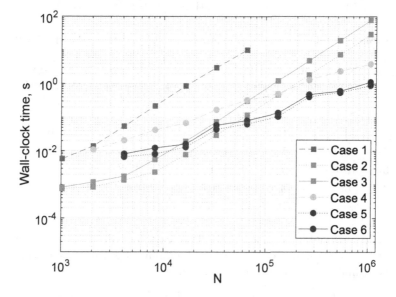

Fig. 1. Wall-clock time of MVP calculation for $p_{fmm} = 12$ using different modules

4.2 Tests for Regular Bubble Cluster

All numerical results were computed for a bubble in water ($\rho = 1000\,\text{kg/m}^3$, $\gamma = 0.073\,\text{Pa/m}$, $\kappa = 1.4$) under atmospheric pressure ($p_0 = 10^5\,\text{Pa}$) and zero gravity. To demonstrate the results, we used the dimensionless coordinates $x' = x/a_0$, $y' = y/a_0$, $z' = z/a_0$, where a_0 is the initial minimum bubble radius. To test the performance of the program code, we used a regular cubic cluster consisting of $M = n_b \times n_b \times n_b$ bubbles of initial radius $a_0 = 10\,\mu\text{m}$ with a distance between bubbles $d = 4a_0$ (Fig. 2), in an acoustic field of frequency $\omega/(2\pi) = 200\,\text{kHz}$ and amplitude $P_a = p_0$. The discretization of the surface of each bubble is $\Delta N = 642$.

Based on the MVP tests for Laplace kernels described above and depending on the problem size and the necessary accuracy, we used Modules 2, 3, and 4

Table 2. Wall-clock time of the program code for 100 time steps and $n_b = 4$

Case	Number of MVP calls	Average number of GMRES iterations	Average time for 1 MVP, s	Average time for BEM solver, s	Average time for 1 step, s
1	851	5	0.15	2.11	8.69
2	690	4	0.1	1.32	8.03
3	691	4	0.038	0.53	0.56
4	543	2	0.023	0.31	0.35
5	690	4	0.083	1.1	1.13
6	543	2	0.04	0.45	0.48

Table 3. Wall-clock time of the program code for 100 steps and $n_b = 12$

Case	Number of MVP calls	Average number of GMRES iterations	Average time for 1 MVP, s	Average time for BEM solver, s	Average time for 1 step, s
1	3070	9	3.5	62	235.05
2	1011	7	2.2	35.25	208.46
3	1016	7	1.1	17.93	18.48
4	693	4	0.5	6.68	7.21

with single and double precision on GPU to calculate all MVP in the BEM. To evaluate the performance of the program code and the accuracy of the bubble dynamics simulations, we considered six cases:

- Case 1: MVP calculation in Module 3 with accuracy parameters $p_{fmm} = 16$, $\epsilon^{GMRES} = 10^{-6}$;
- Case 2: MVP calculation in Module 3 with accuracy parameters $p_{fmm} = 12$, $\epsilon^{GMRES} = 10^{-5}$;
- Case 3: MVP calculation in Module 4 with accuracy parameters $p_{fmm} = 12$, $\epsilon^{GMRES} = 10^{-5}$, double-precision GPU computing;
- Case 4: MVP calculation in Module 4 with accuracy parameters $p_{fmm} = 8$, $\epsilon^{GMRES} = 10^{-4}$, single-precision GPU computing;
- Case 5: MVP calculation in Module 2 with accuracy parameter $\epsilon^{GMRES} = 10^{-5}$, double-precision GPU computing;
- Case 6: MVP calculation in Module 2 with accuracy parameter $\epsilon^{GMRES} = 10^{-4}$, single-precision GPU computing.

Tables 2 and 3 show the profiling of the program code for 100 time steps (from iterations 101 to 200) and bubble cluster sizes $n_b = 4$ and $n_b = 12$, respectively. Tests show that, starting from 100 iterations, the GMRES accelerates and requires fewer iterations for convergence. For 64 bubbles (Table 2), Cases 3 and

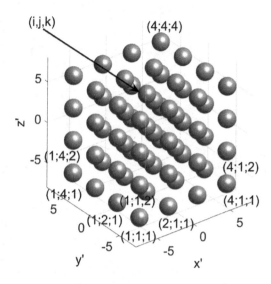

Fig. 2. Benchmark of a regular bubble cluster for $n_b = 4$

4 reached a greater acceleration than Cases 5 and 6; for this reason, calculations by the direct algorithm on GPU were not performed for larger clusters. From Tables 2 and 3 we can see that the GMRES convergence for larger clusters is worse (the average number of iterations increases). The GMRES converges identically in Cases 2 and 3. Note that the modules for calculating geometric characteristics in Cases 1 and 2 are implemented on CPU, which significantly increases the wall-clock time for one step as the number of computational points increases. The geometry calculation modules in the other cases are implemented on GPU, whose computational cost typically does not exceed 10% of the BEM solver cost. The greatest acceleration was observed in Case 4. In this case, however, we lost in accuracy of the calculations due to single-precision GPU computing, fewer elements in the far field expansion, and the GMRES accuracy. For example, in case of 4096 bubbles, which corresponds to $N = 2\,629\,632$, one FMM call takes about 2 s, 4 iterations are required for GMRES convergence, and 1 time step takes 24 s. The wall-clock time required for the BEM code is shown in Fig. 3 in a logarithmic scale. We can see from the figure that the complexity of the BEM code is determined by the complexity of the MVP (i.e. is linear when using the FMM), which shows the good scalability of the FMM. Note that it is more efficient to use Module 2 for $N \le 20\,000$.

We conducted an analysis of the dynamics of regular bubble clusters. Figures 4 and 5 depict sections of clusters consisting of 64 (Fig. 2) and 1728 bubbles, on the plane $Ox'y'$ at the end of one period of oscillations of the acoustic field. We see from the figures that the corner bubbles of the cluster deform significantly. As the cluster size increases, the deformation of the bubbles decreases. An analysis of the cluster volume (Fig. 6) and the bubble center mass shows

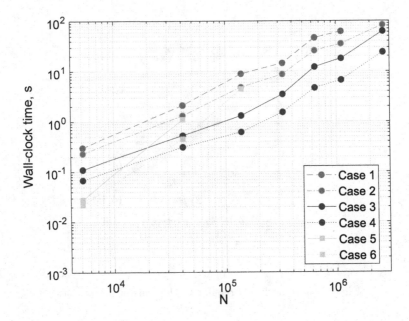

Fig. 3. Code performance, average wall-clock time for the BEM solver

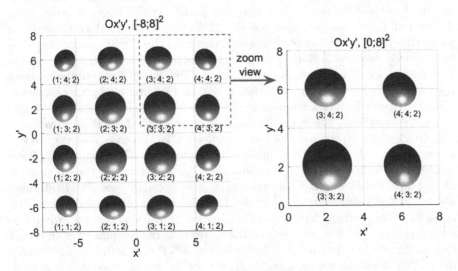

Fig. 4. Dynamics of a regular cluster with 64 bubbles. Second row from the bottom (see Fig. 2)

that the bubbles become less mobile as the cluster size increases and their volume changes slightly. Note that the bubbles move on average toward the cluster center during the oscillation period.

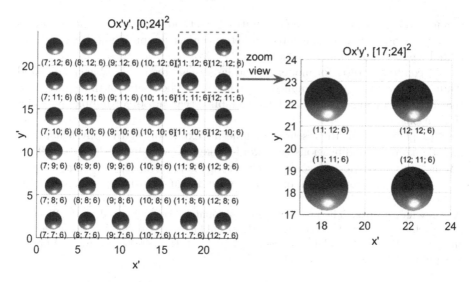

Fig. 5. Dynamics of a regular cluster with 1728 bubbles. Sixth row from the bottom

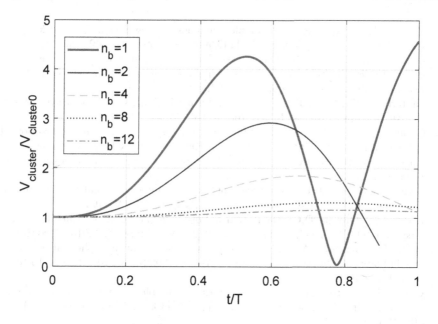

Fig. 6. Dynamics of the volume of regular clusters

5   Conclusions

We developed and tested on personal supercomputers efficient algorithms that enable the direct simulation of bubble systems with dynamic deformable interfaces discretized by millions of boundary elements. The approach combines the boundary element method, the fast multipole method, and the parametric spherical filter. Codes were implemented on heterogenous computing clusters and validated against well-known solutions.

We conducted tests for moderate and large-scale regular bubble clusters and studied the accuracy/performance of the method. The results of this study show that the developed software can be a valuable research tool for bubble dynamics simulations.

Our plans for future research include an extension of the physical model and algorithmic modifications that would take into account neglected effects of strong bubble interaction (coalescence, strong shape oscillations, fragmentation, etc.), as well as a comparison of computational results with experimental data.

Acknowledgments. The program code development was supported by the Skoltech Partnership Program. The reported study of bubble cluster dynamics was funded by the Russian Science Foundation (research project No. 18-71-00068). The FMM library was provided by Fantalgo, LLC (Maryland, USA).

References

1. Plesset, M.S., Prosperetti, A.: Bubble dynamics and cavitation. J. Fluid Mech. **9**, 145–185 (1977). https://doi.org/10.1146/annurev.fl.09.010177.001045
2. Akhatov, I., Gumerov, N., Ohl, C.-D., Parlitz, U., Lauterborn, W.: The role of surface tension in stable single-bubble sonoluminescence. Phys. Rev. Lett. **78**(2), 227–230 (1997). https://doi.org/10.1103/PhysRevLett.78.227
3. Lauterborn, W., Kurz, T.: Physics of bubble oscillations. Rep. Prog. Phys. **73**(10), 88 (2010). https://doi.org/10.1088/0034-4885/73/10/106501
4. Gumerov, N.A., Akhatov, I.S.: Numerical simulation of 3D self-organization of bubbles in acoustic fields. In: Ohl, C.-D., Klaseboer, E., Ohl, S.W., Gong, S.W., Khoo B.C. (eds.) Proceedings of the 8th International Symposium on Cavitation, Singapore (2012)
5. Canot, E., Achard, J.-L.: An overview of boundary integral formulations for potential flows in fluid-fluid systems. Arch. Mech. **43**(4), 453–498 (1991)
6. Blake, J.R., Gibson, D.C.: Cavitation bubbles near boundaries. Ann. Rev. Fluid Mech. **19**, 99–123 (1987). https://doi.org/10.1146/annurev.fl.19.010187.000531
7. Best, J.P., Kucera, A.: A numerical investigation of nonspherical rebounding bubbles. J. Fluid Mech. **245**, 137–154 (1992). https://doi.org/10.1017/S0022112092000387
8. Oguz, H.N., Prosperetti, A.: Dynamics of bubble growth and detachment from a needle. J. Fluid Mech. **257**, 111–145 (1993). https://doi.org/10.1017/S0022112093003015
9. Chahine, G.L., Duraiswami, R.: Dynamical interactions in a multibubble cloud. ASME J. Fluids Eng. **114**, 680–686 (1992). https://doi.org/10.1115/1.2910085
10. Zhang, Y.L., Yeo, K.S., Khoo, B.C., Wang, C.: 3D jet impact and toroidal bubbles. J. Comput. Phys. **166**, 336–360 (2001). https://doi.org/10.1006/jcph.2000.6658
11. Bui, T.T., Ong, E.T., Khoo, B.C., Klaseboer, E., Hung, K.C.: A fast algorithm for modeling multiple bubbles dynamics. J. Comp. Physics. **216**, 430–453 (2006). https://doi.org/10.1016/j.jcp.2005.12.009

12. Lee, M., Klaseboer, E., Khoo, B.C.: On the boundary integral method for the rebounding bubble. J. Fluid Mech. **570**, 407–429 (2007). https://doi.org/10.1017/S0022112006003296

13. Itkulova, Yu.A., Abramova, O.A., Gumerov, N.A., Akhatov, I.S.: Boundary element simulations of free and forced bubble oscillations in potential flow. In: Proceedings of IMECE 2014, Canada, Montreal, vol. 7 (2014). https://doi.org/10.1115/IMECE2014-36972

14. Itkulova(Pityuk), Yu.A., Abramova, O.A., Gumerov, N.A., Akhatov, I.S.: Simulation of bubble dynamics in three-dimensional potential flows on heterogeneous computing systems using the fast multipole and boundary element methods. Numer. Methods Program. **15**, 239–257 (2014). [In Russian]

15. Saad, Y., Schultz, M.H.: GMRES: a generalized minimal residual algorithm for solving nonsymmetric linear systems. SIAM J. Sci. Stat. Comput. **7**, 856–869 (1986). https://doi.org/10.1137/0907058

16. Gumerov, N.A., Duraiswami, R.: Data structures, optimal choice of parameters, and complexity results for generalized multilevel fast multipole method in d dimensions. Technical report CS-TR-4458. University of Maryland, College Park (2003). https://doi.org/10.1016/B978-0-08-044371-3.X5000-5

17. Gumerov, N.A., Duraiswami, R.: Fast Multipole Methods for the Helmholtz Equation in Three Dimensions. Elsevier, Oxford (2005)

18. Gumerov, N.A., Duraiswami, R.: Comparison of the efficiency of translation operators used in the fast multipole method for 3D Laplace equation. Technical report CS-TR-4701. University of Maryland, College Park (2005)

19. Gumerov, N.A., Duraiswami, R.: FMM accelerated BEM for 3D Laplace and Helmholtz equations. In: Proceedings of BETEQ-7, Paris, France, pp. 79–84 (2006)

20. Gumerov, N.A., Duraiswami, R.: Fast multipole methods on graphics processors. J. Comput. Phys. **227**(18), 8290–8313 (2008). https://doi.org/10.1016/j.jcp.2008.05.023

Author Index

Printed in the United States
By Bookmasters